CHINESE FOSSIL VERTEBRATES

CHINESE FOSSIL VERTEBRATES

Spencer G. Lucas

Columbia University Press
New York

Columbia University Press
Publishers Since 1893
New York   Chichester, West Sussex

Library of Congress Cataloging-in-Publication Data

Lucas, Spencer G.
Chinese fossil vertebrates
p.   cm.
Includes bibliographical references and index.
ISBN 978-0-231-08483-3
1. Vertebrates, Fossil–China
I. Title
QE841.L83   2001
566.0951—dc21   2001042435

To the memory of Minchen Chow

Contents

Preface

I first visited China in 1980, not long after the Cultural Revolution ended. The country had just come out of more than a decade of scientific isolation, and I was one of the first vertebrate paleontologists to then reach China. The goals of my visit were purely research—to study fossils of Paleogene mammals, especially pantodonts, as part of the work for my doctoral dissertation. I found the fossils illuminating, the country fascinating, and made fast friends with several colleagues. I left China wanting to return, and have since done so four more times. Those visits allowed me to travel throughout much of the country, conducting both field and museum research. I also studied vertebrate fossils that were early collected in China and now housed at the American Museum of Natural History (New York) and the Paleontological Museum of the University of Uppsala (Sweden). In my travels, I met, shared information with, and, in some cases, collaborated with many Chinese vertebrate paleontologists and many non-Chinese vertebrate paleontologists who also worked in China and/or studied Chinese vertebrate fossils collected by others. Four of them made indelible impressions on me, namely Minchen Chow, Pei Wenchong, Birger Bohlin and Otto Zdansky, men whose contributions to the vertebrate paleontology of China dwarf those of most others. My travels, research, and colleagues taught me much. I amassed a large library of Chinese technical literature and I even managed to learn a bit (just a bit) of Chinese.

I set out to write this book, in part, because I believed that I knew quite a bit about the vertebrate paleontology of China, and therefore that I was qualified to write a book about the subject. China's Silurian-Devonian fishes and Pleistocene mammals humbled me in this belief. However, with apologies to Lucretius, I would say that by writing we ourselves learn. Indeed, realizing this brings home the principal reason I wrote this book: because I have long needed such a book. In it, I organize temporally the vertebrate fossil record of China. The book thus reflects my primary interests, which are in vertebrate biostratigraphy and biochronology. It provides the reader with a comprehensive, chronologically-ordered review of China's vertebrate fossil record. Additionally, this book includes a history of vertebrate paleontological studies in China and an entrée to some important issues of systematics, evolutionary history, paleoecology, taphonomy, and functional anatomy best elucidated by China's vertebrate fossils.

This book proved to be a vast undertaking that pulled together nearly two decades of my research experience and a morass of literature. First, and foremost, I must thank my Chinese friends and colleagues who taught me so much about the fossil record of their great country. In particular, I thank Chang Meeman,

Chen Peiji, Ding Suyin, Huang Xueshi, Li Chuankuei, Li Jianjun, Li Jinling, Luo Zhexi, Ma Ancheng, Miao Desui, Qi Tao, Qiu Zhuding, Su Dezao, Sun Ailin, Tong Yongsheng, Wang Jingwen, Zhai Renjie, Zhen Shuonan, Zheng Jiajian, and Zhou Shiwu. Richard Reyment, Jürgen Schöbel, and Solweig Stuenes made my work in Sweden possible, as did Niall Mateer, who also taught me much about the Cretaceous of China. Malcolm McKenna and Richard Tedford helped me to study the collections of the American Museum of Natural History. The National Geographic Society, National Science Foundation, and Swedish Natural Science Research Council funded much of my research on Chinese vertebrate fossils.

John Estep and Randy Pence expertly executed many of the illustrations in the book, and Mary Bratzler typed the manuscript in its various drafts. Yami Lucas provided diverse help and encouragement. Luo Zhexi, Niall Mateer, Sue Turner, and an anonymous reviewer made constructive comments on the text that improved it. Ed Lugenbeel brought Columbia University Press to me, and me to the Press. Holly Hodder helped me to continue this association. Finally, I owe a special debt of gratitude to the late Minchen Chow (Zhou Mingzhun), to whom I dedicate this book. He, more than any other scientist, made possible my studies of Chinese fossil vertebrates.

— Spencer Lucas

CHINESE FOSSIL VERTEBRATES

Chapter 1

Introduction

China is the world's third largest nation. Its vast land area contains extensive exposures of sedimentary rocks, many of nonmarine origin. Serious scientific study of China's vertebrate fossil record began in the last century. This record extends back to the Early Cambrian, nearly 550 million years. Today, Chinese vertebrate fossils represent one of the most extensive and important records of vertebrate evolution. This book provides a comprehensive review of Chinese fossil vertebrates, reviewed in temporal (stratigraphic) order.

Political Divisions of China

The People's Republic of China (hereafter, simply referred to as China) has a land area of nearly 9.6 million km^2, which is 6.5 percent (one-fifteenth) of the world's land surface. One of the world's largest countries, China is divided into 30 political units under the direct control of its Central Government in Beijing (see figure 1-1). Three of these units are cities (Beijing, Tianjin, and Shanghai), five are autonomous regions (Xinjiang, Ningxia, Nei Monggol or Inner Mongolia, Xizang or Tibet, and Guangxi), and the remaining units are provinces. The Chinese Central Government claims dominion over Taiwan, but in this book Taiwan is not included in China.

Geological Setting

China has a very complex geology (see figure 1-2) that encompasses great thicknesses of sedimentary rock of Phanerozoic age (e.g., Lee 1939; Yang et al. 1986; Meyerhoff et al. 1991). Equally complex is the plate tectonic history of China prior to the late Mesozoic.

Most workers recognize that during the Paleozoic and much of the Mesozoic, what is now China belonged to several microplates (or blocks) (see figure 1-3). The south China block encompasses most of southern China—the region south of the Qinling fold belt. The north China block is north of the Qinling fold belt, extending east of the Qilian Mountains. The Tarim block is north of the Kunlun fold belt, west of the north China block and south of the Tien Shan. It corresponds rather closely to the Tarim basin of western China. A variety of smaller blocks have also been identified (see figure 1-3), though their identity and boundaries are less certain than those of the larger blocks.

Figure 1-1 *This map of China shows the 30 major political divisions (cities, provinces, and autonomous regions) of the country.*

The locations and configurations of these blocks during the Paleozoic and Mesozoic are constrained by paleomagnetic evidence. For the south China block, many paleomagnetic data are available, though not all are reliable. However, these data are few and far between for the other microplates of Paleozoic China. This explains why there is so much uncertainty and controversy over the Paleozoic and early Mesozoic plate tectonics and continental assembly of China. The Paleozoic configurations of China used in this book are those of Z. Li et al. (1993), whereas the early Mesozoic configurations are those of Golonka et al. (1994). Fossil vertebrates of Silurian, Devonian, Permian, and Triassic age from China provide some constraints on the configuration of the Chinese microplates, an important subject discussed in the appropriate ensuing chapters.

Since the Jurassic, sedimentary deposition across China has been almost entirely nonmarine. Prior to that time, extensive marine deposits accumulated in China, especially over the south China block. Nevertheless, significant Paleozoic nonmarine deposits are known from China, beginning in the Devonian rocks. The wide extent and preservation of nonmarine sediments in China partly accounts for its outstanding vertebrate fossil record. Much nonmarine sedimentation in China was focused in about 20 sedimentary basins (see figure 1-4). These basins contain most of China's vertebrate fossil record, especially of pre-Cenozoic vertebrates.

Vertebrate Biochronology

This book employs concepts of vertebrate biochronology—the use of fossil vertebrates to discriminate intervals of geologic time—earlier advocated by Lucas (1990, 1991, 1993a, b, c, 1996a, c, 1998a, b). The basic unit of vertebrate biochronology (indeed, of all biochronology) is the biochron. A biochron (Williams 1901) is simply an interval of geologic time that corresponds to the duration of a taxon. Each vertebrate taxon has a corresponding biochron. Biochronological organization of the Chinese vertebrate fossil record can be achieved by identifying widespread, distinctive, and relatively short-lived vertebrate taxa as the name-bearers of biochrons. For example, the endemic Chinese

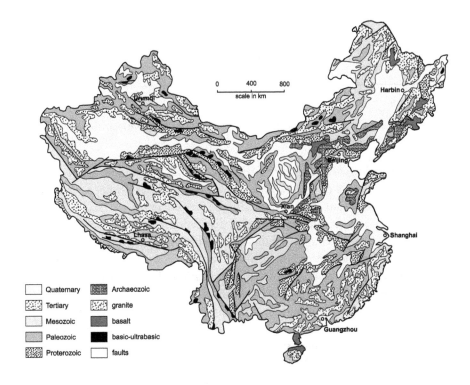

Figure 1-2 *Even a generalized geologic map of China (after Hsieh 1973) reveals the complex geology of the country.*

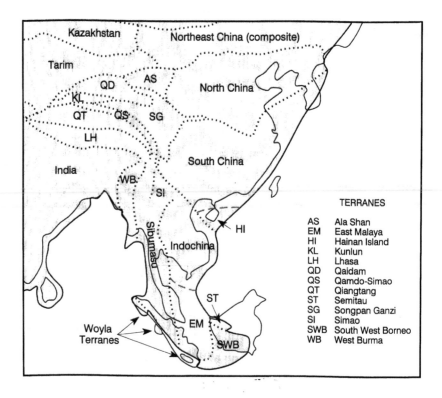

Figure 1-3 *Eastern Asia was assembled in Mesozoic time from the approximately 20 microplates (terranes) shown here (after Metcalfe 1994).*

agnathan *Hanyangaspis* is characteristic of Chinese early middle Silurian strata, so we may speak of a *Hanyangaspis* biochron (see Chapter 3). Recognizing such biochrons throughout the entire Chinese vertebrate-fossil record is useful because it not only clarifies the temporal distribution of the vertebrate taxa, but also sets up a biochronological framework to be tested and refined.

Biochronology is the use of fossils to recognize intervals of geologic time, whereas biostratigraphy is simply the identification of distinctive bodies of rock based on their fossil content. The biochron is thus recognized (operationalized) by its biostratigraphic counterpart, the range zone (see figure 1-5).

Biochrons refer to a single taxon, from the species to (in theory) kingdom level. Vertebrate paleontologists have also distinguished geologic time based on aggregations of taxa, or what they term *faunas*. Biostratigraphically, these time units correspond to assemblage zones of vertebrate fossils. These biochronological units have been termed land-mammal *ages* or land-vertebrate *ages*, although they are not ages in the formal stratigraphic sense (they lack a stratotypical stage). I have called them land-vertebrate faunachrons (Lucas 1993a). A faunachron is the time equivalent to the duration of a fauna. In effect, it is the temporal equivalent of an assemblage zone of vertebrate fossils. Note that this use of the term fauna is a paleontological one that corresponds with the term local fauna used to refer to "a group of fossils local in both time and space" (Taylor 1960: 10), and thus does not equate with the neozoological use of the term (Tedford 1970).

Land-vertebrate faunachrons have been proposed for the Mesozoic and Cenozoic vertebrate-fossil record of China. These faunachrons provide a useful biochronological organization of the Chinese vertebrate faunas that is followed in this book.

Some Features of This Book

Before reading this book you should be alerted to some of its special features. This book provides a technical review of Chinese fossil vertebrates, so it contains

Figure 1-4 *About 20 major sedimentary basins can be identified in China (after Hsü 1989).*

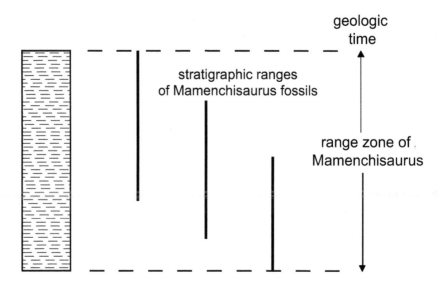

Figure 1-5 *The combined stratigraphic ranges of fossils of the Jurassic dinosaur Mamenchisaurus establish a range zone (biostratigraphic unit). The time during which Mamenchisaurus lived is its biochron.*

an extensive bibliography of all references cited in the text. The text uses the Pinyin romanization of Chinese place names and surnames, except in those cases where confusion might result (for example, the famous Chinese paleontologist C. C. Young is Yang Jungjian in Pinyin, but remains C. C. Young in this book). Note that Rich et al. (1994) provide a useful Chinese character–Pinyin–English (and reverse) dictionary of vertebrate paleontological terms. The last chapter of this book is a short, concise overview of the Chinese vertebrate-fossil record.

Chapter 2

History of Vertebrate Paleontological Studies

Fossil vertebrates have been collected in China and studied scientifically for more than a century. This work began with the serendipitous discoveries by early Western explorers and geologists in nineteenth century China and the idiosyncratic purchases of fossil bones from Chinese druggists by colonial envoys and naturalists. The early decades of the twentieth century saw Westerners pursue the vast vertebrate paleontological wealth of China on a grand scale, both as steady, colonial fossil collectors and as members of the fabled Central Asiatic Expeditions of the American Museum of Natural History headed by Roy Chapman Andrews.

The Second World War and Chinese Communist Revolution initiated a new phase of Chinese vertebrate paleontology. A native Chinese vertebrate paleontologist, C. C. Young (Yang Zhungjian) founded an institute of vertebrate paleontology in Beijing as part of the Chinese Academy of Sciences. This is now the Institute of Vertebrate Paleontology and Paleoanthropology (hereafter IVPP) of the Academia Sinica. From the institute, Young developed a large corps of Chinese vertebrate paleontologists who collected and studied everything from fossil agnathans to anthropoids. After Young's death, Minchen Chow and his successors continued to head the IVPP and developed it along the lines Young established.

The Cultural Revolution was a major setback to Chinese vertebrate paleontology, as it was to all science in China. But under Chow's leadership the IVPP emerged into the 1980s ready to develop further the Chinese vertebrate fossil record. The 1980s saw an explosion in knowledge of this record and a renewal of collaboration between Western and Chinese vertebrate paleontologists after a nearly half-century hiatus.

The current status of vertebrate paleontology is changing as rapidly and dramatically as is all of China. Now, vertebrate paleontology is worked on at numerous Chinese museums and institutes other than the IVPP. Many younger Chinese vertebrate paleontologists have been trained in the West, and some have remained there. Virtually every Western vertebrate paleontologist with a specific interest in the Chinese vertebrate-fossil record has been to China at least once.

It is impossible for me to capture these many changes and put them in proper perspective as they occur. What I can do is review the study of China's fossil vertebrates to the present. This review is in part narrative, as I relate the sequence of events, discoveries, and personalities that are the basis for a history of vertebrate paleontological studies in China. The history is presented as an analysis as it supports the three-stage model of the introduction of science into any non-European nation proposed by Basalla (1967).

Basalla's Model

Basalla argued that there are three overlapping phases (or stages) during the diffusion of Western science into any non-European nation (see figure 2-1). During the first phase, the *non-scientific* (lacking Western science) society provides a source for European science. Westerners visit the new land, survey and collect objects that they take back to the West. The science of the initial phase parallels geographic exploration and the appraisal of natural resources. No scientific knowledge or training is diffused directly to the non-scientific society during this phase.

Basalla (1967) termed the second phase *colonial science*. During this phase, Western scientists live in the non-scientific society as colonials and establish local scientific institutions. The scientific tradition maintains its Western processes and thought, but Western scientists begin to train locals.

The third phase completes the transplantation of the science. The locals struggle to achieve an independent scientific tradition and culture. Local scientists trained locally replace colonial scientists. These local scientists ultimately establish an independent scientific tradition by developing their own institutions, students, and research to a level comparable to that seen in the West.

Chinese vertebrate paleontology fits Basalla's model well, as the following narrative demonstrates: Phase one took place from about 1870 to 1940. Phase two began in roughly 1916 and continued to 1945. Finally, phase three arrived in 1949 with the founding of a fully autonomous People's Republic of China.

Ancient Chinese Observations on Fossil Vertebrates

Needham (1959: 619–621) recounts ancient Chinese written references to fossil vertebrates, which date back as far as 133 B.C., when *dragon bones* were discovered during the digging of a canal. Daoyuan noted the existence of *stone fishes* in about 500 A.D., and the *Yün Lin Shih Phu* of 1133 A.D. offered some

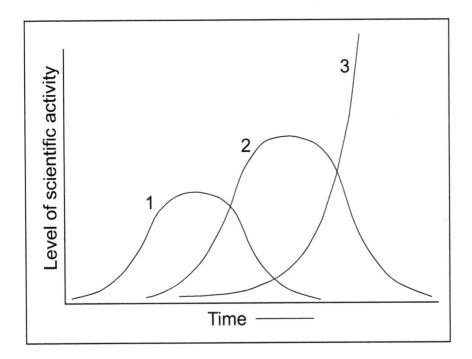

Figure 2-1 *According to Basalla's (1967) model, in the diffusion of Western science into a non-Western society, levels of scientific activity can be divided into three phases.*

remarkably perceptive observations on such fossil fishes (including an analysis of their taphonomy!) quoted at length by Needham (1959: 620). Yet, despite the great antiquity and perspicacity of these Chinese observations on fossil vertebrates—predating by more than a millennium the comments of DaVinci, Hooke, Steno and others who are the starting point of the Western science of paleontology—these observations had no impact on (indeed, were unknown to) Western science.

What proved to be significant to Western paleontology was the ancient Chinese tradition of regarding fossil bones and teeth, mostly of mammals, as dragon's bones, dragon's teeth, or oracle bones. These bones had putative medicinal and mystical powers that ensured their incorporation into the Chinese pharmacopeia by at least 100 B.C. (Needham 1959). Chinese apothecaries and their agents assiduously collected fossils and jealously guarded their provenance. It was largely through the purchase of such dragon bones that Western paleontology first encountered China's vertebrate fossil record during the nineteenth century.

Dragon Bones: Nineteenth-Century Western Paleontology and China's Vertebrate Fossils

During the nineteenth century, British and German diplomats and French Jesuit missionaries spent time stationed in China, mostly in its eastern cities. None of these visitors collected vertebrate fossils, but several purchased dragon bones from local Chinese druggists. Some of these bones, almost all of them of Pleistocene mammals (cave deposits proved a rich source of the Chinese druggists' fossils), found their way back to Europe and were described by Adams (1868), Owen (1870), Gaudry (1872) and Lydekker (1881, 1883, 1891, 1901) (see figure 2-2).

The fieldwork of the middle 1800s that was culminated by von Richthofen's (1877–1912) classic monographs on the geography and geology of China produced the first vertebrate fossils collected in China by a Westerner. Koken (1885) described these specimens. Fossil bones were also collected by the Hungarian Széchenyi in Gansu during 1877–1880 (Széchenyi 1899). However, these explorers made isolated finds, mostly of Neogene-Pleistocene mammals, and their discoveries were only of passing interest.

The richest yields of Chinese fossil vertebrates obtained during the nineteenth century were those purchased from druggists by the German naturalist K. Haberer. Haberer sent these fossils to German vertebrate paleontologist Max

2 cm

Figure 2-2 *This fossil proboscidean tooth was bought from Chinese druggists and described by Richard Owen in 1870 (after Owen 1870).*

Schlosser. Among the many fossils of proboscideans, camels, hipparionids, giraffes, and carnivores Haberer bought, was a single hominid tooth (see figure 2-3). Schlosser's (1903) description of them—most were evidently of Pleistocene age—attracted great interest because Schlosser boldly suggested that the anthropoid tooth suggested an Asian origin of primates. At the American Museum of Natural History in New York, both Henry Fairfield Osborn (1910) and William Diller Matthew (1915b) developed this idea further, proposing that mammals originated in Asia. Indeed, Schlosser's proposal became the original impetus for the extensive collecting of vertebrate fossils in China and Mongolia by the American Museum of Natural History during the 1920s and 1930s.

Johan Gunnar Andersson and the Lagrelius Collection

Johan Gunnar Andersson (1874–1960) (see figure 2-4) was a Swedish geologist hired by the Chinese government as a mining advisor in 1914. Part of his service in this capacity was to travel through the Chinese provinces and inspect mining operations, especially coal mines. During these travels, Andersson encountered numerous fossil localities, primarily in central and eastern China. In 1916, while examining copper deposits in southern Shanxi, Andersson collected mammal fossils on the banks of the Huang He at Yuanjuxian (Andersson 1923). Later that year he discovered large quantities of fossil mammals in the Chang Jiang (Yangtze) Valley. This provoked his curiosity because the great German explorer Richthofen, in his classic monographs on the geography and geology of China (Richthofen 1877–1912), identified the fossil-mammal-bearing rocks as geologically young (Holocene) loess. The fossil mammals Andersson discovered, and his stratigraphic observations, suggested instead that they are actually a thick, complex and fossiliferous sequence of Neogene strata:

> I got interested in this matter in 1916 when I, in crossing the Yellow River between Honan and Shansi, found a highly instructive section exhibiting underneath the loess series of fossiliferous beds, probably young pliocene [sic], we did not know before. The investigations thus made it imperative to me [sic] to try to settle the ages and the climatic conditions of the formation of the loess. This could evidently be done only by studying the fossils contained in the loess and so I set out to collect such. These researches soon brought me to a full understanding of the fact that much of what was earlier been called loess is in fact red clays containing pliocene [sic] *Hipparion* fauna described by Schlosser (J.G. Andersson to R. Chapman Andrews, 19 January 1919).

This moved Andersson to contact an old friend, vertebrate paleontologist Carl Wiman (1867–1944) (see figure 2-4) at the University of Uppsala in Sweden.

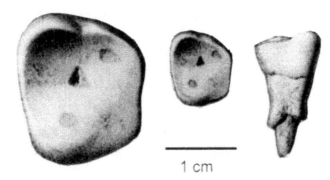

1 cm

Figure 2-3 *This upper molar was purchased by K. Haberer from a Chinese druggist and identified by Schlosser (1903) as ? Homo sp. It is shown in occlusal (left, middle) and posterior (right) views. The occlusal view on the left is twice as large as the occlusal view in the middle. The scale is for the two views on the right.*

Andersson simply wanted to know whether the fossil mammals could be further collected and studied, principally for their biochronological utility, by a trained vertebrate paleontologist. To do this, not only was a vertebrate paleontologist needed in China, but funding was also required to support the work.

Figure 2-4 *The Swedish geologist Johan Gunnar Andersson (1874-1960), on the left, began collecting vertebrate fossils in China early in the nineteenth century. His colleague, Swedish vertebrate paleontologist Carl Wiman (1867-1944), on the right, oversaw the scientific study and publication of these fossils.*

Wiman responded positively to Andersson's request and contacted their mutual friend, Axel Lagrelius (1863–1944), a wealthy printer in Stockholm. Lagrelius provided 45,000 Swedish crowns (about 12,000 U.S. dollars by 1916 exchange rates) to support Andersson's paleontological work in China. However, Wiman could (or would) not go to China, and no other vertebrate paleontologist was available to work with Andersson. So, the arrangement was for Andersson to continue collecting vertebrate fossils as his time and circumstances allowed, and to ship them to Sweden for study.

This worked reasonably well until 1920, especially because Andersson used some of the funds provided by Lagrelius to hire Chinese assistants to collect vertebrate and other kinds of fossils. In 1919, Lagrelius organized a private foundation, the Swedish China Research Committee, better known as the *Kinafond*, to support Andersson's paleontological work. Between 1916 and 1920, Andersson spent about 70,000 Swedish crowns on fieldwork and related costs, mostly in Nei Monggol, Shanxi, and the environs of Beijing. Much of the funding supported collecting fossil plants and shipping them to Sweden.

It is difficult to be exactly certain why in 1920 Andersson induced Wiman to send a vertebrate paleontologist to China. Some of the impetus must have been the sheer wealth of vertebrate fossils (especially of mammals) Andersson and his assistants were uncovering in China, and the technical difficulties they had to solve when collecting them. Another inducement may have been Andersson's first encounter with Roy Chapman Andrews in Beijing in January 1919.

Andrews had already been to China, once in 1912 and again in 1916, collecting recent mammals for New York's American Museum of Natural History. Andrews and Andersson seem to have gotten along well, with Andrews suggesting to Andersson that he could receive ample funding if the American Museum became the recipient of the vertebrate fossils collected by Andersson and his assistants. Andersson refused, regarding Andrews offer as a threat to the Swedish enterprise. Soon after the meeting, Andersson suggested to the *Kinafond* committee that American competition was in the offing; the committee immediately sent additional funds.

Otto Zdansky

Wiman could not go to China, and his only Swedish student—Torsten Ringström, who had undertaken a dissertation on Chinese fossil rhinoceroses collected by Andersson—was dying of cancer. A new student of Wiman, the Swede Birger Bohlin, was untrained and too young. So, Wiman turned to Otto Zdansky, an Austrian who had undertaken postdoctoral work in Uppsala.

Zdansky (1894–1988) (see figure 2-5) was in Uppsala as an exchange student in 1920 after completing a doctoral dissertation in Vienna under the direction of the great Austrian paleontologist Othenio Abel. Wiman offered to send him to China for three years, all expenses paid but without salary. He also promised Zdansky the exclusive right to publish on the fossil vertebrates he would collect in China.

Zdansky left for China on May 1921, sailing from Sweden to Tianjin. He spent the next two to three years (until December 1923) collecting a wide variety of localities in eastern China, mostly of Neogene age (see figure 2-6; Mateer and Lucas 1985). Zdansky's important discoveries include the extensive and now classic Baode *Hipparion* faunas, the Late Jurassic dinosaurs and middle Eocene mammals from the Mengyin district of Shandong, and the first hominid fossil and extensive mammalian fauna from the cave deposit of Zhoukoudian. The hominid discovery is particularly interesting because it has largely been overlooked in histories of the Zhoukoudian discoveries and because it provides a startling insight into Otto Zdansky's scientific personality.

Figure 2-5 *Otto Zdansky (1894–1988) worked for Wiman and Andersson collecting vertebrate fossils in eastern China during the 1920s.*

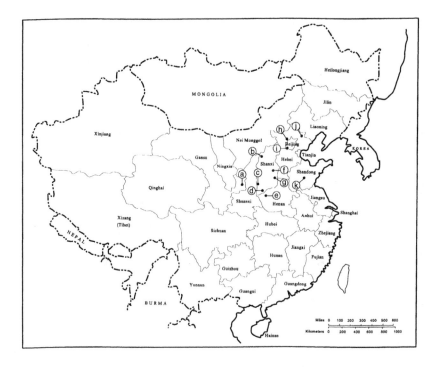

Figure 2-6 *These are the principal areas where Otto Zdansky collected vertebrate fossils in China (1921–1923): a) Qingyang Xian, Gansu; b) Baode Xian, Shanxi; c) Xiangning Xian, Shanxi; d) Yuanju Xian, Shanxi; e) Xinan Xian and Mianzhi Xian, Henan; f) Wuxiang Xian, Shanxi; g) Zhangzhi Xian, Shanxi; h) Huailai Xian and Xuanhua Xian, Hebei; i) Zhoukoudian, Beijing Shi; j) Zhengde Xian, Hebei; k) Xintai Xian, Shandong.*

The first excavation undertaken by Zdansky in China was at Zhoukoudian. Andersson first became aware of this deposit near Beijing in 1918 through J. McGregor Gibb, a chemistry professor at Beijing University. The deposit McGregor told Andersson about was called "chicken-bone hill" (Jigushan), a pillar-like remnant of red clay in a limestone quarry on a hill just west of the village of Zhoukoudian. Andersson, Zdansky, and Walter Granger (chief paleontologist of the Central Asiatic Expeditions of the American Museum of Natural History, see figure 2-7) visited Jigushan a few weeks after Zdansky arrived in China and found Pleistocene bird bones pouring from the pillar. Fortunately, a local quarryman informed them of a nearby place where much larger bones could be found. This proved to be an enormous fissure-filling in limestone that opened on the top of a vertical limestone wall about 10 m high, later known as the "main cave," "locality 1" or simply the "Peking Man cave" at Zhoukoudian (see figure 2-7).

Figure 2-7 *The initial excavation at Zhoukoudian in 1921 entered a ledge on top of the main, collapsed cave (photograph courtesy of the late Birger Bohlin). Compare to figure 12-13.*

Zdansky worked this fissure-filling for several weeks, uncovering the upper 1.3 meters of the deposit. Among the numerous remains of fossil mammals Zdansky collected, he found an isolated upper molar tooth, *unmistakably hominid*, in either a layer of slightly banded brown clay (33 cm thick) or the underlying layer of banded dark-brown clay (24 cm thick). However, instead of immediately announcing what, in retrospect, was one of the most important fossil hominid discoveries ever made, Zdansky simply put the tooth into his pocket, took it back to his dwelling and packed it among pig teeth from the excavation. Thus, having virtually assured its obscurity, he shipped it back to Uppsala. As he explained later (Zdansky, oral communication, 1982), there were two reasons why he took this course of action. The first was that Zdansky was not overly impressed with the isolated hominid tooth (see later comments). The second was that he feared that if he told Andersson about it, Andersson would usurp his right to excavate and study the material from Zhoukoudian.

Andersson, who did not know about the hominid molar from Zhoukoudian, regarded the site as one of many potential sources of Pleistocene mammals in China (Andersson 1922). Andersson's main interest, and the major reason he had brought Zdansky to China, was the excavation of the so-called *Hipparion* deposits farther south, in Henan. Consequently, at the beginning of the fall of 1921, after only a little over a month of excavation at Zhoukoudian, Andersson told Zdansky to terminate his work and proceed to Mianzhi-xian in Henan. Thus began nearly two years of travel for Zdansky (until the summer of 1923) that took him to Henan, Shanxi, Shandong, and Gansu to collect Jurassic, Eocene, and Pliocene vertebrates (Mateer and Lucas 1985).

By the summer of 1923, Zdansky's money from the *Kinafond* was virtually exhausted. He was thus forced to return to Beijing, and he reopened the excavation at Zhoukoudian. At the same time, Andersson sought to secure additional funds for Zdansky, although the latter's enthusiasm for China had waned. Zdansky had spent two years in an alien culture living under primitive conditions and working arduously. His desire to return home to Europe was understandable. Thus, in December of 1923, Zdansky returned to Austria by rail via Siberia, never again to return to China.

After a brief visit with his family in Vienna (his father had just died), Zdansky returned to Uppsala in January 1924. Unable to find a regular professorial position, he worked in the evenings in the Paleontological Museum of Uppsala University as a preparator of fossil vertebrates. During the days, Zdansky undertook the study of the vast collections of fossil mammals he had acquired in China. Wiman studied the fossil reptiles collected by Zdansky in China. At the same time, Andersson continued his paleontological and archaeological studies in various parts of China, but he never once pursued the excavation at Zhoukoudian.

In May of 1926, Gustav V, the Crown Prince of Sweden, departed on a trip around the world via North America and the Far East, which included a stop at Beijing. Here, Gustav intended to make contact with Andersson and observe firsthand some of the scientific results and goodwill the *Kinafond* had promulgated. The recently formed Geological Society of China planned a reception for the Crown Prince, and Andersson felt it would be appropriate if some of the most notable paleontological discoveries made by Zdansky were first announced at this reception. Therefore, Andersson wrote Wiman in Uppsala, who approached Zdansky, then in the throes of completing his large, and now classic, monograph on the Pleistocene mammal fauna from Zhoukoudian (Zdansky 1928). No doubt to Wiman's surprise, Zdansky wrote a short paper on two hominid teeth from Zhoukoudian, the molar he discovered in 1921 and a premolar he uncovered while unpacking the collection in Uppsala. He identified these teeth simply as "?*Homo* sp.," exercising admirable restraint in an area of taxonomy frequently confused by egotism and sensationalism. Zdansky ended his paper with the following conclusions:

> Granted the human origin of the teeth, there arises the question of their relation to the living and prehistoric races of man. As the reader will infer, I am very sceptical towards a great deal of prehistoric-anthropological literature,

and I am indeed convinced that the existing material provides a wholly inadequate foundation for many of the various theories based on it. As every fresh discovery of what may be human remains is of such great interest not only to the scientist but also to the layman, it follows only too naturally that it becomes at once the object of the most detailed—and, in my opinion, too detailed—investigation. I decline absolutely to venture any far-reaching conclusions regarding the extremely meagre material described here, and which, I think cannot be more closely identified than as ?*Homo* sp. (Zdansky 1927: 284)

Besides Andersson, two other scientists with paleontological interests were present at the reception for the Crown Prince: Davidson Black (1884–1934) and Pierre Teilhard de Chardin (1881–1955). Clearly, Andersson must have shared in the excitement created by the announcement of Zdansky's discovery on 22 October 1926, more than five years after Zdansky first found the hominid tooth at Zhoukoudian. However, he must also have felt chagrined that Zdansky had not informed him of the hominid teeth when they were first discovered. Had Zdansky done this in 1921 (or in 1923), while in China with money from the *Kinafond* still available, Andersson could have focused all, or at least most, efforts on the further excavation of Zhoukoudian. It was apparent that Andersson lacked the resources to do this on his own, and this coupled with the great enthusiasm Davidson Black expressed in the reopening of the Zhoukoudian site, made some sort of cooperative, and therefore not a wholly Swedish-controlled, excavation of Zhoukoudian unavoidable.

Black, Bohlin, and Zhoukoudian

Davidson Black (1884–1934) was a Canadian anatomist who had taken the position of Professor of Neurology and Embryology at the Peking Union Medical College in 1917. It was fortunate for Andersson that he and Black were good friends, for this made it easy for the two to strike up a cooperative agreement for the new excavation at Zhoukoudian. Black immediately rushed to print with a short article in *Nature* (Black 1926) extolling the importance of Zdansky's discovery. He then approached the Rockefeller Foundation in New York (which had previously supported renovation of the Peking Medical College where Black taught) for funding to support the new excavation at Zhoukoudian.

Andersson quite logically suggested that Zdansky should direct the new excavation. However, Zdansky had just accepted a position as lecturer in paleontology at the Egyptian University in Cairo and had no desire to return to China. Andersson then insisted that a Swede supervise the excavation, having in mind Birger Bohlin (see figure 2-8), a student of Wiman's in Uppsala who had just completed a Ph.D. on Chinese fossil giraffes (Bohlin 1926). Black

Figure 2-8 *Birger Bohlin (1898–1992) came to Zhoukoudian in 1927 to supervise the excavations (photograph courtesy of the late Birger Bohlin).*

accepted Andersson's choice. In 1927, Bohlin arrived in China and reopened the excavation at Zhoukoudian.

For two years, 1927–1928, Bohlin supervised a much more extensive excavation at Zhoukoudian than Zdansky had been able to undertake. However, by the end of 1928, the first installment of Rockefeller Foundation money was exhausted. At the same time, Bohlin was offered the position of geologist/paleontologist on a scientific expedition to Central Asia organized by Sven Hedin (1865–1952), the celebrated Swedish explorer and geographer. Thus, in early 1929, Bohlin left Zhoukoudian and, with his departure, Swedish involvement in the excavation of this famous locality ended.

The rest of the story of Zhoukoudian can be told here in summary, although it violates the chronological order of this narrative (see Jia and Huang 1990, for a detailed history). The 1928 discoveries at Zhoukoudian included many more teeth and two jaws of the hominid that Black (1926) had christened *Sinanthropus pekinensis*. In order to find continued funding and support for the excavations (6000 m^3 of rock had already been removed), the Cenozoic Research Laboratory was organized as a branch of the Geological Survey of China. Davidson Black headed the Research Laboratory, which was headquartered in the Peking Union Medical College.

In December 1929, Pei Wenchong, a young Chinese scientist, discovered a skullcap of "*Sinanthropus*" in the renewed excavation effort. This energized Black's efforts at the site, which continued until 1934, when he died suddenly of a heart attack at the age of 49.

The French Jesuit priest and paleontologist Pierre Teilhard de Chardin took over the Zhoukoudian excavations after Black's death. A year later, Franz Weidenreich replaced Black as Professor of Anatomy and supervised the excavations until from 1935 to 1937, when the military halted them. At that time, a wide range of hominid fossils, including 14 skulls, had been collected at Zhoukoudian (see figure 2-9).

In December of 1941, the Peking Man fossils left Beijing on the SS *President Harrison* under the care of the U.S. marines. Attacked by a Japanese warship (war between Japan and the U.S. has just been declared), the *President Harrison* ran aground, and the Japanese captured the marines. The Peking Man fossils have never been seen again, leading to wild speculation about their whereabouts (e.g., Janus and Brashler 1975), though it seems most likely that they rest at the bottom of the Pacific.

End of the Sino-Swedish Paleontological Program

Andersson left China in 1926, after almost 12 years of work there. He returned to Sweden to become, temporarily, Professor or Geology at the University of Stockholm. He then assumed the Director of the Museum of Far Eastern Antiquities position in Stockholm, a museum primarily founded upon Andersson's enormous collections from China. Although Andersson returned to China later, he made no further collecting efforts there.

Under Andersson's direction, the Sino-Swedish venture had been funded by the Chinese Government and largely by private funds from Sweden. Despite his energy and success in securing funds, Andersson always operated on a rela-

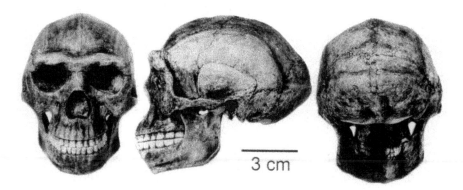

Figure 2-9 *This reconstructed skull of Sinanthropus pekinensis (the Peking Man), now referred to as Homo erectus, is from Weidenreich's (1943) monograph.*

tively small budget. Clearly, it was his great efficiency that enabled the collection of such vast numbers of specimens.

Sven Hedin (1865-1952), a renowned Swedish explorer and geographer, was independently wealthy and also had access to extensive resources similar to those commanded by Roy Chapman Andrews (see figure 2-10). In 1927, Hedin was commissioned by the Chinese Railroad Authority to organize an expedition to investigate Inner Mongolia and Xinjiang in order to improve communications with these far-flung regions. Hedin wanted a palaeontologist as part of his team exploring the interior of China, and Bohlin became the natural choice because of his presence in China.

At the end of 1928, when the first installment of Rockefeller Foundation money for the excavation of Zhoukoudian was gone, it was a reasonable decision for Bohlin to accept the offer from Hedin. Ironically, the great discoveries of hominid fossils at Zhoukoudian initiated by Pei Wenchong's discovery of a skullcap in 1929 occurred after Bohlin's departure for Central Asia.

Bohlin carried out his responsibilities as paleontologist for Hedin's expedition with remarkable success, making large collections of fossil vertebrates and plants in Nei Monggol, Gansu, and Tsaidam during the years 1928–1933. Most of the specimens he collected were shipped to Uppsala, but they were later returned to China during the late 1940s.

Political upheaval in China during the late 1930s, especially in the western provinces, and the pending Japanese invasion of Manchuria, were not conducive to foreign exploration, and Hedin had to terminate his expedition. The dedication of Hedin and his colleagues, several of whom disappeared into the nether regions of desolate Central Asia and were presumed lost, only to reappear years later with a wealth of scientific data, are the stuff of legends. The immense contribution made by the 54 volumes of the *Reports of the Scientific Expeditions to the Northwestern Provinces of China under the Leadership of Dr. Sven Hedin* amply testifies to the scope and success of Hedin's expeditions.

The disturbances of the late 1930s were, in part, a renewal of the problems that had started the Boxer Rebellion in 1899. The establishment of a weak, but fiercely nationalistic government in Nanjing clearly spelled trouble for all foreign enterprises in China and caused a mass exodus of foreigners from the country. Thus ended for some time any foreign scientific research and exploration in China. The tragic loss of the Zhoukoudian hominid fossils during the Japanese invasion well emphasizes the chaos then prevalent in China (Shapiro 1971). Indeed, the Sino-Swedish cooperative exploration of China was over by the late 1930s, ending the longest foreign involvement in palaeontological and geological collecting and research in China.

The Central Asiatic Expeditions

Roy Chapman Andrews (1889–1963) (see figure 2-10) first went to China in 1912 on behalf of the American Museum of Natural History to collect recent mammals for its Department of Mammalogy, in which Andrews was employed. Andrews was very successful in this regard, and in 1915 he proposed a series of expeditions to China over ten years to explore its zoology. In 1916–1917, the First Asiatic Expedition went to Yunnan and Tibet to recover large numbers of animals for the American Museum's collection.

After a stint in the U.S. military during World War I, Andrews returned to Asia for the Second Asiatic Expedition, which went to Mongolia in 1919. However, earlier that year, in January 1919, Andrews met J. Gunnar Andersson in Beijing and became aware of the latter's success in collecting fossil mammals in eastern and northern China. At the meeting, Andrews offered to fund Andersson's collecting efforts if the fossils were turned over to the American

Figure 2-10 *Roy Chapman Andrews (standing third from left), Walter Granter (to Andrews' right), and other members of the Central Asiatic Expedition of 1923 are seen here in front of one of their fleet of Dodge trucks (photograph courtesy of the American Museum of Natural History).*

Museum of Natural History. Andersson refused, and parlayed Andrews' offer into increased funding from the *Kinafond* by suggesting to members of the committee that American competition for Chinese fossils was in the offing.

Early in 1920, Andrews approached Henry Fairfield Osborn (1857–1935), world-renowned vertebrate paleontologist and President (Director) of the American Museum of Natural History, with plans for a new expedition, this one to northern China and Mongolia, to collect fossil mammals. Andrews was well aware that Osborn (following Schlosser 1903) and his brilliant and accomplished curator of fossil mammals, William Diller Matthew (1871–1930), believed the origin of mammals and of humankind lay in Asia (Osborn 1910; Matthew 1915b). Andrews proposed to uncover the fossil evidence to back this belief, and, not surprisingly, Osborn heartily endorsed the expeditions proposed by the younger man. Ironically, Andrews (1926: 4–7) later claimed the idea of collecting mammal fossils came to him through his interest in the natural history of the extant mammals he had been collecting in China, a claim later repeated by Preston (1986: 117), among others. It seems more likely that Andrews got the idea from his knowledge of Andersson's discoveries. To his credit, though, Andrews conceived of a far greater and more diverse effort than the Swede was attempting. He wanted to assemble a team of natural scientists—geologists, paleontologists, cartographers, mammalogists, botanists, entomologists, etc.—to explore vast areas with the aid of gasoline-powered vehicles (see figure 2-10).

To that end, Andrews raised more than half a million U.S. dollars from some 600 individuals and institutions, and secured a fleet of Dodge automobiles. Five Central Asiatic Expeditions followed (in 1922, 1923, 1925, 1928 and 1930), headquartered in Beijing and mostly focused on what is now the Mongolian Republic, though some collecting efforts also took place in China, in Nei Monggol, Gansu, Sichuan, and Yunnan. Andrews (1926, 1929a, b, 1932) provided detailed narrative accounts of the expeditions, and their story has been well summarized by McKenna (1962) and Preston (1986). Here, I briefly review the Chinese facets of the Central Asiatic Expeditions.

Only a minority of the vertebrate fossils collected came from China. Most of these came from Nei Monggol and were fossils of Late Cretaceous dinosaurs and Eocene mammals, especially from the now classic deposits at Iren Dabasu and Irdin Manha. Miocene mammals were also discovered at Tunggur in Nei Monggol, and this remains one of the most important Neogene fossil mammal localities in China (see Chapter 11). Neogene and Pleistocene mammals were collected from fissure fills in Sichuan and in Yunnan as well.

For logistical reasons, Andrews, his wife, and most of the expedition's scientists and their families, lived in Beijing year-round throughout most of the expeditions' duration. They were quartered in the house of a recently deceased

Britisher, Dr. G. E. Morrison. Walter Granger (1872–1941), American Museum curator and fossil-mammal collector extraordinaire, was the expeditions' chief paleontologist (see figure 2-11). Two Columbia University geologists—Charles Berkey and William Morris—undertook studies of the geological context of the vertebrate fossils discovered. They were assisted by trained collectors and technicians of the American Museum staff (see figure 2-10) as well as a diverse array of hired Chinese and Mongol drivers, teamsters, packers, cooks, and other field personnel.

The First Central Asiatic Expedition of 1922 worked only in Mongolia. The Second Expedition of 1923 worked also in Nei Monggol at localities first found in 1922 on the return route from Mongolia to Beijing. The Third Expedition of 1925 followed suit, but in 1926–1927 war in northern China limited fieldwork to the south, in Sichuan and Yunnan (see figure 2-11). The Fourth Central Asiatic Expedition of 1928 worked only in Nei Monggol. The last expedition (the fifth), in 1930, worked in eastern Nei Monggol accompanied by the Chinese vertebrate paleontologists C. C. Young and Pei Wenchong and by Pierre Teilhard de Chardin.

Andrews overcame tremendous logistical difficulties to pull off the five expeditions amidst the political and economic chaos of China during the 1920s and 1930s. He constantly had to negotiate with and bribe a shifting array of

Figure 2-11 *Walter Granger (seated left) is seen here with field associates at Yenchingkou, Sichuan, 1922 (photograph courtesy of the American Museum of Natural History).*

local warlords and power brokers to outfit and insure safe passage for his expe-ditions. His motives, in the eyes of some Chinese, were also not above suspi-cion. A particularly large setback occurred in 1923 after the fabled discovery of dinosaur eggs at the Flaming Cliffs of Shabarakh Usu in Mongolia. That win-ter, in the United States, as a fund-raising stunt, Andrews auctioned off an egg for $5,000. When the Chinese and Mongolians found out about it, they con-cluded that the Central Asiatic Expeditions were plundering their countries of priceless treasures.

Indeed, Andrews had caused suspicion from the very start, having stated in an article in *Asia* magazine (December 1920) that "China has no national insti-tution where natural history objects can be studied and exhibited by modern methods and where the scientific work of her own people can be encouraged and directed." The Chinese (and Andersson) were not amused by this, or by Andrews' planned Central Asiatic Expeditions, about which they had not been consulted. Ultimately, the American Museum's expeditions ended because of political problems—those of strife-ridden China during the 1930s—com-pounded by Andrews' ultimate inability to engender trust and cooperation within the Chinese scientific establishment.

The effect of the Central Asiatic Expeditions on the vertebrate paleontology of China is inestimable. Not only did the scientific discoveries—volumes of technical literature have been written about them—provide the first look at many important facets of China's vertebrate fossil record (especially its Late Cretaceous dinosaurs and Eocene mammals, just to mention the two most prominent). But, the legendary expeditions revealed the rich vertebrate fossil record of China and Mongolia being unearthed in an atmosphere of high adventure. The allure of Asian vertebrate fossils subsequently inspired many youngsters to vertebrate paleontology as a career, and later to China and Mon-golia as a collecting field.

The 1920s–1930s: Foreign Vertebrate Paleontologists Living in China

A larger number of foreign vertebrate paleontologists, or foreigners whose research overlapped with vertebrate paleontology, lived in China during the 1920s and 1930s than at any other time. Other than the vertebrate paleontolo-gists just discussed in conjunction with the Sino-Swedish and Central Asiatic Expeditions, there were Americans, a Canadian, and two French researchers.

Foremost among the Americans was Amadeus Grabau (1870–1946), an invertebrate paleontologist and stratigrapher who came to China in 1920 as Professor of Geology at Peking University. Grabau spent the remainder of his life in China, studying its stratigraphy and fossil invertebrates, training Chinese geologists, and interacting extensively with Andersson, Andrews, Black and the other foreign vertebrate paleontologists working in China (Johnson 1985).

In 1915, Emile Licent, a French Jesuit priest and entomologist, went to Tianjin, where he established a natural history museum at the Jesuit college. Licent collected fossil mammals, primarily Neogene horses from Gansu, and in 1923 was joined by a trained paleontologist and fellow French Jesuit, Pierre Teilhard de Chardin (1881–1955).

Teilhard de Chardin came to China to collect and study fossils, especially of late Cenozoic mammals and of human prehistory. He lived for long periods of time in China, from 1923 to 1946, mostly in Beijing, and worked mostly in Nei Monggol, Gansu, and at Zhoukoudian and other localities in the Beijing environs. Other than the work at Zhoukoudian, the research on Nihewan Pleistocene mammals he published with French paleontologist Jean Piveteau was probably the most important vertebrate paleontological research he undertook in China (Teilhard de Chardin and Piveteau 1930). Teilhard de Chardin interacted extensively with the young, aspiring Chinese geologists and vertebrate paleontologists. He much influenced Pei Wenchong (1904–1982), who earned a doctorate in France during the 1930s and went on to become one of China's premier paleoanthropologists.

The Canadian anatomist Davidson Black has already been discussed. His successor was the German anatomist Franz Weidenreich. Weidenreich lived in Beijing from 1935 to 1941, supervising the Zhoukoudian excavations from 1935 to 1937. His classic monograph (Weidenreich 1943) on the skull of "*Sinanthropus*" still provides essentially all of the direct scientific observations gleaned from the collections that disappeared in 1941.

C. C. Young and Minchen Chow

C. C. Young (Yang Zhungjian) (1897–1979) (see figure 2-12) can rightfully be called the founder of Chinese vertebrate paleontology. Young undertook his doctorate in Germany during the 1920s under the direction of Max Schlosser. The monograph he published of his dissertation (Young 1927), on the fossil rodents of northern China, was the first scientific article on vertebrate paleontology published by a Chinese scientist.

Young was born June 1, 1897, in Shaanxi. From 1917–1923 he attended Beijing University, graduating with a masters degree in geology. He then proceeded to Munich University to study vertebrate paleontology, received his

Figure 2-12 C. C. Young (1897-1979) was the father of Chinese vertebrate paleontology.

doctorate in 1927, and returned to China in 1928 to take a position with the Geological Survey.

Young published nearly 500 scientific articles during his long career. After initial work on Neogene and Quaternary fossil mammals and stratigraphy, Young later specialized in fossil reptiles, and became an internationally recognized authority. Most of Young's work was descriptive; he named more than 200 new species of fossil vertebrates.

The Japanese invaded and annexed Manchuria in 1937. World War II began then in China, ending in 1945 and followed by an internal struggle that led, in 1949, to the establishment of the People's Republic of China. Of the foreign vertebrate paleontologists living in China, all had left by 1941 except Teilhard de Chardin. Vertebrate paleontological research ground to a virtual halt in China, to re-emerge after the war under the aegis of Young. Young thus bridged two gaps. He not only brought vertebrate paleontology in China from the pre-war to post-war eras, but he also was the first Chinese scientist to bring the Western science of vertebrate paleontology to China.

Furthermore, Young trained and mentored a large number of Chinese vertebrate paleontologists. Foremost among them was Minchen Chow (Zhou Mingzhun) (1918–1996), who was to succeed him as dean of Chinese vertebrate paleontologists (Miao 1996).

Born in Shanghai, Chow received an M.A. from Miami University, Ohio, and a Ph.D. in geology at Lehigh University after the Second World War. In 1950, Chow worked in the Bighorn basin of Wyoming with Princeton vertebrate paleontologist Glenn L. Jepsen. Thus began his lifelong interest in mammalian paleontology.

In 1952, Chow returned to China and began working with Young at the institute in Beijing that was later to become the Institute of Vertebrate Paleontology and Paleoanthropology. He spent the rest of his career associated in one way or another with the IVPP, culminating in his appointment as director.

Chow published more than 100 scientific articles and five monographs, mostly on fossil mammals. Perhaps his greatest single scientific contribution was the discovery of Asia's oldest Cenozoic mammals, those from the early Paleocene of Guangdong. Even more important was his mentorship of an entire generation of Chinese vertebrate paleontologists who came to the forefront after the Chinese Cultural Revolution.

The IVPP

The Cenozoic Research Laboratory of the Geological Survey of China, founded in 1929, continued to operate in that capacity until 1953 when it became an independent research unit of the Chinese Academy of Sciences (Academia Sinica). At that time its name was changed to the "Vertebrate Paleontology Laboratory." In 1954, Zhoukoudian was established as its official field station, and a museum was built there. From 1951 to 1954, the Cenozoic Research/Vertebrate Paleontology Laboratory was headquartered in Nanjing as part of the Paleontological Institute of the Academia Sinica. The laboratory became an independent unit in 1954, but it remained in Nanjing until January 1960, when it moved to its own facility in a northern suburb of Beijing. In September of 1957, it was renamed the Institute of Vertebrate Paleontology, and in January 1960, it received its current name, the Institute of Vertebrate Paleontology and Paleoanthropology—IVPP for short.

In the 1950s, the Cenozoic Research Laboratory, under various names, had a staff of about 20. By the mid-1960s, the IVPP staff numbered 150 scientists and technicians. Three field stations—Zhoukoudian, Lantien, and Tai Yuan (Shansi)—were attached to the IVPP. In 1974, the IVPP moved to its present location in Beijing near the Zoological Park. The former facility remained a museum on the outskirts of Beijing, while research and collections were at the main institute. In the fall of 1994, a newly constructed facility opened to replace the old IVPP building. This new building houses a staff of about 250 scientists and technicians. A centralized collection facility (previously, each scientist took care of his/her collections) that houses about 200,000 catalogued vertebrate fossils and a 1500 m^2 exhibition area are part of the new IVPP building. It remains the major research facility for vertebrate paleontology in China.

New Collaboration

The Chinese Cultural Revolution lasted for a decade (1966–1976) and cut off Chinese science, including vertebrate paleontology, from the rest of the world. In the late 1970s, this isolation ended, and Chinese vertebrate paleontology experienced a true renaissance, which continues today.

The IVPP once again became a thriving research center, headed by Minchen Chow. Chinese vertebrate paleontologists worked throughout the country and traveled abroad to study, conduct research, and participate in scientific meetings. Western and Chinese vertebrate paleontologists collaborated extensively. Joint field expeditions in China teamed local vertebrate paleontologists with American, British, Canadian, French and German colleagues. The most celebrated and extensive of these expeditions was the Sino-Canadian Dinosaur Project of 1987–1990. Canadian and Chinese vertebrate paleontologists joined forces to collect Jurassic and Cretaceous dinosaurs (and other vertebrates) across northern China (Dong 1993). Some of the important results of this research were published in special issues of *Canadian Journal of Earth Sciences* in 1993 (volume 30, numbers 10–11) and 1996 (volume 33, number 4). The Sino-Canadian Dinosaur Project (Grady 1993) stands as the most prominent of many recent collaborative research efforts between Chinese and western vertebrate paleontologists.

The Three-Stage Model

The above history shows that Basalla's three-stage model for the diffusion of a Western science into a non-European nation can be applied to the history of vertebrate paleontology in China (see figure 2-1). From 1870 until the 1920s, only foreign vertebrate paleontologists collected and studied Chinese vertebrate fossils, sending their collections to the West. The 1920s and 1930s saw foreign vertebrate paleontologists living in China and beginning to train Chinese vertebrate paleontologists. This corresponds to Basalla's second stage. Between the 1930s and early 1950s C. C. Young established an independent Chinese vertebrate paleontology that thrives today, fitting well with the third and final phase of Basalla's model.

Chapter 3

Cambrian-Silurian

During the early Paleozoic, China consisted of at least six separate microplates, which, by Silurian time, were north of the eastern part of Gondwana and mostly north of the paleoequator (see figure 3-1). These microplates are the Tarim, north China, south China, Indochina, Sibumasu (or Shan-Thai), and Tibetan blocks. Some of these microplates consisted of one or more smaller microplates (or terranes), although there is some debate about exactly how many microplates were present. For example, some paleontologists have divided the south China block into separate Hunan and Yangtze terranes, though the general similarity of vertebrate faunas in these terranes supports their inclusion as a single block (G. Young 1990, S. Wang 1993, Z. Li et al. 1993).

Late Cambrian-Early Ordovician phosphatic fragments from North America, Svalbard, and Greenland were long considered to be the oldest fossil record of vertebrates (Repetski 1978). However, recent discovery of Early Cambrian agnathans in southern China pushes that record back in time (Shu et al. 1999).

The Ordovician record of vertebrates is of ostracoderms from a wide geographic range of localities (e.g., Blieck et al. 1991). These jawless fishes evolved into a diverse array of Silurian and Devonian descendants. It is among these descendants that we find many of China's early vertebrates, those of Early Silurian age.

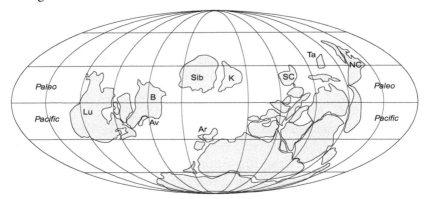

Figure 3-1 *This Silurian paleogeography shows the distribution of the Chinese microplates (after Z. Li et al. 1993). Abbreviations are Ar = Arabia, Av = Avalonia, B = Baltica, K = Kazakstan, Lu = Laurentia, NC = north China, SC = south China, Sib = Siberia, Ta = Tarim.*

Cambrian and Ordovician rocks are widely distributed in China. Cambrian strata are nearly ubiquitous, and it has been claimed that the most complete sequences of Ordovician strata in the world are in China (W. Chang 1962). The recently discovered aganthans from the Lower Cambrian of Yunnan (Shu et al. 1999) not only are the oldest known vertebrate fossils, but they are a harbinger of further discoveries of vertebrate fossils in Chinese Cambrian strata. Also, a possible indication of the presence of vertebrates in the Chinese Ordovician is the problematic *Fenhsiangia* from the Lower Ordovician of Hubei. This animal has a tubular, phosphatic exoskeleton and may be related to early vertebrates (Long and Burrett 1989).

Silurian rocks also have a wide distribution in China (see figure 3-2). However, they are most notably absent across parts of western and northern China, reflecting a major episode of regional uplift and erosion during Silurian time. When examining the distribution of Silurian strata and fossils in China, five or six sedimentological-paleobiogeographic areas ("domains") have been recognized (e.g., Yu et al. 1984; Mu et al. 1986). Known Silurian vertebrate fossils are almost exclusively from two domains, which are in southern China: the Yangtze region of the southern stable domain and the central Hunan-Qingfeng region of the southern mobile domain (see figure 3-2). These domains were part of the south China block during the Silurian.

Silurian Vertebrate-Producing Strata

The oldest Chinese strata that yield vertebrates are of marine origin and are among a range of Llandoverian ages (see figure 3-3). Among the oldest is the marine Rongxi Formation, which is widely exposed in northwestern Hunan, southwestern Hubei, southeastern Sichuan, and northeastern Guizhou. The Rongxi Formation yields agnathans and acanthodians and has a thickness of about 258 m, and consists mainly of variegated purple-red, yellow-green and gray-green silty mudstone, sandy shale, and siltstone. Red beds characterize both the lower and uppermost parts of the formation. Some beds are of marine origin and yield fossils of trilobites and crinoids. The better dated underlying and overlying units indicate a late Llandovery age (Mu et al. 1986).

The lower Wengxiang Formation of southeastern Guizhou is about 100 m thick. It consists of gray-green and yellowish green shale, sandy shale, and sandstone interbedded with arenaceous and bioclastic limestones. The base of the formation is a limestone- and quartzite-pebble conglomerate with a maximum thickness of 7 m. The lower Wengxiang Formation is extremely fossiliferous, yielding "acanthodians" and an especially diverse brachiopod fauna of Llandovery age.

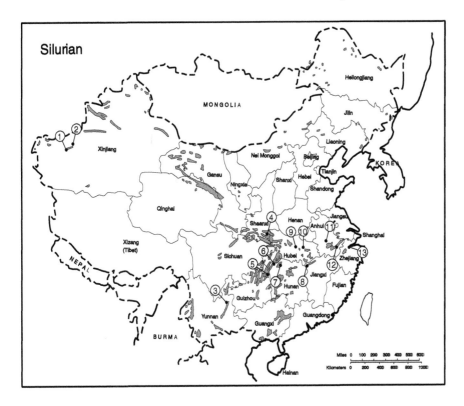

Figure 3-2 *Silurian rocks are widely distributed in China as are the principal Silurian vertebrate-fossil localities: 1 – Bachu, Xinjiang; 2 – Kalpin, Xinjiang; 3 – Qujing, Yunnan; 4 – Ziyang, Shaanxi; 5 – Xiushan, Sichuan; 6 – Sangzhi and Baoqing, Hunan; 7 – Dayong, Hunan; 8 – Xiushui, Jiangxi; 9 – Jingshan, Hubei; 10 – Wuhan, Hubei; 11 – Chaoxian and Wuwei, Anhui; 12 – Ningguo, Anhui; 13 – Changxing, Zhejiang.*

The lower part of the Xiushan Formation gradationally overlies the Rongxi Formation throughout its outcrop belt. It is 217 m thick and consists of quartzitic siltstone, sandy shale, and fine-grained shale in its lower part overlain by sandy limestone and calcareous sandstone in its upper part. The lower part of the Xiushan Formation yields a marine invertebrate fauna of brachiopods, trilobites, nautiloids, and conodonts that suggest a late Llandovery age.

Early middle Silurian vertebrates of China come mostly from the Guoding-shan Formation of Hubei. The correlative Fentou Formation of Hubei, Anhui, and Jiangsu is about 200 m thick and is divided into three parts. The lower part is grayish-yellow, fine-grained quartzitic sandstone with lenses of intraforma-tional mud chips. It yields fin spines of the putative acanthodian *Sinacanthus*, and trilobites. The middle part is grayish-yellow or green, argillaceous siltstone, silty argillite, some interbedded fine-grained quartzite, and a few fossiliferous

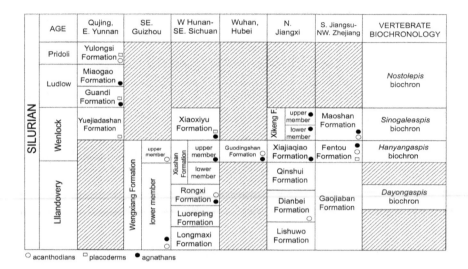

AGE		Qujing, E. Yunnan	SE. Guizhou	W Hunan-SE. Sichuan	Wuhan, Hubei	N. Jiangxi	S. Jiangsu-NW. Zhejiang	VERTEBRATE BIOCHRONOLOGY
SILURIAN	Pridoli	Yulongsi Formation						Nostolepis biochron
	Ludlow	Miaogao Formation						
		Guandi Formation						
	Wenlock	Yuejiadashan Formation		Xiaoxiyu Formation		Xikeng F. upper member / lower member	Maoshan Formation	Sinogaleaspis biochron
			upper member	upper member	Guodingshan Formation	Xiajiaqiao Formation	Fentou Formation	Hanyangaspis biochron
	Llandovery		Wengxiang Formation lower member	Xiushan Formation lower member		Qinshui Formation		
				Rongxi Formation		Dianbei Formation	Gaojiaban Formation	Dayongaspis biochron
				Luoreping Formation				
				Longmaxi Formation		Lishuwo Formation		

○ acanthodians   □ placoderms   ● agnathans

Figure 3-3 *This correlation of the Silurian vertebrate-producing strata of China (after J. Pan 1992) organizes China's Silurian vertebrate fossils into four biochrons.*

yellowish-green shales. Fossils of trilobites, brachiopods, and cephalopods, as well as "acanthodian" spines (*Sinacanthus*), come from the middle part of the Fentou Formation. *Kiangsuaspis nankingensis* Pan from this unit at Fentou, Jiangsu, was supposed to be a cyathaspid agnathan, but now is recognized as a ceriocarid crustacean (J. Pan 1984, 1986).

The upper part of the Fentou Formation is gray-greenish yellow argillaceous siltstone with local beds of mud-chip conglomerate. It yields a brachiopod fauna. Invertebrate fossils of the Fentou Formation suggest an early Wenlock age (Mu et al. 1986).

The oldest nonmarine vertebrate-producing strata in China belong to the Maoshan Formation of Anhui, Zhejiang, and Jiangsu. The Maoshan contains agnathan and "acanthodian" fossils and is mostly sandstone that consists of purple, thick-bedded and crossbedded strata above whitish-gray sandstone beds at its base. About 100 m thick, the Maoshan Formation gradationally overlies the Fentou Formation and is most likely of late Wenlock age.

The lower member of the Xikeng Formation has produced the most diverse late Wenlock vertebrate assemblage from China. The entire Xikeng Formation of Jiangxi is 629 m thick, and its lower part is thick to thin-bedded, quartzitic sandstone, purplish-red sandstone, yellow-green siltstone and sandy shale. The Huixingshao and Maoshan formations, also of mostly nonmarine origin, are correlative.

The Yulongsi Formation of Yunnan is mostly a series of mudstones and siltstones with black shales at its base. It mostly contains marine fossils, including brachiopods, nautiloids, and conodonts, at several levels, but also contains a few brackish and freshwater forms. The 330-m-thick Yulongsi Formation yields placoderms and acanthodians and is of Pridoli age.

The underlying Miaogao (Miaokao) Formation of eastern Yunnan is about 335 m thick. It is composed of thin-bedded and alternating layers of nodular limestone and shale. The Miaogao Formation produces aganthans and a diverse invertebrate fauna of trilobites, brachiopods, corals, bivalves, bryozoans, conodonts, gastropods, and ostracods of Ludlow age.

The Guandi (Kuanti) Formation of eastern Yunnan is about 200 to 500 m thick. It is beneath the Miaogao Formation and consists mostly of yellow-green argillaceous siltstone, shale, and interbedded gray limestone (lower part) overlain by purple-red to yellow silty shale and mudstone (upper part). Its diverse marine invertebrate fauna, of Ludlow age, includes brachiopods, gastropods, nautiloids, corals, ostracods, trilobites, and conodonts.

Early Silurian–*Dayongaspis* Biochron

The oldest Chinese Silurian vertebrates are agnathans, "acanthodians" and chondrichthyans. Their fossils come from marine shale and mudstone of the upper part of the Wujiahe Formation in Ziyang County, Shaanxi and the Rongxi Formation of Dayong County in western Hunan (see figure 3-2). These strata are broadly assigned a Llandovery (Early Silurian) age. "Acanthodian" remains of Early Silurian age have also been described from the lower member of the Wengxiang Formation in Kaili County, Guizhou. N. Wang et al. (1998) recently described the chondrichthyan *Xinjiangichthys* from the Lower Silurian of the Tarim basin in Xinjiang.

The Chinese Early Silurian agnathan is *Dayongaspis hunanensis* Pan & Zeng (see figure 3-4), the sole member of the Dayongaspidae, a family of polybranchiaspiforms named by J. Pan and Zeng (1985). The cephalic shield of *Dayongaspis* is nearly triangular and thus of typical galeaspid shape. The eyes are near the anterior margin of the shield, just behind the large, circular naso-hypophysial opening. The galeaspid *Konoceraspis grandoculus* Pan also is found with *Dayongaspis*.

Putative acanthodian spines collected with *Dayongaspis* resemble *Sinacanthus*, better known from younger Silurian strata (see 3-5). Indeterminate galeaspid fossils of Early Silurian age are known from the Wujiahe Formation of Shaanxi, and indeterminate galeaspid fossils have been reported from the Wengxiang Formation in Guizhou and the Xiushan Formation of western Hunan.

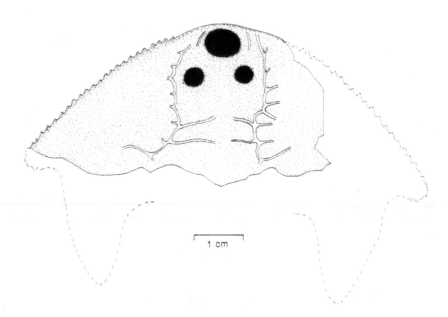

1 cm

Figure 3-4 *In this dorsal view of the cephalic shield of the Early Silurian agnathan Dayongaspis (after J. Pan 1992) note the naso-hypophyseal opening (= median dorsal orifice) in front of and between the two orbits.*

Early Middle Silurian—*Hanyangaspis* Biochron

Chinese early middle Silurian (early-middle Wenlock) vertebrates have a much broader distribution and are significantly more diverse than those of early Silurian age. They mostly come from shallow water marine transgressive facies in the Yangtze region (Hubei–northern Anhui). The endemic Chinese agnathan *Hanyangaspis* Pan & Liu (see figure 3-5G) is characteristic, as are sinacanthid "acanthodians" and the oldest Chinese placoderms.

Key vertebrate-producing formations of early middle Silurian age are:

1. The Guodingshan Formation in Wuhan County, Hubei has produced *Hanyangaspis guodingshanensis* Pan & Liu as well as the "acanthodians" *Sinacanthus wuchiangensis* Pan, *S. triangulatus* Pan & Liu, *S. fancunensis* Liu, and *Neosiacanthus planispinatus* Pan & Liu.

2. The "Shamao" Formation of Jingshan County, Hubei has yielded *Hanyangaspis* sp. and *Sinacanthus* sp.

3. In Hubei, the Fentou Formation contains *Sinacanthus* cf. *S. wuchangensis*.

4. The Qiaotou Formation in northern Jiangxi yielded *Sinacanthus* sp.

5. The Fentou Formation near Nanjing yielded *Sinacanthus* sp.

6. In northern Anhui, the Fentou Formation yielded *Hanyangaspis chaohuensis* (Wang, Xia & Chen), the oldest Chinese arthrodiran placoderms, *Neosiacanthus wanzhongensis* Wang, Xia & Chen and *N. shizikouensis* Wang, Xia & Chen.

7. The lower part of the Maoshan Formation in Zhejiang produced fossils of *Changxingaspis* and *Meishanaspis* (N. Wang 1991).

Hanyangaspis is the archetype of a group of agnathans, the Hanyangaspidida, characterized by a subterminal exonasal opening, a short region of the dorsal shield anterior to the pineal and parapineal foramen, two medio-transversal commisures in the dorsal shield, and a small number (maximum of seven) of

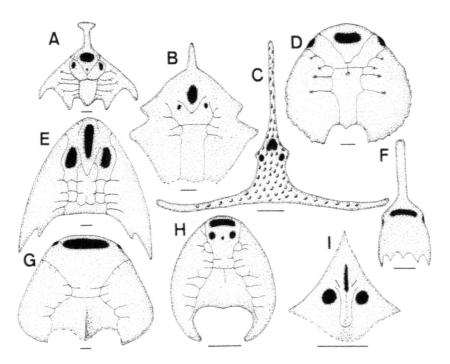

Figure 3-5 *These are dorsal views of the cephalic shields of representative Chinese Silurian-Devonian eugaleaspids: A, Sanchaspis megalarostrata. B, Nanpanaspis microculus. C, Lungmenshanaspis kiangyouensis. D, Cyclodiscapsis ctenus. E, Sinogaleaspis shankonensis. F, Sanqiaspis zhaotongensis. G, Hanyangaspis guodingshanensis. H, Changxingaspis gui. I, Tridenaspis magnoculus. Scale bars = 1 cm (after S. Wang 1993).*

pairs of branchial chambers. Four genera are hanyangaspidids: Early Silurian *Dayongaspis* (see figure 3-4) and middle Silurian *Hanyangaspis, Xiushuiaspis,* and *Changxinaspis.* The group is currently considered an endemic Chinese evolutionary radiation of cephalaspids of Silurian age, though S. Wang (1993) mentions undescribed Early Devonian hanyangaspidids from Guangxi.

Sinacanthus is a form genus for a particular morphology of supposed acanthodian spine from the Silurian and Lower Devonian of China (see figure 3-6). Similar spines are known from Australia (Turner, 1986). The *Sinacanthus* spine has a wide base with a triangular cross section that rapidly tapers to the tip. The spine is strongly curved and is ornamented by numerous, continuous longitudinal ridges that are sharp, lack nodes, and mostly extend to the spine tip, though some terminate at the spine margins. The *Sinacanthus* spines most closely resemble those of some better known climatiid acanthodians (Denison, 1979). However, M. Zhu (1998) has recently concluded that the histology of *Sinacanthus* spines indicates they are of chondrichthyans, not acanthodians

Neosiacanthus is another genus known only from spines. J. Pan (1986) and J. Pan and Dineley (1988) suggested these are the spinal plates of arthrodires. S. Wang (1993), however, has argued that these spines do not co-occur with any other arthrodire fossils and more resemble acanthodian spines than parts of an arthrodire. Here, *Neosiacanthus* is thus considered a form genus for acanthodian spines, though it may be chondrichthyan.

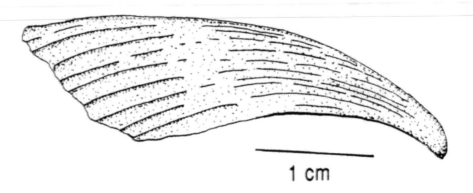

1 cm

Figure 3-6 *Fin spines of Sinacanthus like this one from the Silurian-Devonian of China were long thought to be those of acanthodians, but histological studies indicate they are chondrichthyan (after Denison 1979).*

Late Middle Silurian–*Sinogalaeaspis* Biochron

China's youngest middle Silurian vertebrates are from nonmarine strata that overlie early-middle Wenlock marine rocks. Because these youngest middle Silurian strata are of nonmarine origin, they lack the marine fossils (especially brachiopods, trilobites, crinoids, and corals) by which precise age assignments are made. Mu et al. (1986) argued that these nonmarine rocks probably were deposited relatively rapidly, so it is unlikely they represent more than just a portion of late Wenlock time. For this reason, these rocks are assigned a late middle Silurian age. The agnathan *Sinogaleaspis* Pan & Wang (see figure 3-5E) is characteristic and restricted to the late Wenlockian nonmarine strata, so this time interval is referred to here as the *Sinogaleaspis* biochron.

The lower member of the Xikeng Formation at Xikeng, Jiangxi has produced the most diverse assemblage of the *Sinogaleaspis* biochron. All taxa are agnathans: *Sinogaleaspis shankouensis* Pan & Wang, *S. xikengensis* Pan & Wang, *Xiushuiaspis jiangxiensis* Pan & Wang, and *X. ganbeiensis* Pan & Wang.

In Zhejiang, the Maoshan Formation has produced similar agnathans, *Sinogalaeaspis zhejiangensis* Pan and *Xishuiaspis* sp. The Maoshan Formation in Jiangsu has produced the "acanthodian" *Sinacanthus* sp., and in Anhui the "acanthodian" *Sinacanthus fancunensis* Liu is also known from the Maoshan Formation.

The correlative Xiaoxiyu Formation of northwestern Hunan has produced an indeterminate arthrodire plate and the endemic Chinese agnathan *Fugaleaspis* sp., a genus also (and better) known from the lower Devonian of Yunnan and Guangxi (see next chapter). In Sichuan, the Huixingshao Formation has produced *Eugaleaspis xiushanensis* Liu and an indeterminate birkeniid, the first anaspid found in China (J. Pan and Dineley 1988).

Of the late Wenlock Chinese vertebrates, the hanyangaspidid *Xishuiaspis* and the "acanthodian" form genus *Sinacanthus* are also found in lower Wenlock strata (see above). Characteristic of the upper Wenlock is the eugaleaspid agnathan *Sinogaleaspis*. This genus belongs to the Eugaleaspidiformes, a group of endemic Silurian-Devonian galeaspids from China—the four genera *Eugaleaspis*, *Yunnanogaleaspis*, *Sinogaleaspis,* and *Meishanaspis* (see figure 3-5). *Sinogaleaspis* and *Meishanspis* are of Wenlock age, *Eugaleaspis* ranges from Wenlock to Early Devonian, and *Yunnanogaleaspis* is of Early Devonian age.

The eugaleaspids are characterized by a long exonasal opening, a long prepineal region of the dorsal shield, and a supraorbital sensory line that contacts the medio-transversal commissure. Significantly, the oldest eugaleaspids—*Sinogaleaspis* and *Meishanaspis*—are the most derived members of the group, suggesting that their fossil record is very incomplete (see figure 3-7).

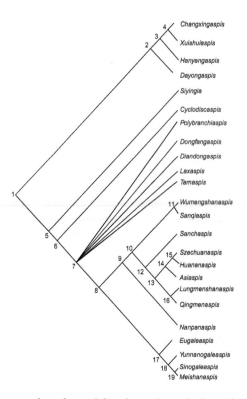

Figure 3-7 *Phylogenetic hypothesis of the relationships of galeaspid agnathans (from N. Wang 1991). Character states that correspond to the numbered nodes are: (1) dorsal surface of the cephalo-thoracic shield consists of a single exoskeletal plate; a large dorsal exonasal opening; galeaspid pattern of sensory lines in dorsal face of the cephalo-thoracic shield; (2) subterminal exonasal opening; short prepineal region of dorsal shield; two medio-transverse commissures in the dorsal face of shield; (3) very broad exonasal opening; (4) large thoracic part of the dorsal shield; orbital openings behind the exonasal opening; (5) one medio-transversal commissure; development of supraorbital sensory line system; (6) posterior margin of the dorsal shield slightly elevated to form a low median dorsal spine; (7) v-shaped pineal canals of supraorbital sensory lines, or pineal canals extending backwards to meet the medio-transversal commissure; (8) dorsal shield with a developmental rostral process or slit-shaped exonasal opening; (9) centrifugal pseudo-cornual process on the posterolateral or lateral margin of the dorsal shield; (10) development of pectoral cornual process; (11) posterior spine-shaped pectoral cornual process; (12) lateral slender pectoral cornual process; (13) pectoral cornual process projecting from the latero-posterior margin of dorsal shield; (14) sickle-shaped pectoral cornual process; (15) length of the main part of the dorsal shield greater than its breadth; (16) latero-dorsal openings at the dorsal shield; (17) longitudinal slit-like exonasal openings, supraorbital sensory lines in contact with the medio-transversal commissure; (18) dorsal shield with intero-pectoral cornual process in posteromesial lateral corner, which is close to and smaller than the pectoral cornual process; (19) pineal and parapineal foramen level with the posterior margin of orbits.*

Late Silurian–*Nostolepis* Biochron

Until recently very little was known of the Late Silurian (Ludlow and Pridoli) vertebrates of China. The upper part of the Guandi Formation in Qujing County, eastern Yunnan, recently yielded the earliest antiarch, *Silurolepis platy-dorsalis* Zhang & Wang. *Silurolepis* has a median ventral plate in the trunk shield, which precludes its assignment to the Sinolepidae. The Guandi Formation also yielded the ichthyoliths *Thelodus sinensis* and *Naxilepis gracilis*, as well as an indeterminate yunnanolepid (N. Wang and Dong 1989). Thelodonts are agnathans known mostly from isolated scales (see figure 3-8) that have a broad distribution in Silurian–Devonian rocks (Halstead and Turner 1973, Turner and Tarling 1982). *Thelodus* is a probable thelodontid scale with a broad blunt

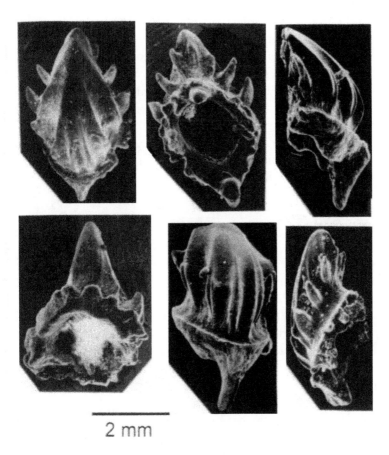

2 mm

Figure 3-8 *These are various views of a tiny thelodont scale, Turinia pagoda from the Middle Devonian of Yunnan (courtesy of S. Turner).*

base, "neck" at the base of the crown and a blunt, pentagonal crown with a peaked, faceted occlusal surface. *Naxilepis* is an actinopterygian scale.

N. Wang and Dong (1989) described an ichthyolith assemblage from the Miaogao Formation in eastern Yunnan of late Ludlow age. These microvertebrates are: the endemic acanthodian *Hanilepis wangi*; the endemic actinopterygians *Kawalepis comptus* and *Naxilepis gracilis*; the thelodontid *Thelodus sinensis*; and the cosmopolitan genera *Gomphonchus, Nostolepis, Ligulalepis,* and *Poracanthodes* (see figure 3-9).

China's youngest Silurian vertebrates are from the Pridoli Yulongsi Formation of Yunnan. They are an indeterminate arthrodire and the acanthodians *Nostolepis* sp. and cf. *Nosotolepis stinata*.

Nostolepis (see figure 3-9D) is the characteristic acanthodian of the Chinese late Silurian. The genus was based originally on scales, which are ornamented with converging or parallel ridges or strong ribs. These scales also have a peculiar histology of mesodentine penetrated by a system of radial, concentric, and ascending canals. The complete fish *Nostolepis* is known as a climatiid with a broad distribution, especially in Eastern Europe, Spitsbergen (Svalbard), Scandinavia, India, Iran, Australia, and Western Europe (Denison 1979). Characteristic features include a head covered with tesserae and a histology similar to that of the scale; tooth spirals with transverse, leaf-shaped teeth, fin spines ornamented with nodose ridges and lacking an inserted base, and the presence of paired intermediate spines. *Nostolepis* is one of the few cosmopolitan vertebrates known from the Chinese Silurian

China's Oldest Vertebrates

China's oldest fossil vertebrates are recently discovered Lower Cambrian agnathans from Yunnan (Shu et al. 1999). After a long gap, the succeeding Chinese fossil vertebrates are of Early Silurian age. These are agnathans (the galeaspid *Dayongaspis*) and the "acanthodian" form genus *Sinacanthus*, both first appearing in Llandovery strata. China's oldest thelodont agnathans are a bit younger, first occurring in Wenlock strata. The oldest placoderm from China is *Silurolepis*, an antiarch first found in Wenlock strata. China's oldest actinopterygian and chondrichthyan records are microvertebrates of Silurian (Ludlow) age.

It is worth comparing the timing of these first occurrences of these vertebrates with their first records elsewhere, partly to develop some perspective on the global significance of some of China's earliest vertebrates (see table 3-1). This comparison highlights some important conclusions:

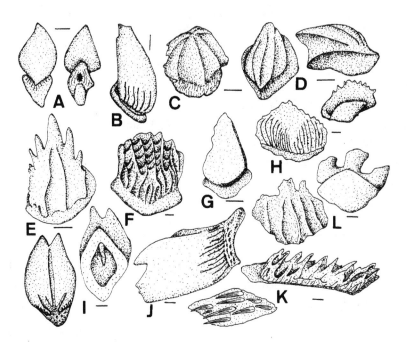

Figure 3-9   *These are representative ichthyoliths from the Upper Silurian Miaogao Formation: A–C, Thelodonti: A, Turinia asiatica; B, Kawalepis comptus; C, Thelodus sinensis. D–G, Acanthodii: D, Nostolepis sp.; E, Hanilepis wangi; F, Poracanthodes qujingensis; G, ischnacanthid gen. indet. H–J, Chondrichthyes; H, Gualepis elegans; I, Changolepis tricuspidatus; J, Peilepis solida. K–L, Actinopterygii. K, Naxilepis gracilis; L, Ligulalepis yunnanensis (after S. Wang 1993). Scale bars = 0.1 mm.*

1. The galeaspids were an evolutionary radiation of agnathans totally endemic to the south China microplate that first appeared in the early Silurian (e.g., J. Pan 1992; Long 1995). They are further discussed in the next chapter.

2. Thelodonts and chondrichthyans first appear in China well after their first appearance elsewhere. Only in the early 1980s were Chinese Silurian thelodonts and chondrichthyans first reported, based on microvertebrate remains. The prospect for older Silurian records of thelodonts and chondrichthyans remains high.

3. Antiarchs and osteichthyans appear at the same time in China as their oldest records elsewhere.

Table 3-1 Comparison of oldest vertebrate records from the Chinese Cambrian-Silurian and the oldest records of these taxa outside of China

First Record	China	Elsewhere
Taxon:		
Agnatha	Early Cambrian	Late Cambrian
Galeaspida	Early Silurian	Early Silurian
Acanthodii	Late Silurian	Late Silurian
Thelodonti	"Middle" Silurian	Late Ordovician
Antiarchi	"Middle" Silurian	Middle Silurian
Osteichthyes	Late Silurian	Late Silurian
Chondrichthyes	Early Silurian	Early Silurian

4. Chinese Early Silurian "acanthodian" spines (form genus *Sinacanthus*), long thought to be the oldest known record of acanthodians, are actually of chondrichthyans.

5. The oldest fossil vertebrates known are Early Cambrian agnathans from China.

Silurian Vertebrate Biochronology

The text above identified four biochrons that correspond to the time represented by the Chinese Silurian vertebrate fossil assemblages (see figure 3-3). Further refinement of this scheme is desirable; each biochron is equal to about one epoch of the Silurian, or about 5 to 10 million years on the Harland et al. (1990) numerical time scale. The vertebrate biochronology proposed here for the Chinese Silurian thus is coarse, but has the potential for further development as a useful tool for correlation.

Silurian Vertebrate Paleobiogeography

China's Silurian vertebrate fossils are essentially confined to the south China block (see figure 3-2), so the biogeographical affinities of these vertebrates should parallel the paleogeographic affinities of this microplate. During the Silurian, south China was an isolated microplate (see figure 3-1). The latitude of this block was low, as suggested by the widespread Silurian limestones, dolomites and reefs on the south China block (Nie 1991). A relatively low paleolatitude for Silurian south China is also supported by the presence of a prolific shallow-water marine invertebrate fauna, especially the profuse coral growth.

The Silurian vertebrates of China emphasize the clear isolation of the south China microplate. Endemic galeaspids and "acanthodians" dominate the Early-Middle Silurian vertebrate-fossil assemblages of China. The first hint of cosmopolitanism comes in the Late Silurian when some wide-ranging chondrichthyans appear in the Chinese vertebrate-fossil record.

G. Young (1981) first drew attention to the vertebrate endemism that characterized south China during the Silurian-Devonian, although Y. Liu (1965) had mentioned it much earlier. G. Young (1981) proposed a separate south China vertebrate province to recognize this isolated center for the radiation and diversification of many early fish groups (see figure 3-10). Recent attempts to further divide this province into realms, for example, by J. Pan and Dineley (1988) or N. Wang (1991), are not convincing. They are little more than attempts to discriminate some biochronological or paleoecological pattern in the still burgeoning record of Chinese Silurian vertebrates

The isolation and vertebrate endemism of south China during the Silurian is striking. All galeaspids (four genera), all antiarchs (one genus), and most "acanthodians" (three genera) are endemic, as are one of the two thelodont genera. This endemism persisted well into the Devonian and will be discussed at greater length in the next chapter.

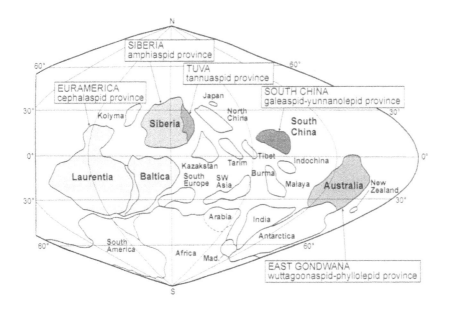

Figure 3-10 *Vertebrate provinces of the Early Devonian include a separate south China province (after G. Young, 1981).*

Chapter 4

Devonian

The Devonian paleogeography of China resembles the Silurian with one major exception—the counterclockwise rotation of the Tarim, south China and north China blocks so that Tarim is northwest of south China and north China is north of south China (see figure 4-1). The Silurian endemism of most Chinese vertebrates continued into the Devonian, but by Middle-Late Devonian time many cosmopolitan vertebrates evolved to live side-by-side with many endemics.

Devonian rocks are found in three main regions of China (see figure 4-2) (Yang et al. 1981; H. Hou and Wang 1985). North of the Inshan-Tienshan Mountains (latitude 41°–42°N) are shallow-water marine clastics and volcanic rocks. These rocks were deposited on the north China block and have not yet produced any vertebrate fossils. Western Chinese Devonian rocks are located between the Inshan-Tienshan and the Tsinling-Kunlun Mountains (see figure 4-2) and are mostly terrestrial red beds. Late Devonian vertebrates have been recovered on the slopes of the Qilianshan. South of the Kunlun-Tsinling-Mountains is where most of China's Devonian rocks and vertebrate fossils are found. These rocks were deposited on the south China block. They represent an array of marine environments of deposition, mostly shallow water platform carbonates.

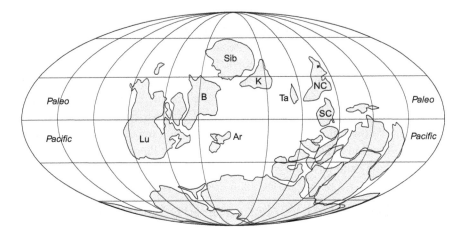

Figure 4-1 *This Devonian paleogeography shows a configuration of the Chinese microplates broadly similar to that of the Silurian (after Z. Li et al. 1993). Abbreviations as in figure 3-1.*

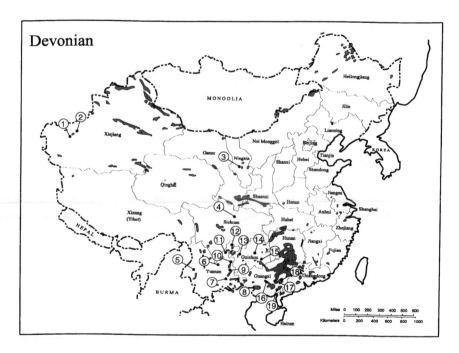

Figure 4-2 *Devonian rocks and principal vertebrate fossil localities are widely distributed in China. Localities are: 1 = Kalpin, Xinjiang; 2 = Bachu, Xinjiang; 3 = Zhongwei, Ningxia; 4 = Jiangyou, Sichuan; 5 = Shidian, Yunnan; 6 = Wuding, Yunnan; 7 = Yiliang, Yunnan; 8 = Guangnan, Yunnan; 9 = Qujing, Yunnan; 10 = Zhanyi, Yunnan; 11 = Zhaotong, Yunnan; 12 = Yiliang, Yunnan; 13 = Hezhang, Guizhou; 14 = Wudang, Guizhou; 15 = Duyun, Guizhou; 16 = Liujing, Guangxi; 17 = Guiping, Guangxi; 18 = Xinlong, Guangxi; 19 = Bobai, Guangxi.*

The Devonian has long been referred to as the "age of fishes," and China's Devonian vertebrate-fossil record confirms this eponym. No terrestrial vertebrates (tetrapods) are known from the Devonian of China. The vertebrate record here is wholly of fishes: agnathans, placoderms, acanthodians, chondrichthyans, and osteichthyans. This is one of the most significant Devonian fish records on earth, and China's Devonian fishes have impacted our understanding of early vertebrate evolution in a major way. The endemism of many of China's Devonian fishes is also of great significance to Devonian paleogeography and paleobiogeography.

Devonian Vertebrate-Producing Strata

Most of China's Devonian vertebrate fossils come from its thick and complex Lower Devonian stratal successions in southern China (figures 4-2 and 4-3).

The most complete and fossiliferous section is in the Qujing basin of Yunnan. The section begins here with 30 to 70 m of black shale, the Miandiancun Formation, which conformably overlies the Upper Silurian Yulongsi Formation. The overlying Cuifengshan Formation is composed of four members (in ascending order):

1. Xishancun Member, 320 m of yellow sandstone and green shale

2. Xitun Member, 300 m of gray shale

3. Guijiatun Member, 300 to 360 m of red sandstone and shale

4. Xujiachong Member, 870 m of red and yellow sandstone and shale

The Miandiancun Formation and all four members of the Cuifengshan Formation produce fossil vertebrates of Early Devonian age, and I consider this the typical succession of Early Devonian vertebrates from China.

The Devonian section at Liujing, Guangxi has been advocated as one of the stratotype sections of the Devonian in south China (Yang et al. 1981). Two formations here—Lianhuashan and Nagaoling—contain Early Devonian vertebrates (see figure 4-3). The Lianhuashan Formation disconformably overlies Cambrian strata and consists of three members (in ascending order): (1) basal Lingli Member, 70 to 100 m of sandstone and conglomerate; (2) Hengxian Member, 110 m of purplish-red mudstone and dolomitic limestone; and (3) Liukankou Member, 150 m of red sandstone. The Hengxian and Liukankou members yield fossil fishes.

Middle Devonian (Eifelian-Givetian) strata that contain fossil vertebrates are much less extensive in China than Lower Devonian vertebrate-producing strata. The most fossiliferous strata are in eastern Yunnan, where they overlie the most fossiliferous Lower Devonian strata of the Qujing basin. The Chuandong Formation is 80 to 90 m of yellow sandstone that conformably overlies the top of the Cuifengshan Formation. It yields fossils of the placoderms *Xichonolepis, Bothriolepis* and *Wudinolepis*. The overlying Sanshuanghe Formation is 200 m of yellow sandstone and gray dolomite that has yielded fossils of *Bothriolepis*. The youngest Middle Devonian strata in the Qujing basin belong to the Haikou Formation, 15 to 100 m of yellow sandstone, quartzite, and shale, which contain fossils of *Bothriolepis, Hunanolepis,* and *Quasipetalichthys*.

In Guangxi, Middle Devonian vertebrates come from the Donggangling Formation. Along the flank of the Tsinling Mountains in Ningxia, *Bothriolepis* is known from the Middle Devonian upper part of the Shiaxiagou Formation, 180 to 200 m of purple sandy shale and sandstone. In western Yunnan, the Malutang Formation and overlying Heyuanzhai Formation contain rare specimens of the thelodont *Turinia*.

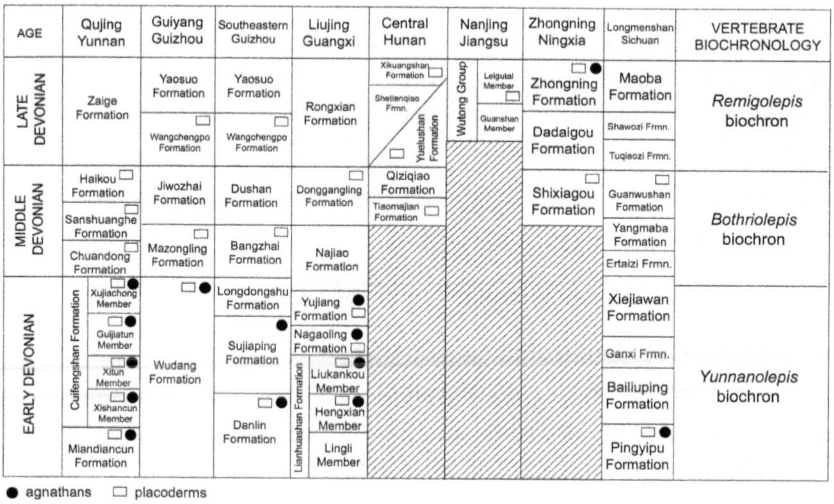

AGE	Qujing Yunnan	Guiyang Guizhou	Southeastern Guizhou	Liujing Guangxi	Central Hunan	Nanjing Jiangsu	Zhongning Ningxia	Longmenshan Sichuan	VERTEBRATE BIOCHRONOLOGY
LATE DEVONIAN	Zaige Formation	Yaosuo Formation / Wangchengpo Formation	Yaosuo Formation / Wangchengpo Formation	Rongxian Formation	Xikuangshan Formation / Shetianqiao Frmn. / Guanshan Formation / Yuelushan Formation	Wutong Group (Leigutai Member / Guanshan Member)	Zhongning Formation / Dadaigou Formation	Maoba Formation / Shawozi Frmn. / Tuqiaozi Frmn.	*Remigolepis* biochron
MIDDLE DEVONIAN	Haikou Formation / Sanshuanghe Formation / Chuandong Formation	Jiwozhai Formation / Mazongling Formation	Dushan Formation / Bangzhai Formation	Donggangling Formation / Najiao Formation	Qiziqiao Formation / Tiaomajian Formation		Shixiagou Formation	Guanwushan Formation / Yangmaba Formation / Ertaizi Frmn.	*Bothriolepis* biochron
EARLY DEVONIAN	Cuifengshan Formation (Xujiachong Member / Guijiatun Member / Xitun Member / Xishancun Member) / Miandiancun Formation	Longdongshu Formation / Wudang Formation	Sujiaping Formation / Danlin Formation	Yujiang Formation / Nagaoling Formation / Liukankou Member / Hengxian Member / Lingli Member (Lianhuashan Formation)				Xiejiawan Formation / Ganxi Frmn. / Bailiuping Formation / Pingyipu Formation	*Yunnanolepis* biochron

● agnathans   □ placoderms

Figure 4-3 *This correlation of Devonian vertebrate-producing strata of China (after J. Pan 1992) organizes the fossil vertebrates into three placoderm biochrons.*

Upper Devonian (Frasnian-Fammenian) vertebrate-producing strata are more widespread in China than those of Middle Devonian age. These Upper Devonian rocks are in eastern Yunnan, southern Guizhou, central Hunan–northern Guangzhou, southern Jiangsu, southern Jiangxi, and Ningxia (see figure 4-3).

In eastern Yunnan, the Yidade Formation (Frasnian) has yielded fossils of *Panxiosteus* and *Eastmanosteus*. To the east, in southern Guizhou, the marine Daihua Formation (Fammenian) is 94 m of banded and nodular limestone that produces a variety of ichthyoliths, including the cosmopolitan *Thrinacodus* and symmoriids. In central Hunan, Frasnian and Fammenian vertebrate-bearing strata are superposed. The Frasnian strata are the Shetiangqiao Formation, 580 m of mudstone, siltstone, and siliceous shale (lower part) overlain by argillaceous and thin-bedded limestone (upper part). Vertebrate fossils are of *Bothriolepis* and *Changyanophyton*. The overlying Xikuangshan Formation is about 430 m thick and consists of three members, only the uppermost (Aojiechong Member) of which yields vertebrates.

Near Nanjing in Jiangsu, the nonmarine Wutong Group is 50 to 180 m of yellow quartzite and gray shale that contain the most diverse known assemblage of Late Devonian fishes from China. The continental Upper Devonian Zhongning Formation of Ningxia is red-bed shales and sandstones that have yielded the antiarch *Remigolepis*. In Jiangxi, at Yudu, the nonmarine Xiashan Formation is 300 m of sandstone and quartzite that yields fossil plants and *Bothriolepis*.

There is an important distinction between marine and nonmarine vertebrate producing strata of Devonian age in China. The fish faunas of these different facies will be discussed below.

Early Devonian–*Yunnanolepis* Biochron

The antiarch placoderm *Yunnanolepis* (see figure 4-4) typifies the Early Devonian vertebrate faunas of China, so I refer to this time interval as the *Yunnanolepis* biochron. *Yunnanolepis* and closely related antiarchs, the Yunnanolepidae, are endemic to the Lower Devonian of south China (Y. Liu 1963). These forms superficially resemble other antiarchs but have distinctive structures in the trunk shield. These include lack of the brachial process, axial fossa, and axial foramen seen in other antiarchs; in their place, *Yunnanolepis* has a deep pectoral fossa posterior to the spinal plate. Furthermore, unlike other antiarchs, *Yunnanolepis* has separate intero-lateral and spinal plates

A host of endemic agnathans coexisted with *Yunnanolepis* in China during the Early Devonian (see figure 3-5). These agnathans are assigned to two.

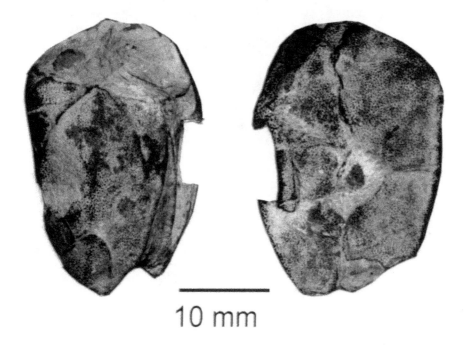

10 mm

Figure 4-4 *This trunk shield and armor of* Yunnanolepis parvus *is seen in dorsal (right) and ventral (left) views. This specimen is from the Lower Devonian at Qujing, Yunnan.*

orders, the Eugaleaspida and Polybranchiaspida. The eugaleaspids are endemic to the Chinese Lower Devonian.

Eugaleaspids have a shield that covers the head and anterior part of the trunk, as in osteostracans. The shield in *Eugaleaspis* is triangular, whereas that of *Polybranchiaspis* is heart-shaped. The head shields (see figure 4-5) are composed of a single bone, lack a pineal opening, and have a median dorsal orifice (nasohypophysial opening) anterior to the orbits, as well as a ventral oral opening. Numerous, polygonal, ornamented scales cover the body. Polybranchiaspids had rather similar skulls, though there is obviously a wide range of shield shapes in the Chinese Early Devonian agnathans (see figure 3-5).

Next to agnathans, antiarch placoderms are the most abundant Early Devonian fishes from China, and these are the oldest and most primitive antiarchs.

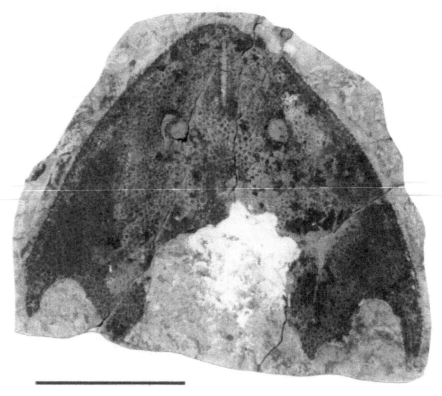

2 cm

Figure 4-5 *This dorsal view of the skull of the Devonian eugaleaspid Nochelaspis shows it to be a single bone with a slit-like naso-hypophyseal opening between the orbits.*

The characteristic Early Devonian *Yunnanolepis* (see figure 4-4) has already been mentioned, and is a totally endemic form, as are the other yunnanolepids. This diverse group includes the genera *Qujinolepis, Phymolepis, Zhanjilepis, Eoantiarchilepus, Grammaspis, Tsuifengshanensis, Orientolepis, Lianhuashanolepis, Macrothyraspis, Sinoszechuanaspis, Qingmenaspis, Pentathyraspis, Microhoplonaspis, Konoceraspis, Hyperaspis,* and *Kwangsilepis* (G. Zhang 1978; J. Pan 1992; J. Pan and Lu 1997). Several hundred well-preserved head shields and trunk shields of yunnanolepids are known.

Other Early Devonian placoderms from China include *Szelepis* and *Parawilliamsaspis*, the only two dolichothoracids known from China (Y. Liu 1979). *Szelepis*, from the Cuifengshan Formation at Qujing, Yunnan is the better known, being represented by molds of the head shield and a thoracic spine. *Kueichowlepis* from the Wudang Formation of Guizhou (J. Pan et al. 1975) is a brachythoracid. *Livosteus sinensis* Wang, from the Lower Devonian Jiucheng Formation of Yunnan, is a coccosteid taxon also known from Latvia.

Petalichthyid placoderms of the Chinese Lower Devonian (e.g., M. Zhu 1991; M. Zhu and Wang 1996; S. Ji and Pan 1997) are *Diandongpetalichthys, Holopetalichthys, Xinanpetalichthys, Neopetalichthys, Guangxipetalichthys,* and *Sinopetalichthys* (see figure 4-6). The latter is rather similar to *Macropetalichthys* from Europe and North America, though all Chinese Early Devonian petalichthyid genera are endemics. Indeed, they represent a minor evolutionary radiation of "quasipetalichthyids" that took place only in south China. *Asiacanthus,* from the Cuifengshan Formation of Yunnan, was originally thought to be an acanthodian, but is now agreed to be based on a spinal plate of an indeterminate placoderm (Denison 1978, 1979). Thelodonts are known from isolated scales referred to the genus *Turinia.*

The sarcopterygian fish *Youngolepis* (M. Zhang and Yu 1981; M. Zhang 1982, 1991; M. Zhu and Fan 1995) is known from the Xishancun and Xitun members of the Cuifengshan Formation at Qujing, Yunnan. M. Zhang's (1982) extremely detailed study of the cranial anatomy of this fish (see figure 4-7) reveals it to have many "rhipidistian" features and to be most similar to *Powichthys* (Jessen, 1975 1980) from the Lower Devonian of the Canadian Arctic. M. Zhang and Yu (1981) suggested that *Youngolepis* and *Powichthys* constitute a distinct rhipidistian group separate from the porolepiforms, osteolepiforms, and other crossopterygians. *Achonolepis* (from the same locality as *Youngolepis*) may also belong to this group. The youngolepids are among the oldest and most primitive sarcopterygians.

The same locality that yielded *Youngolepis* is also the site of the earliest dipnoan, *Diabolepis speratus* (M. Chang and Yu 1984). Another early dipnoan is *Sorbitorhynchus* from the Dale Formation in Guangxi (Campbell and Barwick 1990). Whether or not *Diabolepis* is a dipnoan or "protodipnoan" has been

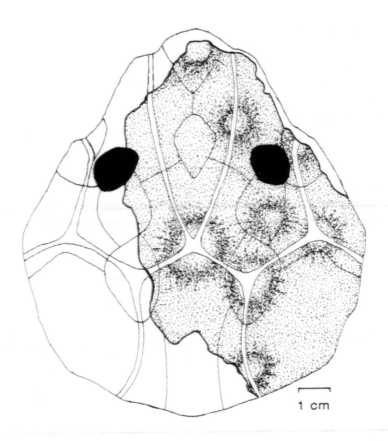

Figure 4-6 *The head shield of the Lower Devonian placoderm Sinopetalichthys (after J. Pan et al. 1975) is seen here in dorsal view.*

debated at some length and is discussed below. M. Chang and Wang (1995) recently described *Erikia jarviki*, another Lower Devonian dipnoan, from Guangnan, Yunnan.

Acanthodians and chondrichthyans of the Chinese Early Devonian are known mostly from microvertebrate remains—ichthyoliths, which are teeth, scales, and dermal denticles. These are best known from the Ganxi and Xie-jiawan formations at Longmenshan, Sichuan (Turner et al. 1995) and the Xitun Member of the Cuifengshan Formation at Qujing. N. Wang (1997) described thelodonts from the Xishancun and Xitun members of the Cuifengs-han Formation, erecting the new genus *Parathelodus*. N. Wang (1997) consid-ered *Parathelodus* transitional between Silurian *Thelodus* and Early-Middle Devonian *Turinia*, so he suggested the age of the Xishancun and Xitun Forma-tions could be as old as Late Silurian. Other ichthyoliths (see figure 3-9) have been assigned to various taxa, including the chondrichthyan form genera *Gual-*

epis, Peilepis, and *Changolepis* and the acanthodian form genera *Gomphonchus, Nostolepis, Cheiracanthoides,* and *Machaeracanthus* (N. Wang 1984). They closely resemble contemporaneous ichthyolith assemblages from North America and elsewhere, and indicate acanthodian and chondrichthyan cosmopolitanism during the Early Devonian.

Early Devonian Paleocommunities

Chinese Early Devonian fishes are highly endemic to southern China. Three characteristic assemblages, which have received various names (see table 4-1), can be recognized. Regardless of what they are called, the three assemblages actually represent three biostratigraphically separate assemblages of Lockhovian, Pragian and Emsian age, here referred to (oldest to youngest) as A, B, and C. The A assemblage is best known from the lower part of the Xishancun

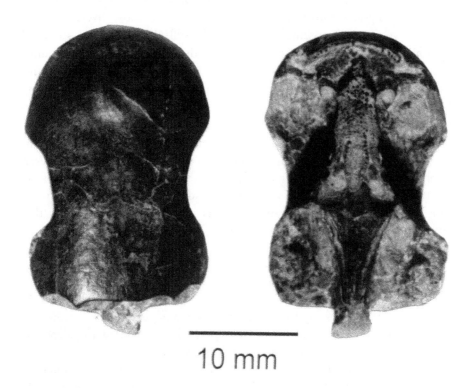

10 mm

Figure 4-7 *Dorsal (left) and ventral (right) views of a tiny and exquisitely preserved skull of the sarcopterygian Youngolepis praecursor from the Lower Devonian of Yunnan.*

Table 4-1 Early Devonian Vertebrate-Fossil Assemblages of China

Wang	Liu	Pan & Dineley	Age
C *Sanchaspis megalarostrata—Qingmenaspis microculus*	*Sanchaspis—Asiaspis*	*Eugaleaspis xujiachongensis—Sanchaspis megalarostrata*	Emsian
B *Yunnanolepis chi—Youngolepis praecursor*	*Yunnanolepis—Qujinolepis*	*Eugaleaspis changi—Nanpanaspis microculus*	Pragian
A *Polybranchiaspis liaojiaoshanensis—Laxaspis qujingensis*	*Polybranchiaspis—Laxaspis*	*Yunnanogaleaspis major—Dongfangaspis qujingensis*	Lockhovian

Member of the Cuifengshan Formation at Qujing, Yunnan. The fishes of this time interval are almost exclusively galeaspid agnathans and *Yunnanolepis*. The B interval is best represented by the assemblage of the overlying Xitun Member of the Cuifengshan Formation at Qujing. This is the most diverse Chinese Early Devonian fish assemblage, but it is still dominated by galeaspid agnathans and *Yunnanolepis*, and also has a diversity of yunnanolepid antiarchs, thelodonts, chondrichthyans, acanthodians, rhipidistians, and dipnoans. Assemblage C is best known from the overlying lower part of the Xujiachong Member of the Cuifengshan Formation at Qujing. Galeaspids and *Yunnanolepis* are still dominant, and petalichthyids make their first appearance.

This division of the Chinese Early Devonian fish faunas into three assemblages might, at first glance, appear to be of biochronological value. However, it fails to account for sampling and facies biases which are obvious when the great diversity of assemblage B is contrasted with that of assemblages A and C. The entire Cuifengshan Formation consists of nonmarine siliciclastics that range from yellow sandstones and shales (Xishancun Member) to red-bed sandstones and shales (Xujiachong Member) to gray mudstones (Xitun Member). Differing facies and varied sampling efforts probably explain the differences between the three assemblages, not temporal differences, which are real but remain untested.

S. Wang (1991) presented a fruitful attempt to analyze assemblage composition among China's Early Devonian vertebrates. He recognized 10 paleocommunities, each based on a distinctive fossil assemblage and its associated lithofacies (see figure 4-8). This analysis distinguishes nonmarine from marine assemblages and clearly indicates galeaspids and antiarchs of the Chinese Early Devonian were euryhaline animals. According to S. Wang (1991), an extensive Middle Devonian (Givetian) transgression in southern China fundamentally altered this paleocommunity structure, leading to the virtual disappearance of the galeaspids and a rise to dominance of placoderms, crossopterygians, and chondrichthyans.

Middle Devonian–*Bothriolepis* Biochron

The cosmopolitan antiarch placoderm *Bothriolepis* appears in China during the Middle Devonian (Eifelian) and is one of the most common vertebrates of the Chinese Middle Devonian. Therefore, I refer to this interval as the *Bothriolepis* biochron.

Bothriolepis is particularly well known from thousands of specimens collected in the Upper Devonian Escuminac Formation of the Gaspé Peninsula in eastern Canada (e.g., Denison 1978). Normally a Late Devonian genus, *Bothriolepis* occurs in a wide range of Chinese Middle Devonian localities, in Hunan, Jiangxi, Guangdong, Guangxi, Yunnan, and Ningxia (Chi 1940; K.Chang 1963, 1965; K. Pan 1964; J. Pan et al. 1978; J. Pan and Wang 1980). Like other antiarchs, *Bothriolepis* has a very long trunk shield, dorsally located orbits and narial openings, a terminal mouth that opens ventrally, pectoral fins completely enclosed in bone, and a dorsoventrally flattened body, among other features.

Early Devonian vertebrate assemblages from China are dominated by galeaspid agnathans, but those of the Middle Devonian are placoderm dominated. In addition to *Bothriolepis*, other antiarchs are: *Hohsienolepis hsintuensis* Pan from Xindu, Henan; *Dianolepis liui* Chang, *Wudinolepis weni* Chang, and *Microbrachius sinensis* Pan from Qujing, Yunnan; and *Xichonolepis qujingensis* Pan & Wang from the Haikou Formation of Yunnan.

Other placoderms—arthrodires and petalichthyids—are much less common than antiarchs. Arthrodires are mostly from the lower part of the Haikou Formation in Yunnan: *Jiuchengia longoccipita* Wang & Wang, *Yinostius major* Wang & Wang, *Kunmingolepis lucaowanensis* Liu & Wang, *Exutaspis megista* Liu & Wang, and *Yangaspis linningensis* Liu & Wang. *Kianyousteus youii* Liu is an anthrodire from the Kuanwushan Formation of Sichuan. These genera are Chinese endemics, as are some of the antiarchs. The petalichthyids are: *Hunanolepis tieni* Pan & Tzeng from the Tiaomachien Formation of Hunan, the Dahepo Formation of Guangdong, and the Haikou Formation of Yunnan; and *Quasipetalichthys haikouensis* Liu and *Eurycaraspis incilis* Liu from the Haikou Formation of Yunnan and the Shixiagou Formation of Ningxia. Indeed, fossils of *Bothriolepis, Hunanolepis,* and *Quasipetalichthys* dominate Middle Devonian vertebrate fossil assemblages of China in nonmarine facies (J. Wang 1991). Other elements are rare: thelodonts (*Turinia pagoda* Wang,

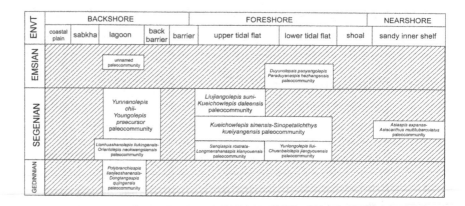

Figure 4-8 *Early Devonian vertebrate paleocommunities of south China ranged from freshwater lagoonal to shallow marine shelf environments (after S. Wang 1991).*

Dong & Turner from Yunnan; see figure 3-8) and the sarcopterygian *Heimenia* from Yunnan.

Most of China's Middle Devonian vertebrate record is from nonmarine red beds similar to the "Old Red Sandstone facies" of Europe and Greenland. Although most of the Chinese Middle Devonian fishes are endemic, *Bothriolepis* and others are cosmopolitan. Recent discoveries of marine vertebrate microfossils of Middle Devonian age (e.g., S. Wang et al. 1986) include the cosmopolitan thelodont *Turinia* and further demonstrate the Middle Devonian breakdown of the vertebrate endemism in south China that is the hallmark of the Early Devonian.

Late Devonian–*Remigolepis* Biochron

By the Late Devonian, most of the jawless fishes are extinct, and antiarchs remain the dominant vertebrates of the Chinese Late Devonian. *Remigolepis* (see figure 4-9), however, is the most widespread placoderm (an antiarch). It is best known from the Zhongning Formation of Ningxia, where numerous specimens have been recovered and assigned to six species: *R. zhongningensis* Pan & Wang, *R. major* Pan, *R. microcephala* Pan, *R. xixiaenensis* Pan, *R. zhongweiensis* Pan, and *R. xiangshanensis* Pan. *Remigolepis* also occurs in the Xikuangshan Formation of Hunan (J. Pan and Dineley 1988).

Remigolepis is broadly similar to *Bothriolepis* and other antiarchs but has numerous unique features, including more separate posterior plates in the trunk shield and pectoral fins lacking a joint. *Sinolepis* is another Late Devonian Chinese antiarch. It is known from the Wutong Group of Jiangsu (*S. macrocephala* and *S. wutungensis* Liu & Pan) and the Zhongning Formation of Ningxia (*S. szei* Pan). Other Chinese Late Devonian antiarchs are *Asterolepis sinensis* Pan from the Wutong Group and *Jiangxilepis longibrachius* Zhang & Liu from the Sanmentan Formation in southern Jiangxi. *Bothriolepis* is widespread in the Upper Devonian of Hunan and Jiangxi.

Arthrodires are less common than antiarchs, but include the dinichthyid *Dunkleosteus yunnanensis* Wang from the Yidade Formation (possibly late Middle Devonian) in Yunnan. The specimen is very fragmentary (isolated nuchal and suborbital plates; see figure 4-10) but clearly belongs to this genus of giant predator (body length more than two meters) also known from the Upper Devonian of North America, Europe, and North Africa. *Changyanophyton hupeiense* Sze from Hubei, originally described as a fossil plant, is a placodermof uncertain affinities (J. Pan 1992). Another placoderm of uncertain affinities is *Huaningichthys* from Yunnan (N. Wang and J. Wang 1999). No petalichthyids have been reported from the Chinese Upper Devonian, and very few agnathan fossils have been recovered (J. Pan et al. 1987).

As in the Middle Devonian, recent efforts to collect marine microvertebrate fossils from Upper Devonian strata in China are vastly augmenting diversity. S. Wang and Turner (1985) described microvertebrate remains from the Daihua Formation of Guizhou that consist mostly of chondrichthyans assigned to the form genera *Phoebodus, Petalodus, Protacrodus, Thrinacodus,* "*Diplodus*" and

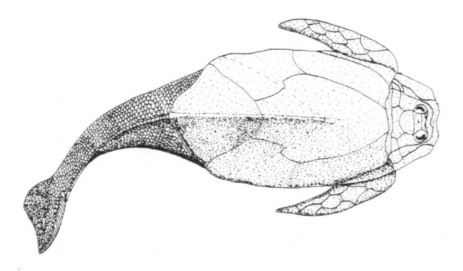

Figure 4-9 *This reconstruction of the characteristic Late Devonian antiarch Remigolepis (after Burrett et al. 1990) shows the fish in dorsal view. Note the heavily armored head, thorax, and pectoral appendages.*

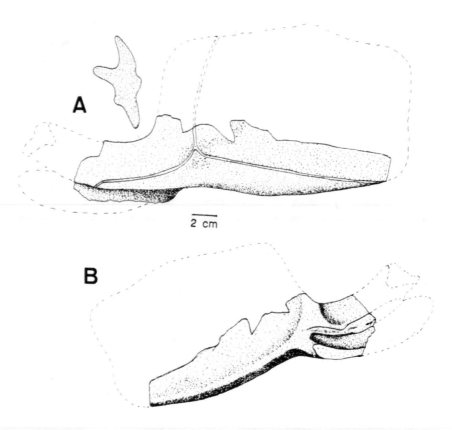

Figure 4-10 *The isolated left suborbital plate of Dunkleosteus yunnanensis is a rare Chinese fossil of the giant placoderm (after J. Wang, 1982).*

Ctenacanthus. S. Wang (1993) noted that scales and teeth of actinopterygians and crossopterygians occur in the Xikuangshan Formation of central Hunan, but these have not been described.

Song and Zhang (1991) recently reported the lungfish *Chirodipterus* from the Shetianqiao Formation in Hunan. *Chirodipterus* is also known from Europe, North America, Australia, and Iran, so its discovery in China further attests to the cosmopolitanism of the Chinese Late Devonian fish fauna.

Systematics of Devonian Agnathans

There has been great disagreement as to the affinities of the endemic Devonian Chinese agnathans to other Agnatha. The "superclass" Agnatha, the jawless fishes, is divided into nine classes (see figure 4-11). Chinese paleontologists have mostly assigned *Eugaleaspis* and *Nanpanaspis* to the Osteostraci, and other Chinese Devonian agnathans to the Heterostraci (e.g., Y. Liu, 1965, 1975; J. Pan et al., 1975, 1978). Tarlo (1967), however, argued that the eugaleaspids and polybranchiaspids should be united in a higher category Eugaleaspida, of equal rank to the other classes of agnathans and allied with the Anaspida, Osteostraci and Petromyzontida as cephalaspidimorphs, an argument later developed and endorsed by Janvier (1975), Halstead et al. (1979) and Janvier and Blieck (1979). J. Pan and Wang (1981) went further to raise the eugaleaspids of Halstead to a group of equal rank to the Cephalaspidimorphi and Pteraspidimorphi. The newly described agnathans from the Lower Cambrian of Yunnan, *Myllokunmigia* and *Haikouichthys*, do not fit readily into any of the agnathan classes (Shu et al. 1999).

A close relationship between eugaleaspids and polybranchiaspids is well established by their uniquely shared features, including a one-piece bony shield with a large, median dorsal orifice in front of the orbits and bone structure with hollow cavities covered with ornamental tubercles or "blisters" (see figure 4-12). This bone structure is quite unlike the "honeycomb" bone structures of heterostracans, although the Chinese agnathans do resemble heterostracans in lacking a pineal opening. The single-plate shield, ventral oral opening and dorsal nasohypophysial opening of the Chinese agnathans are similar characteristics to cephalaspids,

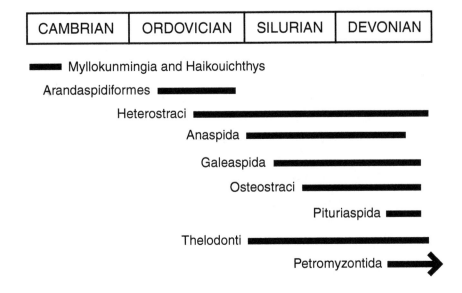

Figure 4-11 *The early-middle Paleozoic temporal ranges of different agnathan classes (modified from Long 1993) extends from the oldest known vertebrates to the still-living petromyzontidans (lampreys).*

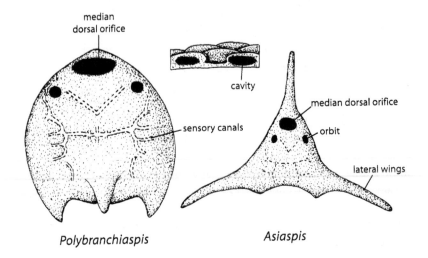

Figure 4-12 *Some key features of the skulls of galeaspid agnathans (after Long 1993). The middle view is a microscopic cross section through the bone. Note that the median dorsal orifice is also called the naso-hypophyseal opening.*

though they do not easily fit into any group of cephalaspids. It thus seems most reasonable to recognize the Chinese Devonian agnathans as a distinct group of cephalaspids, the Galeaspida (Long 1995).

Not all Chinese Devonian agnathans are necessarily galeaspids. *Hanyangaspis* has a head shield made of multiple plates, as does *Latirostraspis* (J. Pan et al. 1975). These forms may be heterostracans. And, as already noted, thelodont scales are now known from the Chinese Devonian (e.g., N. Wang 1984).

Diabolepis and Lungfish Phylogeny

M. Chang (= Zhang) and Yu (1984) described *Diabolichthys speratus* as a new dipnoan (lungfish) based on several skulls and jaws from the Cuifengshan Formation at Qujing, Yunnan. The genus name was later found to be preoccupied and was replaced by *Diabolepis* (M. Zhang and Yu 1987).

The cranial material of *Diabolepis* is exquisitely preserved (see figure 4-13) and shows a mosaic of characters long considered characteristic of lungfish and of primitive rhipidistians. Like later lungfish, *Diabolepis* has an extensive palatal dentition and a very distinctive pattern of dermal bones of the skull roof. However, unlike other lungfish, *Diabolepis* lacks a totally fused braincase and has marginal teeth on separate premaxillaries. These and other features are similarities of

Diabolepis to rhipidistians. M Chang and Yu (1984) nonetheless argued that *Diabolepis* is more closely related to dipnoans than to rhipidistians, and thus should be regarded as a lungfish or as the closest relative (sister taxon) of the lungfishes (e.g., Janvier 1996).

An extremely significant feature of *Diabolepis* is its external nasal openings (choanae), which are located ventrally at the anterior margin of the mouth (see figure 4-13). One of the main reasons why paleontologists long identified rhipidistians as the ancestors of tetrapods is because these were thought to be the only fishes with choanae truly homologous with tetrapod choanae (e.g., Jarvik 1980, 1981). In contrast, Rosen et al. (1981) argued that lungfishes have true choanae, but not rhipidistians, so that the Dipnoi are the sister taxon of Tetrapoda. *Diabolepis*, however, does not support this suggestion because the posterior external nasal opening is lateral to the premaxillary, not medial as in tetrapods. *Diabolepis* from the Devonian of China thus has played an important role in deciphering the phylogenetic ancestry of tetrapods.

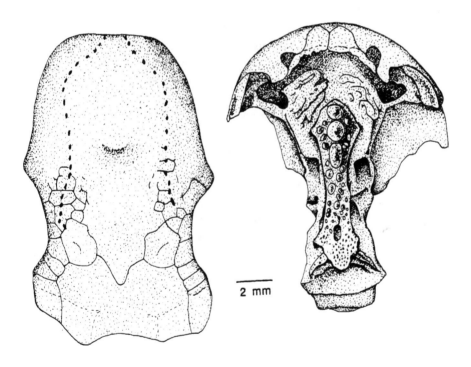

Figure 4-13 *This tiny skull of Diabolepis is seen in dorsal (left) and ventral (right) views (after Zhang and Yu 1987).*

Devonian Vertebrate Biogeography

G. Young (1981) named the diverse and highly endemic Early Devonian fish fauna of south China the yunnanolepid-polybranchiaspid biogeographic province (see figure 3-10) (J. Pan and Dineley [1988] renamed it the eugaleaspid-polybranchiaspid-yunnanolepid realm, a redundancy best forgotten.) This province has an endemic fauna (some 50 genera belonging toendemic higher taxa) of galeaspid agnathans and yunnanolepid placoderms. Most localities are on the south China block, stretching from eastern Yunnan to Guangdong. The idea that the south China block consisted of two separate terranes, the Yangtze and Huanan (Hsü et al. 1988), is not supported by vertebrate evidence. Northeastern Vietnam also seems to have belonged to the same Early Devonian province as south China (Thanh and Janvier 1987, 1990).

The endemism of this province begins to break down during the Middle Devonian with the appearance in south China of the antiarch *Bothriolepis* and other cosmopolitan taxa. By Late Devonian time, it seems south China was closely connected to eastern Gondwana, and its vertebrate endemism is virtually unrecognizable (G. Young 1990, 1993).

What explains the high endemism of south China's Early Devonian fishes? This endemism is part of a global pattern of Early Devonian endemism of both vertebrates and invertebrates, followed by Late Devonian cosmopolitanism. Two explanations seem possible, intrinsic or extrinsic.

Burrett et al. (1990) argued for an intrinsic cause, namely that Late Devonian fishes were better swimmers and thus more capable of dispersing than were Early Devonian forms. In particular, they argued that antiarchs were able to disperse widely along marine coastlines during the Middle-Late Devonian, whereas agnathans could not. G. Young (1993), however, argued that changing global paleogeography might have caused the breakdown of south China's Early Devonian vertebrate endemism, particularly a fusion of south China and Indochina along the Song Ma suture. Perhaps both factors played a role in the vector endemic-to-cosmopolitan that is so well documented by China's Devonian vertebrate-fossil record.

Chapter 5

Carboniferous

By the Late Carboniferous, the Pangean supercontinent had mostly amalgamated (Veevers 1988). However, most of the microplates of China were very loosely connected to Pangea, forming a sort of archipelago at its extreme eastern end (see figure 5-1). A broad Tethys Ocean separated the Chinese microplates from most of Pangea, which lay well to the west. The Tarim and north China blocks were close to the Kazakstan block to the west-northwest, but a wide ocean expanse isolated the Chinese blocks on the south.

Strata and fossils of Carboniferous age are widely distributed and abundant in China (see figure 5-2). The rocks encompass marine and nonmarine facies, both richly fossiliferous. Those of the north China block are intercalated successions of marine and nonmarine strata, whereas on the south China block, deposition was almost entirely marine, except in easternmost China. Yet, despite its rich and varied Carboniferous rock and fossil record, China has produced very few Carboniferous vertebrate fossils (see figure 5-2).

Tetrapods first emerged onto land during the Devonian. Their apparent earliest record is footprints of Lockhovian to basal Frasnian age from the Grampians Group of Victoria, Australia (Warren et al. 1986). By Late Devonian time,tetrapod tracks and body fossils are known from eastern Greenland, Latvia, Scotland, Russia, and Australia (Milner 1993). The Carboniferous has

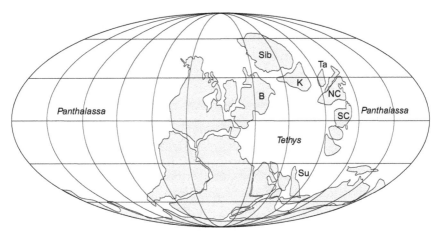

Figure 5-1 *Late Carboniferous paleogeography shows the beginnings of the integration of the Chinese microplates into Pangea (after Z. Li et al. 1993). Abbreviations as in figure 3-1, except Su = Sibumasu.*

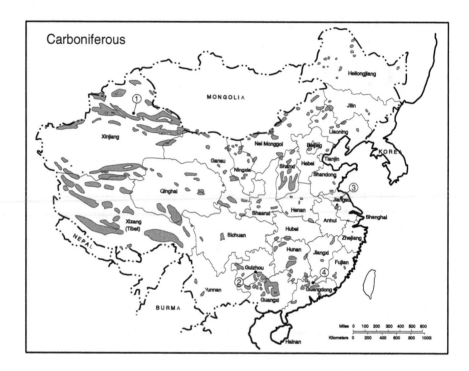

Figure 5-2 *Carboniferous rocks are widely distributed in China (after Yang et al., 1986) but Carboniferous vertebrate fossil localities are few: 1 = Ortu Formation, Xinjang; 2 = Muhua and Dapoushang, Guizhou; 3 = Baoying, Jiangsu; 4 = Dasaiba Formation, Guangdong.*

been referred to as the "age of amphibians" because of the diversity of amphibians known from much of Pangea. Reptiles also first appeared during the Carboniferous. Nevertheless, no tetrapods are known from the Chinese Carboniferous; the earliest Chinese tetrapods are of Permian age (see chapter 6). What are we to make of this?

Given the relative isolation of the Chinese microplates prior to the Permian, it might be tempting to argue that tetrapods arose elsewhere in Pangea and only arrived in China during the Permian when the Chinese microplates fully joined Pangea. However, the early tetrapods (amphibians) were aquatic or amphibious animals. Some of their fossils are from shallow marine rocks, suggesting that they may have been able to cross oceanic barriers. Indeed, the relatively isolated Kazak microplate yields Carboniferous amphibians, *Utegenia* from Kazakstan and *Ariekanerpeton* from Tadjikistan (Ivakhnenko 1987). It thus seems more likely that tetrapods inhabited the Chinese microplates (at least the north China block) during the Carboniferous and simply remain to be discovered.

Carboniferous vertebrates from China thus are fishes, mostly of Early Carboniferous age (see figure 5-2). These vertebrates come primarily from Guizhou, at Muhua (S. Wang and Turner 1985) and Dapoushang (Q. Ji 1989). The vertebrate-bearing rocks at these localities are limestones deposited in deep marine environments that yield conodonts and a dwarfed invertebrate fauna that includes ammonoids, brachiopods, and trilobites.

Carboniferous Vertebrate Occurrences

Microvertebrates, mostly teeth and dermal elements of osteichthyans and chondrichthyans, are found in a number of Chinese Carboniferous marine units, but have been little described (see figure 5-2). An example is the actinopterygian and crossopterygian teeth and scales from shallow marine facies of the Dasaiba Formation at Shaoguan County, northern Guangdong, mentioned briefly by S. Wang (1993).

Published Chinese Carboniferous vertebrates include those from the Wangyou Formation at Dapoushang, Guizhou (S. Wang 1989). The Wangyou Formation is as much as 110 m thick and consists of thin limestones interbedded with dark gray shale. The vertebrate fauna consists of the acanthodian fish *Acanthodes* and the chondrichthyan ichthyolith taxa *Harpagodens ferox*, "*Cladodus*," *Protracodus* and "*Diplodus*."

A relatively new occurrence, of a heliocoprionid chondrichthyan, is from the Ortu Formation in the northern Tien Shan of Xinjiang (see figure 5-3). The Ortu Formation is mixed marine carbonates and clastics that yield an extensive ammonite fauna.

Acanthodes

The archetypal acanthodian fish, *Acanthodes* (see figure 5-4) was one of the last members of the group, occurring principally in strata of Carboniferous-Permian age (Blieck and Goujet reported Late Devonian *Acanthodes* from western Europe: S. Wang 1993). *Acanthodes guizhouensis* Wang at Muhua and Dapoushang in Guizhou are the only occurrences of the genus now known from China. Otherwise, *Acanthodes* was widely distributed in Europe, North America and Australia (Denison 1979).

Acanthodes is a small to moderate-sized acanthodian (about 40–50 cm total body length) with a long slender body. It has long, slender, slightly curved fin spines. The pectoral spine locks into a groove in the scapula (Moy-Thomas and Miles, 1971). This acanthodian seems to have been rather eurytopic and euryhaline because its

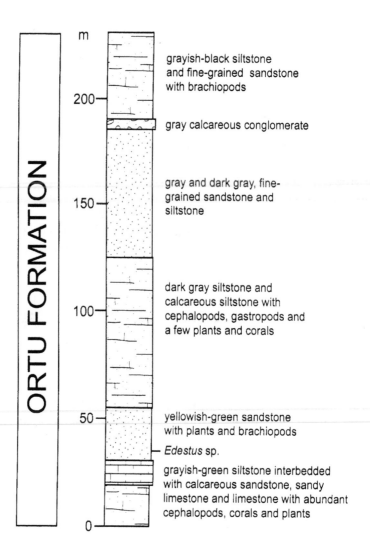

Figure 5-3 *In this stratigraphic section of the Ortu Formation at Qijiagou, Xinjiang (after Z. Cheng et al. 1996), the occurrence of the elasmobranch Edestus is indicated.*

fossils have been found in a wide range of environments, encompassing deep marine limestones (such as the Chinese occurrences in Guizhou) through freshwater ponds (Denison 1979).

Chondrichthyans

Not surprisingly, most of the fishes from the deep marine Lower Carboniferous deposits of Guizhou are chondrichthyans. The genera *Phoebodus, Protacrodus, Stethacanthus, Thrinacodus, Symmorium* and *"Diplodus"* are represented by microvertebrate ichthyoliths (S. Wang and Turner 1985; S Wang 1989). These taxa are also known from underlying Upper Devonian strata.

Phoebodus is the generic name for Paleozoic shark teeth with three principal, widely divergent cusps. On some specimens, two smaller cusps are between the large cusps, and the thick blunt crown base has a distinct nutrient foramen. *Protacrodus* is also a tooth with three broad, rather blunt cusps on a broad and thick, elongate crown base.

Stethacanthus teeth have a broad base with a single dominant pointed and fluted cusp on the crown flanked by four or more smaller, similar cusps. *Thrinacodus* is another phoebodont chondrichthyan tooth type. The teeth are tricuspid and fang-like, with three very curved, pointed cusps supported by a large crown base.

Diplodus is a form genus for teeth of xenacanth sharks. These teeth have a wide, concave base that supports two stocky triangular cusps that diverge at an angle of about 60 to 70° with a small medium cusp in between them.

Teeth referred to *Symmorium* are cladodont, having a large base supporting a row of several pointed cusps. They belong to a fusiform, 200 to 300-cm-long. shark well known from the Devonian-Carboniferous of the United States (Zangerl 1981). *Petalodus* is a genus for selachian teeth that are low crowned and elongate, having one blunt cusp.

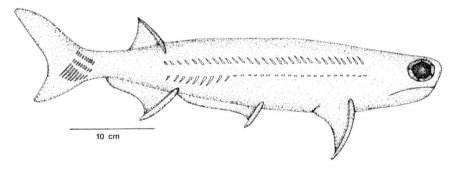

10 cm

Figure 5-4 *Restoration of the Carboniferous-Permian acanthodian Acanthodes (after Zidek 1976).*

Heliocoprionid from Xinjiang

A recent discovery by Cheng Zhengwu of the Chinese Academy of Geological Sciences adds a heliocoprionid shark to China's meager list of Carboniferous vertebrates (Z. Cheng et al. 1996). The specimen is part of the crown base of a tooth from a symphysial tooth whorl. This fragment suggests a relatively large, laterally compressed tooth with a dorso-ventrally elongate base and coarse serrations. Unfortunately, the specimen was found isolated, not as part of a fossil vertebrate assemblage. However, its location, in Xinjiang, on what was the Kazakstan microplate, represents a new Carboniferous record.

Prospectus

So little is known of the Carboniferous vertebrates of China that no analysis or conclusions can really be drawn. The Lower Carboniferous acanthodian and chondrichthyans from Guizhou—the principal described Chinese Carboniferous vertebrates—are representatives of a nearly cosmopolitan marine fish fauna of the Late Devonian-Early Carboniferous (G. Young 1981). These fossils, and other reported, but undescribed, ichthyoliths from the Chinese Carboniferous suggest a rich fossil fish record remains to be recovered here. Chinese coal swamp deposits and their rich floras indicate great potential for the discovery of Carboniferous tetrapods in China.

The Carboniferous now stands as one of the largest gaps in the Chinese vertebrate-fossil record. Further collecting is all that is needed to fill this gap.

Chapter 6

Permian

During the Early to Middle Permian, the north and south China blocks were in tropical latitudes separated from the main Pangean land mass by a nearly closed Tethys marine basin (see figure 6-1). The Tarim block bridged the water gap to the northeastern edge of Pangea (Kazakstan and Siberia) to the north, whereas no apparent direct land connection existed between south China and Gondwana to the south. By Late Permian time, however, the amalgamation of Pangea proceeded by the northeastward drift of the two principal China blocks, opening Tethys to Panthalassa to the east, and joining north and south China to eastern Pangea in the north

During the Permian, early amniotes (reptiles), especially synapsids, became the dominant terrestrial vertebrates. There was also a great diversity of labyrinthodont amphibians, the temnospondyls. In fresh and marine waters, the agnathans (except the cyclostomes) and placoderms had suffered extinction at or before the end of the Devonian. Chondrichthyan and primitive actinopterygian fishes dominated the Permian waterways. Much of what we know about Permian vertebrate evolution comes from extensive fossil records in the western United States, the Ural Mountains region of Russia, western Europe, and the Karoo basin of South Africa (e.g., Romer 1973; Anderson and Cruickshank 1978; Olson and Chudinov 1992; Milner 1993; Lucas 1998a). The Chinese

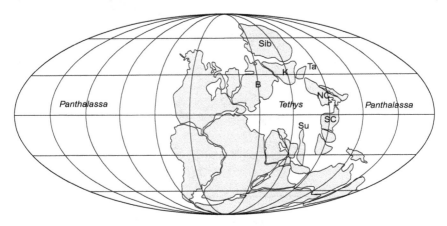

Figure 6-1 *This Late Permian paleogeography (modified from Z. Li et al. 1993 and Golonka et al. 1994) shows a further amalgamation of the Chinese microplates into eastern Pangea than in the earlier Paleozoic. Abbreviations as in figure 3-1.*

record of Permian vertebrate fossils is much less extensive than that of other regions, so it has little impacted our view of Permian vertebrate evolution.

Permian strata, mostly of marine origin, are widely exposed in China (see figure 6-2). But no Early Permian vertebrates have been described from China. Only Middle and Late Permian vertebrates are known, and they represent two distinct, temporally successive assemblages. As in the subsequent Triassic, the Chinese Middle-Upper Permian vertebrate-fossil record comes from two great basins in north China, the Junggur and Ordos basins (see figure 6-2).

Permian Nonmarine Strata in the Junggur and Ordos Basins

The Junggur and Ordos basins are discussed at length in the next chapter because they were such important places for the accumulation of nonmarine strata that entombed most of China's Triassic vertebrate fossil record. Here, a

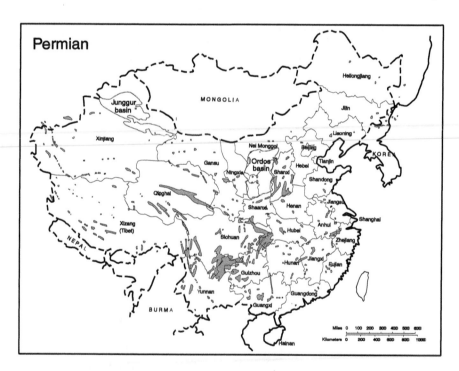

Figure 6-2 *This distribution map of Permian rocks in China shows the location of the Junggur and Ordos basins, which are the primary collecting areas of China's Permian vertebrates (after Yang et al. 1986).*

brief summary focuses on the Middle-Upper Permian vertebrate-bearing strata in these basins.

In the Junggur basin, the Permian vertebrates come from two distinct units, the upper part of the Jijicao Group and the overlying lower part of the Cangfanggou Group (see figure 6-3). The oldest vertebrate-producing strata of the upper part of the Jijicao Group belong to the Lucaogou Formation. This lacustrine deposit consists of black shale and oil shale interbedded with thin-bedded dark gray limestone. China's oldest Permian vertebrates—the anthracosaur *Urumqia* and the palaeoniscid fish *Turfania*—come from the Lucaogou Formation. The formation also contains palynomorphs, bivalves, ostracods, and a few fossil plants and is assigned an early Late Permian age (Z. Cheng, 1980a).

The youngest formation of the Jijicao Group, the Hongyanchi Formation, is a thin unit less than 30 m thick. Thin-bedded and interbedded grayish-black mudstone, sandy mudstone, and siltstone with lenses of fine-grained sandstone and limestone overlying its conglomeratic base. Like the underlying Lucaogou Formation, the Hongyanchi Formation has a gymnosperm-dominated palynomorph assemblage, as well as megafossil plants, ostracods, and bivalves. Indeterminate fossils of palaeoniscid fishes are also known from the Hongyanchi Formation.

The overlying Cangfanggou Group begins with the 270-m-thick Quanzijie Formation. Dark gray mudstone and grayish-green sandstone cap color-mottled conglomerates with muddy siltstone lenses of the lower part of the formation. An extensive palynomorph assemblage from the Quanzijie Formation resembles the palynomorphs from Kazanian (Upper Permian) strata of Russia, and the megafossil plants from the Quanzijie Formation resemble the late Angara flora of the Kuznetsk basin of Russia (Yang et al. 1986). Bivalves, ostracods, and the dicynodont "*Kunpania*" are also known from the upper Quanzijie Formation.

The overlying Wutonggou Formation is 120 to 220 m thick and consists of six, repetitive packages of thick-bedded, dark gray and gray conglomerate, sandstone, siltstone and mudstone. Its palynomorphs, megafossil plants, ostracods and bivalves resemble those of the underlying Quanzijie Formation and thus indicate a Late Permian age (Yang et al. 1986). North of the Tien Shan, the Wutonggou Formation has produced only fragmentary and indeterminate dicynodonts. But, south of the Tien Shan, the dicynodonts "*Jimusaria*" and "*Turfanodon*" come from equivalent strata.

Most of the Late Permian vertebrates from the Junggur basin are from the Guodikeng Formation (see figure 6-4). These are China's youngest Permian vertebrates; all are dicynodonts of the genera "*Jimusaria*," "*Striodon*," and *Diictodon*. The Guodikeng Formation is 140 to 170 m thick. Its upper part is of Early Triassic age and is discussed in the next chapter. Most of the formation

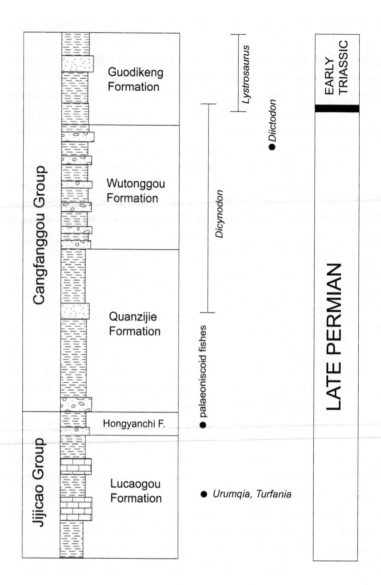

Figure 6-3 *This chart shows the Upper Permian stratigraphic succession and fossil vertebrate distribution in the Junggur basin.*

is interbedded mudstone and sandstone variegated grayish black, yellowish green, and purplish red. Palynomorphs, megafossil plants, ostracods, and bivalves from all but the uppermost Guodikeng Formation indicate a Late Permian age (Yang et al. 1986).

In the Ordos basin, Middle and Late Permian vertebrates are less concentrated and less diverse than those of the Junggur basin. Localities in the Ordos basin are in Gansu, Henan, Hubei, Shanxi, and Nei Monggol. The principal stratigraphic units that yield vertebrates are the Shihezi and Sunjagou Formations in Henan and are similar to the Shiqianfeng Formation of Shanxi. The Shihezi Formation is mostly dark purple, purplish-red, and purple mudstone and siltstone interbedded with gray, grayish green, and grayish white sandstone. It has an average thickness of 100 to 200 m and produces vertebrates discussed below.

The overlying Sunjiagou Formation is 100 to 300 m thick and is mostly dark red and purplish-red mudstone and siltstone intercalated with grayish-green, purplish gray, or grayish white arkosic sandstone. The Sunjiagou Formation and its equivalent, the Shiqianfeng Formation, yield a vertebrate fauna dominated by pareiasaurs that includes the oldest dicynodonts from the Ordos basin. This fauna probably is about the same age as the vertebrates from the Wuttongou and lower Guodikeng Formation in the Junggur basin. Deposition of Upper Permian strata in the Ordos basin was predominantly fluvial, whereas in the Junggur basin lacustrine depositional systems dominated.

Figure 6-4 *This outcrop of the upper part of the Guodikeng Formation at Dalonggkou, Xinjiang encompasses the Permian-Triassic boundary. Note that the strata shown here are overturned toward the right of the photograph.*

Urumqia

China's oldest known tetrapod is the seymouriamorph anthracosaur *Urumqia liudaowanensis* Zhang, from the Lucaogou Formation at Liudaowan, Xinjiang. *Urumqia* is known from a nearly complete skull and lower jaw and part of the anterior vertebral column and forelimb (the holotype: see figure 6-5) and about 30 other specimens (F. Zhang et al. 1984). It has a relatively tall skull with a deep optic notch and an occiput that does not project posteriorly. This neotenic form is a member of the Discosauriscidae, a group of Early Permian seymouriamorphs known from central and eastern Europe, Kazakstan, Russia, and China (Milner 1993; Berman et al. 1997).

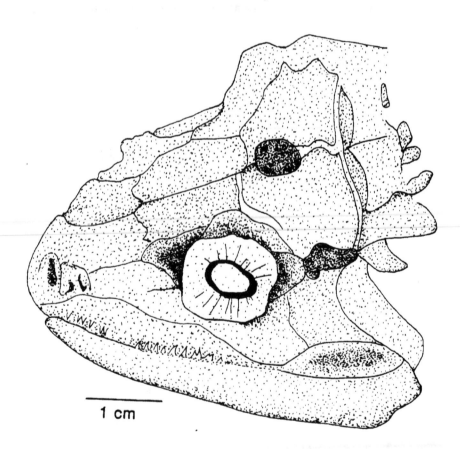

1 cm

Figure 6-5 *The skull and lower jaw of Urumqia (after F. Zhang et al. 1984) is one of several kinds of discosauriscid seymouriamorphs found in Permian strata deposited on the Kazak microplate.*

The discosauriscids were relatively small, aquatic seymouriamorphs. The seymouriamorphs have a peculiar geographic distribution. The larger, terrestrial seymouriids are known from the Lower Permian of the United States (New Mexico and Texas), Germany, and the Upper Permian of Russia. The smaller, aquatic discosauriscids are well known from the Lower Permian of Czechoslovakia, Germany, and France (Werneburg 1988). Other discosauriscids are *Utegenia* from Kazakstan, *Ariekanerpeton* from Tadjikistan, and *Urumqia* from Xinjiang (Ivakhnenko 1987; F. Zhang et al. 1984; Laurin 1996). These localities were part of the Kazakstan microplate during the Permian. The Late Permian age of *Urumqia* is well established, but the other two genera are of less certain age (*Utegenia* has even been assigned a Carboniferous age). Seymouriamorphs must have reached the Kazakstan microplate by the Late Permian, and this suggests a secure connection of the microplate with the European portion of Pangea before that time (Milner 1993; Berman et al. 1997).

Turfania and *Yaomoshania*

The palaeoniscid fish *Turfania taoshuyuanensis* Liu & Ma (see figure 6-6) coexists with *Urumqia* in the Lucaogou Formation. *Turfania* was a long-bodied palaeoniscid about 200 mm long. It shows many characteristic features of palaeonisciforms, including the thick and rhomboidal scales covering the body, long jaws that articulate behind the braincase, triangular dorsal and anal fins, and a deeply cleft heterocercal caudal fin. Of Late Permian age, *Turfania* is one of the last palaeonisciforms and the only well-known representative of a large fish fauna that lived in the deep lake that filled the Junggur basin during part of the Late Permian. *Yaomoshania* is a poorly known contemporary represented by scale rows found in Upper Permian strata of the Jijicao Group near Urumqi; it is a primitive actinopterygian (Poplin et al. 1991).

The Dashankou Locality

J. Li and Cheng (1995b) reported the recent discovery of what may be the best-preserved assemblage of Permian vertebrates known from China. These fossils are from a quarry developed in the upper part of the Xidagou Formation at Dashankou in Gansu. To date, the following taxa have been reported: the dissorophid temnospondyl *Anakamacops petrolicus*, an *Intasuchus*-like temnospondyl, the anthracosaurs *Ingentidens corridoricus* Cheng & Li and *Phratochronis gilianensis* Cheng & Li, the bolosaurid *Belebey vegrandis* Ivakhnenko, a

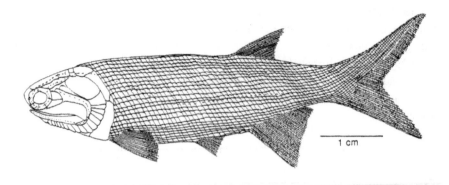

Figure 6-6 *The palaeonisciform fish* Turfania taoshuyuanensis *was a common inhabitant of the Permian lake that occupied the Junggur basin (after H. Liu and Ma 1973).*

captorhinid, the dinocephalians *Sinophoneus yumenensis* Cheng & Li and *Stenocybus acidentatus* Cheng & Li, and the eotitanosuchian *Biseridens qilianicus* Li & Cheng (J. Li and Cheng 1995a, b, 1997a, b; Z. Cheng and Li 1996, 1997). J. Li and Cheng (1995b) argued that this assemblage is of Middle Permian age and correlate to the *Tapinocephalus* zone of the South African Karoo. I consider this assemblage to be the same age as the pareiasaur fauna discussed below.

Pareiasaur Fauna

I use the informal term "pareiasaur fauna" for the vertebrate-fossil assemblage from the upper part of the Shihezi Formation in Jiyuan County, Henan and from the Sunjiagou Formation in Shanxi. This vertebrate fauna includes indeterminate labyrinthodonts; the kotlassiid temnospondyl *Bystrowiana sinica* Young; the pareiasaurs *Shihtienfenia permica* Young & Ye, *Tsiyania simplicidentata* Young, *Honania complicidentata* Young, *Shansisaurus xuecunensis* Cheng, and *Huanghesaurus liulinensis* Gao; the possible gorgonopsian *Wangwusaurus tayuensis* Young; the "procynosuchid" *Hwanghocynodon multienspidus* Young; and the "tapinocephalid" *Taihangshania imperfecta* Young.

The *Bystrowiana* is known from a single vertebra (see figure 6-7) and several plate fragments (Young 1979b). Other coeval temnospondyl specimens are even more fragmentary.

Young (1979b) named *Tsiyania* and *Honania* for isolated teeth. These are clearly of pareiasaurs, but it is doubtful the two genera are valid. Similarly, *Wangwusaurus* is a genus of doubtful validity based on teeth that appear to belong to a gorgonopsian.

Hwanghocynodon and *Taihangshania* are no more clearly established taxa. The former is based on three teeth, two of which are procynosuchid and the other of which is pareiasaurian. The latter is based on supposed tapinocephalid teeth that Sigogneau-Russell and Sun (1979) considered to represent worn or digested pareiasaurian teeth.

Shihtienfenia (see figure 6-8) is the best known Chinese pareiasaur, based on an incomplete skeleton that consists of 20 vertebrae and shoulder and hip girdles. This and other postcranial material of *Shihtienfenia* were collected from the Shiqianfeng Formation at Baode, Shanxi. Nearby at Lishi, Shanxi, several isolated pareiasaur postcrania from the Shiqianfeng Formation provided the basis for *Shansisaurus xuecunensis* Cheng. The only differences between these bones and the bones of *Shihtienfenia* are the more robust humerus of *Shansisaurus*, not a valid taxonomic difference (Sun et al. 1992). Here, *Shansisaurus xuecunensis* is considered a junior subjective synonym of *Shihtienfenia permica*.

A third Shiqianfeng Formation parieasaur is *Huanghesaurus liulinensis* Gao, known from a lower jaw and partial skeleton from Liulin, Shanxi. Sun et al. (1992) correctly synonymize this taxon with *Shihtienfenia*. Thus, *Shihtienfenia*

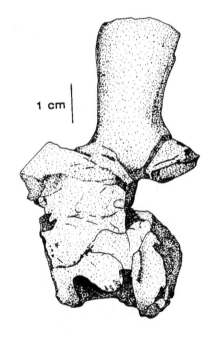

1 cm

Figure 6-7 *Lateral view of dorsal vertebra of Bystrowiana sinica, a rare fossil of a Permian amphibian from China (after Young 1979b).*

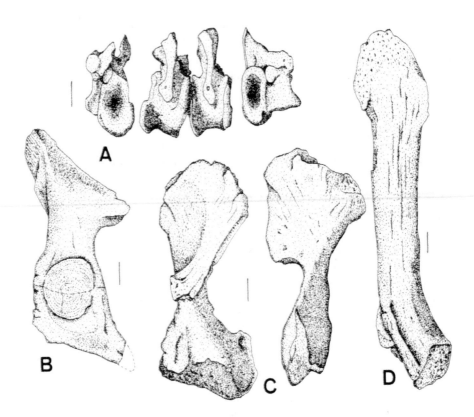

Figure 6-8 *These are selected postcrania of the holotype of the Permian pareiasaur Shitienfenia permica: A, Dorsal vertebrae. B, Innominate. C, Left humerus. D, Scapula. Bar scales = 5 cm. After Young and Ye (1963).*

(= *Shansisaurus*, = *Huanghesaurus*) is the best known, most common and last Chinese pareiasaur. This large animal has many typical pareiasaur features, including teeth with laterally compressed, leaf-shaped crowns, amphicoelous vertebrae, a very long scapula, a strikingly mammal-like pelvis, and very robust limbs. The pareiasaur fauna vertebrates are temnospondyls, a possible gorgonopsian, and, dominantly, pareiasaurs. No dicynodonts are known from Henan.

Because this fauna is largely based on fragmentary material, interpretation of its age is tentative. Z. Cheng (1980b) correlated this fauna with the "*Endothiodon* zone" of the South African Karoo. The term "*Endothiodon* zone" of Broom (1906) was abandoned by Kitching (1970) and subsumed into the *Tropistodoma* assemblage zone by Keyser and Smith (1977). This zone is dominated by fossils of dicynodonts (especially *Tropistodoma* and *Rhachiocephalus*) and gorgonopsians. A surviving cotylosaur is *Pareiasuchus*.

The pareiasaur fauna lacks dicynodonts, is dominated by pareiasaurs, and has only one possible gorgonopsian. It therefore seems quite different from, and older than, the South African *Tropistodoma* zone. A more likely correlation of the pareiasaur fauna is with the older, pareiasaur-rich dinocephalian assemblage zone of Keyser and Smith (1977), which is the lower two-thirds of the old *Tapinocephalus* zone. Nevertheless, given the fragmentary nature of the Chinese pareiasaur fauna vertebrate-fossil assemblage, this correlation must be considered tentative.

Another tentative correlation is to equate the vertebrate faunas of the upper Jijicao Group in the Junggur basin and the pareiasaur fauna vertebrates (Z. Cheng 1980b). Indeterminate temnospondyls and palaeonisciform fishes (e.g., *Turfania*) are the vertebrate faunas of the upper Jijicao Group. These lacustrine vertebrate taxa do not occur in the pareiasaur fauna assemblage. Only stratigraphic position (both the upper Jijicao and pareiasaur fauna vertebrates are beneath *Dicynodon*-biochron-age strata, discussed below) can be used to support broad correlation of the upper Jijicao and pareiasaur fauna vertebrate-fossil assemblages.

Dicynodon Fauna

China's youngest Permian vertebrates are informally referred to here as the *Dicynodon* fauna. In the Junggur basin, this encompasses the fossil vertebrates from the lower Cangfanggou Group, the Quanzijie Formation, Wuttonggou Formation, and the overlying lower-middle Guodikeng Formation. This stratigraphic interval is more than 600 m thick, and vertebrate distribution is patchy and not prolific (see figure 6-3). Subsuming this entire biostratigraphic assemblage into a single "fauna" thus produces relatively coarse temporal resolution. However, further discoveries are needed to allow subdivision of the *Dicynodon* fauna.

Dicynodon fauna vertebrates are all dicynodonts: *Jimusaria sinkiangensis* (Yuan & Young), *Jimusaria taoshuyuanensis* Sun, *Kunpania scopulusa* Sun, *Striodon magnus* Sun, *Turfanodon bogdaensis* Sun, and *Diictodon tienshanensis* (Sun).

Yuan and Young (1934a) first described a dicynodont from China when they coined the name *Dicynodon sinkiangensis* for a complete skull and lower jaw from the Guodikeng Formation. Sun (1973) later transferred this species to her new genus *Jimusaria*, including *Jimusaria taoshuyuanensis*, based on the anterior portions of two skulls and the ventral aspect of a third, all from the Guodikeng Formation at Taoshuyuanzi in the Turpan physiographic basin. Sun (1973: 53) diagnosed *Jimusaria* as follows:

Medium-sized dicynodont; skull slightly triangular in shape; snout small
and narrow; orbit opens dorsally and laterally; interorbital region narrow;
interparietal width = 2/3 width of interorbital region; parietal ridge not
much elevated and parietal bone deep; tusk projects anteriorly and down-
ward; ectopterygoid present; interpterygoid fossa length = 30% of skull
length.

Sun distinguished *J. sinkiangensis* from *J. taoshuyuanensis* by the flat (not
curved) posterior margin of the maxillary process bearing the tusk in the latter.

Jimusaria clearly is synonymous with *Dicynodon*, so Yuan and Young's original generic assignment was correct. *Dicynodon* was a common and widespread
Late Permian dicynodont (see figure 6-9) of medium to large size (skull lengths
range from 100 mm to more than 400 mm). Key features of *Dicynodon* (Cluver and Hotton 1981) include:

- a single pair of maxillary tusks in the upper jaw

- an edentulous lower jaw

- post-orbitals that cover the parietals behind the parietal foramen

- an exposed septomaxilla that does not meet the lacrimal and merges
  smoothly with the outer surface of the snout

- a low boss formed by the nasals above the external nares

- a ventral extension of the palatal rim forms the carniniform process

- a sharp-edged continuous palatal rim with a notch

- a large exposure of the palatine on the palate that contacts the premaxilla

- a short interpterygoid vacuity

- vomers forming a long narrow septum in the interpterygoid fossa

- a small and laterally displaced ectopterygoid

- a labial fossa between the maxilla, palatine, and jugal

- a short contact between the pterygoid and maxilla

- an intertuberal ridge between the basioccipital tubera

- fused dentaries with narrow dentary tables followed by a deep dentary
  sulcus

- a weak coronoid process

- a large mandibular fenestra bounded dorsally by a lateral dentary shelf

Not only do the specimens of "*Jimusaria*" fit well within this morphological
concept of *Dicynodon*, but most of the other Chinese Late Permian dicyn-

odonts fit here as well. *Kunpania scopulusa* (see figure 6-9) is known from part of a skull, a lower jaw, and part of the forelimb from the top of the Quanzijie Formation at Gongbangou, Xinjiang (Sun 1978). This material is of a large dicynodont that cannot be distinguished from *Dicynodon* except by the unusually long mandibular fenestra, lateral shelf and fossa dorsal to that shelf. This difference does not merit generic separation, and *Kunpania scopulusa* is here termed *Dicynodon scopulusa*.

Striodon magnus (see figure 6-9) is based on the posterior part of a skull from the Guodikeng Formation at Dongxiaolongkou, Xinjiang. This is a large dicynodont, with an estimated total skull length of more than 600 mm. It shows no morphological differences from specimens of *Dicynodon*, but the *Striodon* skull is not diagnostic because it lacks the face and rostrum. Here, *Striodon magnus* is regarded as a *nomen dubium*, and its type specimen is identified as *Dicynodon* sp.

Turfanodon bogdaensis is better known than both *Kunpania* and *Striodon*. It is based on a nearly complete skull from the Guodikeng Formation at Taoshuyuan in the Turpan physiographic basin. King (1988: 90) assigned this species to

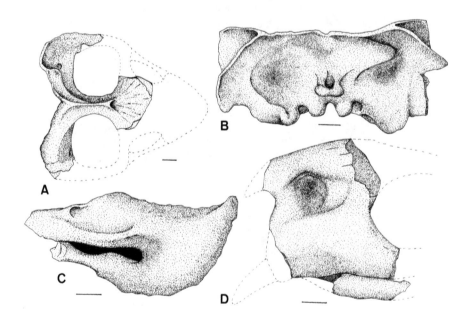

Figure 6-9 *The holotypes of two Permian dicynodonts, Striodon and Kunpania, both synonyms of Dicynodon. A–B, Striodon magnus, dorsal (A) and posterior (B) views of partial skull. C–D, Kunpania scopulusa, right lateral view of lower jaw (C) and left lateral view of part of snout (D). Bar scales = 5 cm. After Sun (1978).*

Dicynodon as *D. bogdaensis,* a decision well grounded in its morphology, which is that of a typical, relatively large (skull length about 300 mm) *Dicynodon.*

The only Late Permian dicynodont from the Junggur basin not assignable to *Dicynodon* is *Diictodon tienshanensis* (Sun, 1973) (see figure 6-10). In the uppermost Permian strata of the Karoo basin of South Africa, *Diictodon* and *Dicynodon* are the two common genera of dicynodonts. They are readily distinguished by skull and lower jaw morphology (see table 6-1). Particularly important features of this distinction are that, unlike *Dicynodon,* *Diictodon* lacks tusks, has a notched palatal rim and has a dentary without a dorsal sulcus (Cluver and Hotton 1981). Sun (1973) originally described *Dicynodon tienshanensis,* but Cluver and Hotton (1977, 1981) reassigned this species to *Diictodon.* The tuskless skull with its notched palatal rim (see figure 6-10) belongs to the genus *Diictodon.*

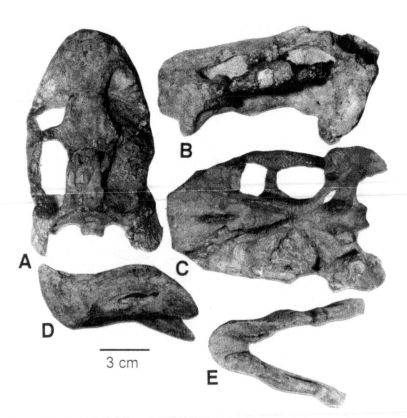

Figure 6-10 *The holotype of the Permian dicynodont Diictodon tienshanensis. A–C, Skull, dorsal (A), left lateral (B) and ventral (C) views. D–E, Lower jaw, lateral (D) and occlusal (E) views.*

Table 6-1 Comparison of Cranial Characters of *Dicynodon* and *Diictodon* (after Cluver and Hotton 1981)

Cranial Characters	*Dicynodon*	*Diictodon*
postcaniniform crest	no	no
palatal rim	continuous	notched
inter-temporal region	narrow	narrow
tusks	present	absent
dorsal sulcus in dentary	yes	no
dentary tables	yes	yes
weak dentary shelf	yes	yes
septomaxilla	exposed	recessed
palatine	large	small

Therefore, in the Upper Permian strata of the Junggur basin, only two dicynodont genera are known—abundant *Dicynodon* and rare *Diictodon*. In the Ordos basin, only one identifiable dicynodont of Taoshuyuanian age is known. This is *Daqingshanodon limbus* Zhu from the Naobaogou Formation at Shiguai, Nei Monggol (see figure 6-11). Y. Zhu's (1989) diagnosis of this taxon mentions many features diagnostic of *Dicynodon*, and does not identify *Daqingshanodon* as a distinct genus, so I synonymize it with *Dicynodon*. The Naobaogou Formation occurrence of *Dicynodon limbus* thus extends the *Dicynodon* fauna into the Ordos basin.

The *Dicynodon* Biochron

Clear recognition of *Dicynodon* in northern China further establishes the cosmopolitanism of this Late Permian dicynodont genus. The distribution of *Dicynodon* establishes a *Dicynodon* biochron of Late Permian age recognized at the following locations (see figure 6-12):

1. Karoo basin, South Africa, where specimens of *Dicynodon* (= *Daptacephalus*) first occur in the Upper *Cistecephalus* Assemblage Zone and are the dominant tetrapod fossils in the *Dicynodon* Assemblage Zone of the Teekloof and Balfour formations (Beaufort Group) (Kitching, 1995).

2. The type of *D. roberti* (Boonstra 1938) is from "Horizon 5" of Boonstra (1953) in the Nt'ware Formation of the Luangwa Valley, Zambia. The

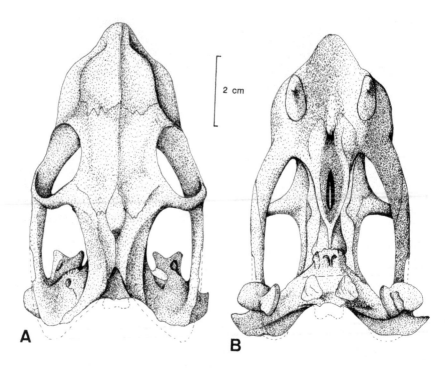

Figure 6-11 *Holotype of the dicynodont Daqingshanodon limbus, dorsal (A) and ventral (B) views of skull. After Y. Zhu (1989).*

skeleton of *Dicynodon* described by King (1981) is also from the Luangwa Valley (Kemp 1975).

3. The *lower bone bed* at Kingori in the Ruhuhu basin of Tanzania (Haughton 1932). Haughton (1932) and Huene (1942) named three species of *Dicynodon* from the Ruhuhu basin.

4. Cutties Hillock quarry, Elgin, Scotland (Newton 1893; King 1988) has produced fossils of *Dicynodon* (= *Gordonia*).

5. Northern Dvina fauna, near Kotlass, Russia, Tatarian zone IV of Efremov (1937) also yielded fossils of *Dicynodon* (Amalitzky 1922; Sushkin 1926).

6. The Quanzijie, Wutonggou, and Guodikeng Formations of the Junggur basin, Xinjiang, China, as reviewed above.

7. The Naobaogou Formation of the Ordos basin, Nei Monggol, China, just discussed.

8. North of the Mekong River in the Luang-Prabang area of Laos (Battail et al. 1995).

The similarity of the dicynodonts from these dispersed localities forms a powerful argument for the assembly of Pangea by Late Permian time. Yet many recent plate-tectonic reconstructions of Late Permian Pangea show a clear marine separation of the north China and Kazakstan blocks from the rest of the supercontinent (e.g., Ziegler 1990; Golonka et al. 1994; Metcalfe 1988, 1996). This implies that the *Dicynodon*-biochron-aged tetrapods dispersed across water bodies to achieve their broad Pangean distribution. Although such dispersal is, of course, possible, it seems more likely the fully terrestrial dicynodonts of the Late Permian dispersed over a single landmass (see figure 6-12), as has been well argued for their successors, the *Lystrosaurus*-biochron tetrapods of the Early Triassic.

Chinese Late Permian vertebrate localities come from two microplates, the Kazakstan (Junggur basin) and north China (Ordos basin) blocks. Perhaps the greatest lesson Chinese Permian vertebrates have to teach us is that these microplates had joined Pangea by Late Permian time. China's youngest Permian vertebrates are part of a dicynodont-dominated land-vertebrate fauna that was present throughout the vast supercontinent.

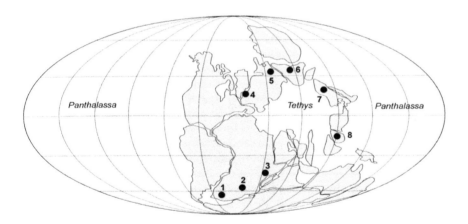

Figure 6-12 *This reconstruction of Late Permian Pangaea shows localities of the Dicynodon biochron. 1 = South Africa, 2 = Zambia, 3 = Tanzania, 4 = Scotland, 5 = Russia, 6 = Junggur Basin, China, 7 = Ordos Basin, China, 8 = Laos.*

Chapter 7

Triassic

During the Triassic Period, the microplates of China joined the easternmost portion of the assembled Pangean supercontinent (see figure 7-1). On the Kazakstan and north China blocks, nonmarine deposition took place, but the south China block remained a site of marine deposition (e.g., D. Qiu 1990). Today, this separation is well delineated by an east-west line drawn through the Kunlun Shan (Xinjiang) and the Dabie Shan (Hubei-Anhui), which essentially separates nonmarine Triassic rocks to the north from marine Triassic rocks to the south.

Chinese Triassic vertebrate fossils are mostly found in the northern part of the country and are almost exclusively of Early and Middle Triassic age. Two ancient sedimentary basins, the Junggur and Ordos, contain most of the Triassic vertebrate-bearing strata in China (see figure 7-2). Other occurrences are confined to isolated specimens, except for the important record of marine reptiles (especially ichthyosaurs and sauropterygians) from marine strata across southern and eastern China. The Early-Middle Triassic succession of vertebrate faunas from the Junggur and Ordos basins resembles correlative faunal successions in the Karoo basin of South Africa and in the Urals of central Russia. However, the virtual absence of a Late Triassic vertebrate fauna from China is one of the most significant gaps in China's Mesozoic vertebrate record.

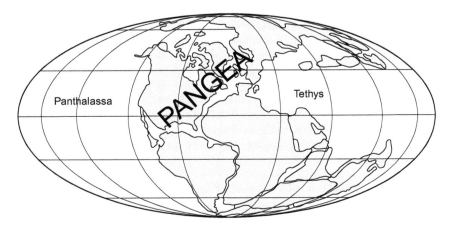

Figure 7-1 *By Triassic time the Chinese microplates of Kazakstan, north China, and Tarim were sutured to the Pangean supercontinent.*

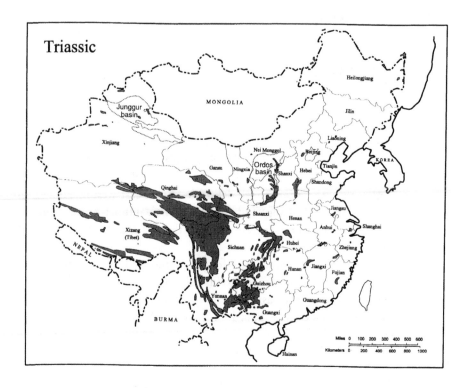

Figure 7-2 *Triassic rocks are widely distributed in China, but mostly concentrated in the south. Most terrestrial vertebrate fossils of Triassic age are found in the Junggur and Ordos basins (modified from Yang et al. 1986).*

Junggur Basin

Today, the Junggur basin of northern Xinjiang is located between the Tien Shan to the south, and the Altai Mountains to the north (see figure 1-4). Roughly triangular in shape, the present Junggur basin has an area of about 140,000 km² and has a sedimentary fill more than 11 km thick. During the Permian and Triassic, no Tien Shan existed, so the southern margin of the Junggur basin extended to the southeast into what is now the Turpan physiographic basin. Similar successions of Upper Permian-Triassic nonmarine strata and vertebrate fossil assemblages are now found both north and south of the Tien Shan (see figure 7-3).

Triassic strata in the Junggur basin belong to five formations (in ascending order):

1. Guodikeng Formation—Only the upper 30 m of the Guodikeng Formation are considered to be Triassic. These rocks contain the lowest occurrence of the dicynodont *Lystrosaurus*, usually taken to mark the base of
   the Triassic. However, they also produce the stratigraphically highest
   specimens of the characteristically Permian dicynodont *Dicynodon* (=
   Jimusaria). This creates an overlap zone that raises real questions about
   the placement of the Permian-Triassic boundary using vertebrate fossils
   (see figure 7-4). The 30 m of the uppermost Guodikeng Formation considered Triassic are mostly purplish-red, silty mudstones and siltstones
   that are finely laminated. They contain abundant conchostracans and are
   obviously lacustrine deposits.

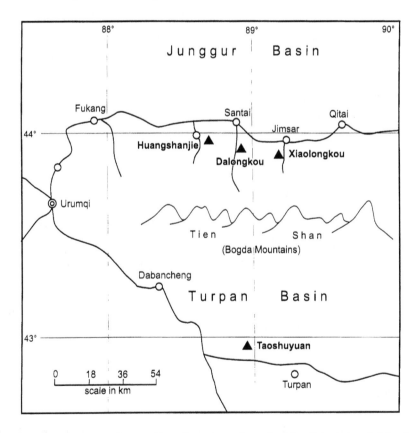

Figure 7-3 *Generalized map of key Triassic vertebrate fossil localities (triangles) in the
Junggur and Turpan physiographic basins of Xinjiang.*

2. Jiucaiyuan Formation—As much as 370 m thick, the Jiucaiyuan Formation conformably overlies the Guodikeng Formation. It consists of purplish and dark-red mudstones with numerous calcrete nodules and grayish-green sandstones of fluvio-deltaic origin. A significant vertebrate fauna dominated by *Lystrosaurus* comes from the Jiucaiyuan Formation and represents most of the Jimsarian land-vertebrate faunachron, discussed below.

3. Shaofanggou Formation—The Shaofanggou Formation is as much as 350 m thick and is a red-bed sequence of mudstones, conglomerates, and sandstones of lacustrine and fluvial origin. No vertebrate fossils have been reported from it.

4. Kelamayi (= Karamay) Formation—The Kelamayi Formation is usually divided into two members, lower and upper. The lower member is as much as 120 m thick and consists of purplish-red sandstones and clayey siltstones interbedded with grayish-green sandstone. These strata are of fluvial origin and contain fossil vertebrates of Ningwuan age. The upper Kelamayi Formation is yellowish-green and grayish-black sandstones, mudstones and shales as much as 380 m thick. An extensive flora of Middle-Late Triassic age (Ladinian-Carnian) is known from these strata. However, only the few vertebrates of the "Fukang fauna" (see below) have been recovered from the upper Kelamayi Formation.

5. Huangshanjie and overlying Haojiagou formations—These units are a coalbearing sequence of grayish-yellow and green mudstones and sandstones as much as 830 m thick. An extensive flora of Late Triassic age is known, but no vertebrates have been recovered.

Triassic deposition in the Junggur basin was largely in lake basins. Most remarkable is the relative continuity of this deposition, especially through the Permian-Triassic boundary interval (figures 6-3 and 7-4).

Ordos Basin

The rectangular-shaped Ordos basin is located north of the great bend of the Huang He and encompasses parts of Nei Monggol, Shanxi, Ningxia, Gansu, and Shaanxi (see figure 1-4). (Indeed, some Chinese authors call this basin the Shaanganning basin, a contraction of the names of the three principal prov-

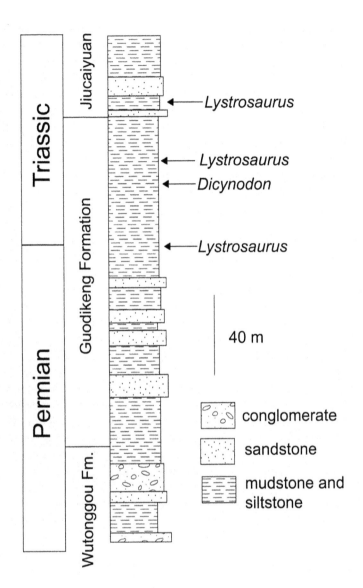

Figure 7-4 *Triassic stratigraphy and vertebrate fossil distribution across the Permian-Triassic boundary in the Junggur basin.*

inces it encompasses—Shaanxi, Gansu, and Ningxia.) The Ordos basin encompasses an area of more than 250,000 km^2 surrounded by mountains, the Qinling to the south, Daqing and Lang to the north, Luliang to the east and Liupan and Helan to the west. Its sedimentary fill is many kilometers thick; the Triassic strata alone are more than 1500 m thick.

As in the Junggur basin, the Triassic strata of the Ordos basin are fluvio-lacustrine red-bed siliciclastics that are coal bearing at their top. Five formations (in ascending order) are present:

1. Liujiagou Formation—As much as 630 m thick, this unit consists of purplish-red and grayish-purple mudstone, siltstones, conglomerates and cross-bedded sandstones. No vertebrates are known, but in Shaanxi, palynomorphs, megafossil plants and conchostracans suggest an earliest Triassic age.

2. Heshanggou Formation—Conformably overlying the Liujiagou Formation, this unit is brick-red, purplish-red, and purplish-gray mudstones and arkosic sandstones as much as 280 m thick. The oldest Triassic vertebrates found in the Ordos basin come from the Heshanggou Formation and are characteristic of the Fuguan land-vertebrate faunachron.

3. Ermaying Formation—Perhaps the best known vertebrate-producing stratigraphic unit in China is the Ermaying Formation. Numerous articles on its vertebrate fossils by C. C. Young, Sun Ailing and others have brought the Ermaying much international attention. As much as 600 m thick, the Ermaying Formation is a complex sequence of red-bed mudstones intercalated with grayish-green and yellow sandstones. Usually the Ermaying is divided into lower and upper members. The lower member is mostly sandstone, whereas the upper member is intercalated sandstones and mudstones. Both members produce distinctive vertebrate fossil assemblages dominated by kannemeyeriid dicynodonts. The lower member contains Ordosian age vertebrates, whereas the upper member produces the vertebrates characteristic of the Ningwuan land-vertebrate faunachron.

4. Tongchuan Formation—This unit conformably overlies the Ermaying and is as much as 600 m thick. The Tongchuan Formation is yellowish-green and greenish-gray sandstones, mudstones, and shales capped by coal beds. It mostly represents lacustrine deposition and includes extensive assemblages of palynomorphs, megafossil plants (Tongehuan flora), conchostracans, ostracods, insects, and bivalves. A single fossil vertebrate, the hybodontid selachian *Hybodus youngi* Liu, is known from the Tongchuan Formation. This genus has a temporal range of Middle Triassic-Late Cretaceous in rocks of mostly marine origin (Cappetta, 1987).

5. Yanchang Formation—The youngest Triassic strata in the Ordos basin belong to the Yanchang Formation, which is as much as 753 m thick. The Yanchang Formation is usually divided into three members: (a) a lower member of massive, grayish-green sandstones and some mudstones; (b) a middle member of thick-bedded grayish-green sandstones and mudstones intercalated in repetitive beds; and (c) an upper member of dark gray mudstones and siltstones with coal beds. No vertebrate fossils are known from the Yanchang Formation, but like the underlying Tongchuan Formation, it yields extensive assemblages of palynomorphs, megafossil plants, conchostracans, ostracods, insects, and bivalves of Late Triassic age.

China's Triassic vertebrate fossils are mostly confined to the Junggur and Ordos basins. It is indeed remarkable how similar the stratigraphic succession and fossil biotas are in these two basins. Clearly, similar tectonic and climatic events drove sedimentation in these two large, closed, drainage basins during the Triassic.

Jimsarian Vertebrates

The oldest Mesozoic vertebrate fauna from China is from the upper part of the Guodikeng Formation and the lowermost Jiucaiyuan Formation (both in the Cangfanggou Group), near Jimsar northeast of Urumqi (Pinyin: Wulumuqi) in western Xinjiang (see figure 7-4). The Jimsarian land-vertebrate faunachron is the time equivalent to these vertebrate fossils (Lucas 1993b, 1996c). These vertebrates are the "*Lystrosaurus* fauna" of northwestern China of some earlier workers (e.g., Sun 1972). Taxa present (Z. Cheng 1980a, table 10) are: the prolacertid protorosaur *Prolacertoides jimusarensis* Young; the eolacertilian *Santaisaurus yuani* Koh; the proterosuchian *Chasmatosaurus yuani* Young; the regisaurid therocephalian *Urumchia lii* Young; and the dicynodont *Lystrosaurus* (see figure 7-5), of which seven species have been named, most of which are not valid (Colbert 1974): *L. youngi* Sun (= *L. curvatus*: Colbert 1974), *L. weidenreichi* Young (a *nomen dubium* based only on postcrania), *L. robustus* Sun, *L. latifrons* Sun, *L. hedini* Young, *L. broomi* Yuan & Young (= *L. murrayi*: Colbert, 1974) and *L. shichanggouensis* Cheng.

Lystrosaurus (see figure 7-5) is the most abundant fossil in this assemblage, a dominance characteristic of age-equivalent vertebrate fossil assemblages outside of China, especially in the Katberg Formation of the Karoo basin in South Africa. *Prolacertoides* is known from the anterior part of a skull. It has a long,

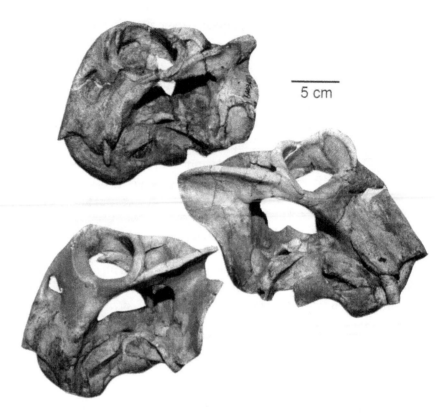

5 cm

Figure 7-5 *These Lystrosaurus skulls from the Lower Triassic of the Junggur basin are the most characteristic fossil vertebrates of the Jimsarian land-vertebrate faunachron.*

pointed snout with long, ellipsoidal external nares. About 20 closely spaced marginal teeth are present, but the pterygoid teeth are rudimentary

The possible Jimsarian eolacertilian, *Santaisaurus*, is better known than *Prolacertoides*, being represented by three incomplete skeletons. The best preserved of these includes a nearly complete skull, lower jaw, and partial postcrania. The short rostrum, large orbit, t-shaped interclavicle, and amphicoelous vertebral centra of *Santaisaurus* suggest inclusion in the Procolophonidae (Romer 1966; Carroll 1988). However, *Santaisaurus* has small, subpleurodont teeth, not the acrodont ("proto-thecodont") teeth of procolophonids. Primarily for this reason, Romer (1956), and Sun et al. (1992) most recently, assigned *Santaisaurus* to the Eolacertilia.

Chasmatosaurus is a well-known, rather crocodile-like proterosuchian that was first described from South Africa (Haughton 1924). In the Jiucaiyuan Formation, the genus is well represented by two skulls and skeletons, one of which is nearly complete (Young 1936b). The Chinese *Chasmatosaurus* is much smaller than the genotypic species from South Africa, *C. vanhoepeni*. It has a slender rostrum with 28 to 30 teeth and posteriorly located choanae. The recurved, serrated (on both edges) teeth show some differentiation along the tooth row. The lower jaw is massive with a thick symphysis.

When originally described, *Urumchia* was assigned a Permian age (Young, 1952). However, its type locality was later determined to be a Lower Triassic horizon of the Jiucaiyuan Formation (Sun 1991). *Urumchia* is known only from the type skull, which greatly resembles that of the therocephalian *Regisaurus* from the Lower Triassic of South Africa (Mendrez 1972). Indeed, *Urumchia* only differs from *Regisaurus* by being larger and having a flat (not pointed) anterior process of the vomer (Sun 1991). These minor differences probably merit species-level separation of the two taxa at most, and thus the Chinese form might more properly be termed *Regisaurus lii* (Young 1952).

The dicynodont *Lystrosaurus* is discussed below. At the base of the Jimsarian interval at Dalongkou, in the uppermost Guodikeng Formation, the "characteristically Permian" dicynodont *Dicynodon* (= *Jimusaria*) *sinkianensis* (Yuan and Young 1934a) co-occurs with *Lystrosaurus* over a stratigraphic interval about 30 m thick of mostly purplish-red silty mudstone (Z. Cheng 1980a; Z. Cheng and Lucas 1993; Lucas 1993b). A similar overlap zone of *Dicynodon* and *Lystrosaurus* is known from South Africa in an approximately 15-m-thick interval at the base of the Palingkloof Member of the Balfour Formation (Smith 1993). *Dicynodon* disappears at the top of the overlap zone in both China and South Africa, and then *Lystrosaurus* is the abundant dicynodont of the vertebrate fauna. In South Africa, major changes in fluvial style (from meandering to incised anastomosed channels) and climate (wetter to drier) accompanied this "replacement" of *Dicynodon* by *Lystrosaurus*. But, in China no evident facies change took place.

Fuguan Vertebrates

Near Fugu, Shanxi, the upper part of the Heshanggou Formation produces a vertebrate fauna (see figure 7-6) that is the basis of the Fuguan land-vertebrate faunachron. Taxa present are the lungfish *Ceratodus heshanggouensis* Cheng, indeterminate capitosauroid labyrinthodonts (Cheng,1980b; Lucas and Hunt, 1993a), the procolophonids *Eumetabolodon bathycephalus* Li and *E. dongshengensis* Li, the proterosuchian *Xilousuchus sapingensis* Wu, and the erythrosuchid *Fugusuchus hejiapensis* Cheng (see figure 7-7), and the ordosiid therocephalian *Hazhenia concava* Sun & Hou.

Ceratodus heshanggouensis is known only from toothplates, which are very similar to those of *Ceratodus donensis* from the Early Triassic Baskunchak Series

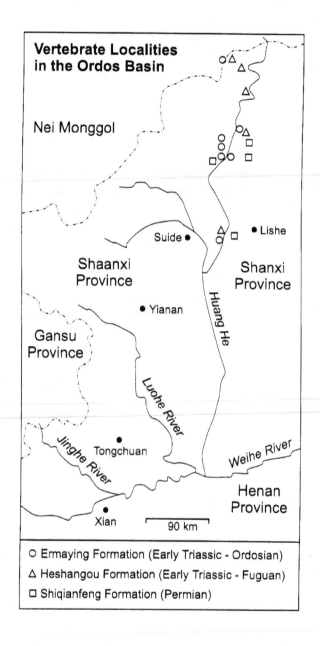

Figure 7-6 *Major fossil vertebrate localities of the Ordosian and Fuguan land-vertebrate faunachrons in the Ordos basin (modified from Z. Cheng 1980b).*

of Russia. Indeed, the Chinese lungfish taxon is almost certainly based on a specimen that Vorobyeva and Minikh (1968) would have identified as *C. donensis donensis* (compare Cheng 1980b, figure 21 to Vorobyeva and Minikh 1968, pl. 14, figure 11).

Cheng (1980b: 122–24, pls. 129–30) described and illustrated skull, girdle, and jaw fragments and vertebral centra from the Heshanggou Formation he identified as benthosuchid and capitosaurid. However, these fossils are not diagnostic of either family, so Lucas and Hunt (1993a) identified them only as capitosauroid.

The procolophonid *Eumetabolodon* is known from numerous skulls collected at localities in Shaanxi and Nei Monggol. Prior to their discovery, only two procolophonid fossils were known from China (the holotypes of *Neoprocolophon* and *Paoteodon*, see below). The triangular skull, large orbits and small number of transversely broad teeth are typical procolophonid features of *Eumetabolodon*. It is similar to *Procolophon* and *Koiloskiosaurus* from the Early Triassic of South Africa and western Europe, respectively. Nevertheless, the short high skull, short snout, anteriorly positioned pineal foramen partly bordered by the frontals, and the long and posteriorly located lower jaw articulation of *Eumetabolodon*, are unique features among procolophonids. The large number (18 total) of skulls of *Eumetabolodon* fall into four size classes and allow tooth replacement during ontogeny to be analyzed. This analysis (J. Li 1983) indicates that the conical postcanine teeth of young procolophonids were replaced by transversely broad, bicuspid teeth later in life.

Xilousuchus is a medium-sized reptile known from parts of the skull and very little postcrania. Originally assigned to the proterosuchians by X. Wu (1981), J. Peng (in Sun et al 1992) suggests it may be an erythrosuchid because of the notch between the premaxilla and maxilla, the strong and distally expanded paraoccipital processes, the maxilla making up part of the margin of the narial opening, the extremely large external nares, and the absence of intercentra.

Fugusuchus (see figure 7-7) is an unquestioned erythrosuchid known principally by its skull. Parrish's (1992) phylogenetic analysis of the erythrosuchids identified *Fugusuchus* as the most primitive member of the family. *Fugusuchus* does have intercentra, but as Parrish observes, some other erythrosuchids apparently also have intercentra. The key primitive feature of *Fugusuchus* is its long upper tooth row that extends back under the orbit. More derived erythrosuchids only have upper teeth anterior to the orbit.

Hazhenia is a therocephalian somewhat similar to Jimsarian *Urumchia*. However, *Hazhenia* is more advanced; note, for example that its secondary palate is formed exclusively by two palatal processes of the maxillaries, and the postcanine teeth are cylindrical with definite crown structure. The extremely

Figure 7-7 *Lateral view of the skull of the erythrosuchid Fugusuchus hejiapensis from the Lower Triassic of the Ordos basin. Note the primitive erythrosuchid feature of teeth extending back under the orbit.*

large lower canines of *Hazhenia* fit into large openings in the front top of the palate when the jaws were closed.

The Fuguan fauna is unusual among Chinese Triassic vertebrate faunas because it lacks dicynodonts. This probably reflects no more than a lack of discovery, not a real absence of dicynodonts during Fuguan time. Z. Cheng (1980b, 1981) referred to the Fuguan fauna as a "labyrinthodont-procolophonid fauna," even though both labyrinthodonts and procolophonids occur in Jimsarian and Ordosian faunas. Furthermore, Z. Cheng (1980b, 1981) correlated the Fuguan with the "*Procolophon* zone" of the South African Karoo. "*Procolophon* zone" is an outmoded term used by Broom (1905) and Watson (1914) but rejected by later workers (Huene 1925; Hotton and Kitching 1963; Keyser and Smith 1977) because *Procolophon* also occurs in the *Lystrosaurus* zone. Without dicynodonts it is difficult to evaluate the global correlation of the Fuguan, but it probably is the same age as the "*Procolophon* zone" of South Africa and thus of Early Triassic (Induan) age. In the Junggur basin of Xinjiang, the Shaofanggou Formation, which overlies the Jiucaiyuan Formation, produces indeterminate labyrinthodonts and *Lystrosaurus* of probable Fuguan age (Cheng 1980a).

Ordosian Vertebrates

The lower Ermaying Formation in the Ordos basin contains a vertebrate fauna, the time equivalent of which is the Ordosian land-vertebrate faunachron, named for the Ordos basin (Lucas 1993b, 1996c). Vertebrate taxa present are: the procolophonid *Paoteodon huanghoensis* Chow & Sun; the proterosuchian *Guchengosuchus shiguainensis* Peng; the euparkeriids *Halazhaisuchus giaoensis* Wu and *Turfanosuchus shangeduensis* Wu; the ordosiid therocephalians *Ordosiodon* (= *Ordosia*: Sigogneau-Russell and Sun, 1979) *linchenyuenensis* Young and *O. youngi* Hou; the therocephalian *Yikezhaogia megafenestrala* Li; and the dicynodonts *Parakannemeyeria xingxianensis* Cheng, *Kannemeyeria* (= *Shaanbeikannemeyeria*) *sanchuanheensis* Cheng and *K. buerdongia* Li.

Chow and Sun (1960) named *Paoteodon* for a fragmentary piece of a maxilla with three teeth from Baode County, Shanxi. They interpreted the fossil as a maxillary fragment from the middle part of the tooth row, but Sun et al. (1992) suggest it is actually the posterior portion of the maxilla.

J. Peng (1991) recently named *Guchengosuchus* for an incomplete skull and partial skeleton from Fugu County, Shaanxi. This erythrosuchid is very similar to *Vjushkovia* (see below) and probably represents an early, distinct species of that genus, *Vjushkovia shiguaiensis* (Peng).

Of the two Ordosian euparkeriids, *Turfanosuchus* is the best known, being represented by a right mandible and partial skeleton. No cranial material of *Halazhaisuchus* is known, and the taxon is based on a partial vertebral column and forelimb. With southern African *Euparkeria* and *Wangisuchus* (see below), the two Ordosian genera make up the Euparkeriidae, a very distinct group of thecodonts. Euparkeriids were small thecodonts (less than 1 m long) with slender limbs, long tails and dermal scutes that ran along the trunk and tail vertebral column. These animals are usually portrayed as bipeds because of their relatively short forelimbs (two-thirds the length of the hind limbs).

The therocephalian *Ordosiodon* is known from cranial and postcranial specimens. It resembles Fuguan *Hazhenia* in many features, but has a shorter snout, smaller anterior teeth and larger postcanines. *Yikezhaogia* from Nei Monggol is a possible therocephalian known from a partial skull, lower jaw and some postcrania. It shows some similarities to other therocephalians but differs from them in having postcanine teeth of identical size and morphology (cylindrical with blunt tips).

Parakannemeyeria (see figure 7-8) has its oldest occurrence in Ordosian strata. *Shaanbeikannemeyeria* clearly is the same genus as the widespread Pangean Early-Middle Triassic dicynodont *Kannemeyeria* (see below).

Most Chinese workers have considered the lower Ermaying Formation (and hence the Ordosian) to be of Middle Triassic age (also see Ochev and Shishkin, 1989, table 2). This seems doubtful, given that *Shansiodon* from the upper Ermaying Formation (Ningwuan) indicates an early Anisian age (see below). I thus consider the Ordosian to be late Early Triassic (Olenekian).

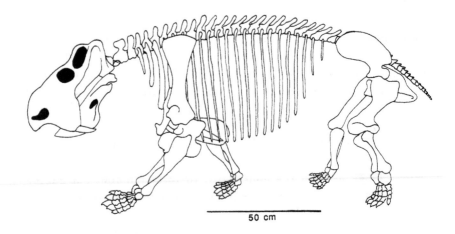

Figure 7-8 *Parakannemeyeria was an Early Triassic dicynodont endemic to China (after Sun 1963).*

Ningwuan Vertebrates

The upper part of the Ermaying Formation in the Ordos basin contains what has been called the "*Sinokannemeyeria* fauna" or "kannemeyeriid fauna" of China (e.g., Sun 1972; Cheng 1981). The Ningwuan land-vertebrate faunachron is the time equivalent to this fauna (Lucas 1993b, 1996c). Ningwu is a city on the Sanggan He in northern Shaanxi near the principal fossil-vertebrate localities. The vertebrate fauna consists of indeterminate labyrinthodonts; the procolophonid *Neoprocolophon asiaticus* Young; the proterosuchian "*Chasmatosaurus*" *ultimus* Young; the erythrosuchids *Shansisuchus shansisuchus* Young (see figure 7-9) and *S. kuyeheensis* Cheng; the ornithosuchid *Fenhosuchus cristatus* Young; the euparkeriid *Wangisuchus tzeyii* Young; the cynodont *Sinognathus gracilis* Young; and the dicynodonts *Shansiodon wangi* Ye, *S. wuhsiangensis* Ye, *S. wupuensis* Cheng, *Sinokannemeyeria pearsoni* Young, *Sino. sanchuanheensis* Cheng, *Parakannemeyeria dolicocephala* Sun, *P. youngi* Sun, *P. ningwuensis* Sun and *P. shenmuensis* Cheng.

A variety of skull and jaw fragments, isolated centra and assorted girdle and limb elements of labyrinthodonts have been reported from the upper Ermaying Formation (Huene 1958; Sun 1972, 1989; Lucas and Hunt 1993b). These fossils cannot be identified more precisely than capitosauroid (Lucas and Hunt 1993a).

Neoprocolophon is known from a single skull from Yushe, Shanxi (Young 1957). This skull closely resembles that of Lower Triassic *Procolophon* from Antarctica and South Africa, but differs in some key features, including the very anterior placement of the quadratojugal "horns" in the Chinese genus.

Shansisuchus (see figure 7-9) is one of the best known erythrosuchids. Hundreds of isolated skull and postcranial bones are known from the upper Ermaying Formation (Young 1964a). This large erythrosuchid (reconstructed body length is about 3 m) has a very large head, a fenestra below the nares, large recurved blade-like teeth, a heavy lower jaw with a thick symphysis, stout limbs of nearly equal lengths, and an overall massive build (see figure 7-9).

Fenhosuchus may be a composite taxon (not actually a single genus) based on many isolated bones, mostly vertebrae and dermal scutes. These fossils are distinct from those of *Shansisuchus* and may represent an ornithosuchid. The possible euparkeriid *Wangisuchus* also may be a composite taxon based on a variety of skull fragments, vertebrae and limb bones (Sun et al. 1992).

Sinognathus is a cynodont known from only a skull and lower jaw from Wuxiang, Shanxi. The skull is generally similar to the well-known South African genus *Thrinaxodon*, especially in its possession of a complete secondary palate.

The Ningwuan saw the zenith of dicynodont diversity in the Chinese Triassic. Three genera were present, the small *Shansiodon* and the much larger *Parakannemeyeria* and *Sinokannemeyeria*.

In the Junggur basin of Xinjiang, the lower part of the Kelamayi (= Karamay) Formation yields a correlative vertebrate fauna of Ningwuan age that consists of the semionotid fish *Sinosemionotus urumchii* Yuan & Koh, indeterminate labyrinthodonts (includes the holotype of the *nomen dubium "Parotosaurus"* [= *Parotosuchus*] *turfanensis* Young: Lucas and Hunt 1993a), the euparkeriid *Turfanosuchus dabanensis* Young, the erythrosuchid *Vjushkovia* (= *Youngosuchus*

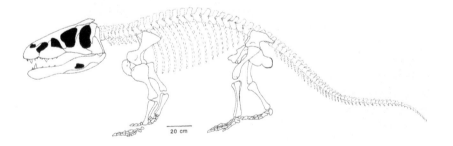

20 cm

Figure 7-9 *The skeleton of the predatory pseudosuchian Shansisuchus (after Young 1964a).*

Sennikov: Parrish 1992) *sinensis* Young, and the dicynodont *Parakannemeyeria brevirostris* Sun.

Sinosemionotus is a member of the Semionotidae, which encompasses some 20 genera of primitive Mesozoic neopterygian fishes. The labyrinthodont specimens, including the holotype of *Parotosuchus turfanensis*, are fragmentary and generically undiagnostic. *Turfanosuchus* and *Parakannemeyeria* were already discussed. *Vjushkovia* was originally described by Huene (1960) for an array of specimens collected in the Donguz Formation of the Russian Urals, strata that belong to the *Eryosuchus* biochron of Ochev and Shishkin (1989) of Middle Triassic (Anisian) age. *V. sinensis* is a small species of *Vjushkovia* known from a skull, lower jaw and partial postcranial skeleton (Young 1973c). This erythrosuchid resembles *Shansisuchus* but is smaller and has three-headed dorsal ribs.

Originally described by Young (1974b) as a traversodontid cynodont, *Traversodontoides wangwuensis* is a bauriid (Sun 1991). It is known from an incomplete skull (see figure 7-10) and some postcrania from the upper Ermaying Formation at Jiyuan, Henan, and probably is of Ningwuan age.

Fukang Fauna

No Late Triassic vertebrate fauna is known from China. However, Chinese paleontologists long assigned a Late Triassic age to a small assemblage of vertebrate fossils from the upper part of the Kelamayi (sometimes spelled Karamay) Formation near Fukang along the northern foot of the Tien Shan in Xinjiang (Young 1978; Su 1978). This "Fukang fauna" consists of three taxa: (1) *Fukangolepis barbaros* Young, a supposed aetosaur based on dicynodont fragments (Lucas and Hunt 1993a); (2) *Bogdania fragmenta* Young, a supposed metoposaurid labyrinthodont based on capitosauroid fragments (Lucas and Hunt 1993b); and (3) *Fukangichthys longidorsalis* Su (see figure 7-11), a palaeoniscid fish similar to but more primitive than Late Triassic *Tanaocrossus* Schaeffer from the Chinle Group of the western United States.

An aetosaur and a metoposaurid would be convincing evidence for a Late Triassic age of the Fukang fauna. However, re-identification as a dicynodont and a capitosauroid make the age determination for the Fukang fauna far less certain. There are Late Triassic dicynodonts and capitosauroids, but these taxa are more characteristic of Early and Middle Triassic tetrapod faunas. The Fukang fauna occurs stratigraphically just above a Middle Triassic tetrapod fauna, but this does not necessarily mean it is of Late Triassic age. It may merely be another tetrapod horizon of Middle Triassic age, slightly younger than the underlying fauna.

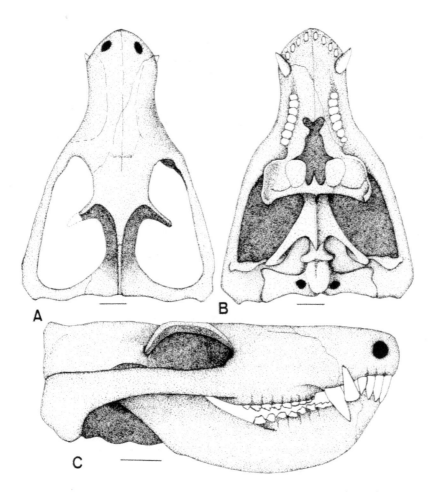

Figure 7-10 *The skull of Traversodontoides wangwuensis, a rare non-dicynodont therapsid from the Chinese Triassic (after Young 1974b), in dorsal (A), ventral (B), and lateral (C) views. Scale bars = 2 cm.*

The fossil fish of the Fukang fauna, *Fukangichthys* (see figure 7-11), also does not provide conclusive evidence of a Late Triassic age. This fish is endemic to the Fukang fauna; its closest resemblance is to a Late Triassic fish from the western United States, *Tanaocrossus* Schaeffer. Points of similarity include the following: (1) elongated dorsal and anal fins; (2) fin rays that do not bifurcate distally; (3) anal fin ray lepidotrichae short, wide, and ornamented with longitudinal, parallel striae; (4) body fusiform to deeply fusiform; (5) caudal fin hemiheterocercal; (6) scales without peg-and-socket articulation and ornamentation; (7) 30 ± 2 vertical scale rows up to the caudal peduncle; (8) anal fin inserted at 24th vertical scale; (9) operculum 1.5 to 2 times larger than and

2 cm

Figure 7-11 *These are two representative specimens of the Triassic semionotid fish Fukangichthys from the Middle Triassic of Xinjiang.*

directly dorsal to suboperculum, preoperculum dorso-ventrally elongate and antero-dorsally expanded; (10) paired post-temporals; and (11) four extra scapulars arranged in a transverse row. Despite these similarities, *Tanaocrossus* is more advanced than *Fukangichthys* in the following features: (1) dorsal fin of *Tanaocrossus* begins adjacent to post-temporals and extends to caudal peduncle, whereas that of *Fukangichthys* is inserted at the 14th vertical scale row; (2) *Tanaocrossus* has at least 10 branchiostegal rays, but *Fukangichthys* has a shorter branchiostegal series of four to six rays; (3) *Fukangichthys* has a typical palaeoniscid skull roof with paired post-temporals, four extra scapulars, paired frontals and paired nasals separated at the midline by the rostral, whereas *Tanaocrossus* has a skull roof with three dermopterotics, a reduced rostrum and nasals meeting along the midline; and (4) *Tanaocrossus* has a maxilla with a triangular posterior expansion, whereas the posterior maxillary expansion of

Fukangichthys is lobate/subrounded. Therefore, based on stage of evolution, an argument can be made that *Fukangichthys* is older than Late Triassic *Tanaocrossus*. No convincing evidence thus exists that the Fukang fauna is of Late Triassic age.

Chinese Triassic Dicynodonts

The above review makes it clear that the Chinese Triassic vertebrate faunas, with a few exceptions, are dicynodont dominated. Indeed, the Chinese Triassic record of dicynodonts is one of the most extensive records of Triassic dicynodonts and bears a brief review.

This record can be divided into three, temporally successive biochrons (oldest to youngest), the *Lystrosaurus, Kannemeyeria,* and *Shansiodon* biochrons. *Lystrosaurus* records begin in the upper 30 m of the Guodikeng Formation in Xinjiang where a few specimens of *Lystrosaurus* co-occur with a specimen of the characteristic Late Permian dicynodont *Dicynodon* (= *Jimusaria*). At the top of the overlap zone, the base of the Jiucaiyuan Formation, *Lystrosaurus* becomes the dominant vertebrate in the Jimsarian fossil assemblage. All the named species of Chinese *Lystrosaurus* are from the Jiucaiyuan Formation in the Junggur basin (see figure 7-5), as follows:

1. *Lystrosaurus hedini* (Young 1935b) is based on a nearly complete skeleton including a skull and lower jaw. This species is very similar to *L. maccaigi* in its long, box-like snout with a flat and ridged facial plane. However, there is a large embayment of the lateral rim of the palate in *L. hedini* that has been the basis for its continued recognition as a distinct species (Cluver 1971; Colbert 1974).

2. *Lystrosaurus broomi* (Young 1939b) is based on an incomplete skull originally identified as *L. murrayi,* a South African species, by Yuan and Young (1934b). Recent workers consider *L. broomi* to be a junior subjective synonym of *L. murrayi* (Cluver 1971; Colbert 1974; King 1988).

3. *Lystrosaurus weidenreichi* (Young 1939b) was based on a partial postcranial skeleton. Because the taxonomy of *Lystrosaurus* is based wholly on cranial characters (Cluver 1971; Colbert 1974; Cosgriff et al. 1982; King 1988), *L. weidenreichi* is a *nomen dubium.*

4. *Lystrosaurus youngi* (Sun 1964) is based on a skull and lower jaw. Later workers regard this species as a junior subjective synonym of *L. curvatus* (Colbert 1974; King 1988).

5. *Lystrosaurus robustus* (Sun 1973) is also based on a skull and lower jaw. The type specimen is part of a quarry sample of skulls of *Lystrosaurus* studied by Sun (1973), who named two species from the sample *L. robustus* and *L. latifrons*. Metric variation in this sample (see table 7-1) hardly merits such a distinction, and I agree with J. Li (1988) that *L. latifrons* is the same species as *L. robustus*. Furthermore, a strong case can be made that the holotype of *L. youngi* is merely a relatively small individual of the *L. robustus* quarry sample, and that the type of *L. broomi* is just a relatively large individual of this same sample (see table 7-1). If these arguments are accepted, then *L. broomi, L. youngi, L. robustus,* and *L. latifrons* are a single species synonymous with *L. murrayi*.

6. *Lystrosaurus shichanggouensis* (Cheng 1980a) is based on a skull, lower jaw and incomplete postcranial skeleton. It most resembles *L. hedini* and may be synonymous.

Viewed conservatively, there were only two or three species of Chinese *Lystrosaurus* during the Jimsarian. After the Fuguan gap in dicynodont distribution (although note the possible occurrence of *Lystrosaurus* in the Shaofanggou Formation of Xinjiang mentioned above), a major change had taken place in the Chinese dicynodont fauna. This is best seen in the Ordosian "kannemeyeriid fauna" of the lower Ermaying Formation

Here, the oldest occurrence of the endemic Chinese dicynodont *Parakannemeyeria* is recorded in *P. xingxianensis* Cheng 1980b. More significant biochronologically is the presence of the genus *Kannemeyeria* (= *Shaanbeikannemeyeria*) in the form of two species, *K. xilougoensis* (Cheng 1980b) from Shaanxi and *K. buerdongia* (J. Li 1980) from Nei Monggol.

Z. Cheng (1980b) named the genus *Shaanbeikannemeyeria* for dicynodonts of large size in which the skulls have a long robust snout, large temporal fossae, high occiput, short basicranium, and prominent maxillary tusks. Cox (1991) correctly noted that the skull of *Shaanbeikannemeyeria* could not be distinguished from that of the Indian genus *Rechnisaurus*. However, *Rechnisaurus* and the Russian genus *Uralokannemeyeria* are best-considered synonyms of *Kannemeyeria* (Keyser and Cruickshank 1979; Lucas 1993b). The presence of *Kannemeyeria* in the lower Ermaying Formation is part of a global distribution of the genus in late Early Triassic (Olenekian) deposits that rivals the earlier cosmopolitanism of *Lystrosaurus*. *Kannemeyeria* is also known from South

Table 7-1 Cranial measurements (in mm) and ratios for skulls of *Lystrosaurus* from a single quarry sample in the Jiucaiyuan Formation (specimens 3242–3248 and 3264–3265) compared with type specimens of three other species of *Lystrosaurus* from the Jiucaiyuan Formation (modified from Sun 1973)

	IVPP Specimen Number											
	3242	3243[a]	3244[b]	3245	3246	3247	3248	3264	3265	35012[c]	39060[d]	8532[e]
1. skull length	148	230	230*	164	240	245*	197	168	245	190	242	135
2. width of frontals	73	130	150	70	131	166	100	82	144	91	130	60
2/1 x 100	50	56.5	65.2	42.7	54.6	67.8	50.7	48.8	58.8	48	53.7	44.5
3. width interparietal crest	31	42	30	25	39	40	33	39	42	43	46	27
3/1 x 100	21	18	13	15.2	16.2	16.3	16.7	23.2	17.1	22.6	19	20
4. angle between face and skull	106°	106°	122°	98°	118°	106°*	102°	116°	118°	—	122°	103°
5. length braincase/length snout x 100	42	52	52	50	52.2	—	56	62.5	43	45.5	60*	100

[a] holotype of *Lystrosaurus robustus* (Sun 1973); [b] holotype of *Lystrosaurus latifrons* (Sun 1973); [c] holotype of *Lystrosaurus hedini* (Young 1935b); [d] holotype of *Lystrosaurus broomi* (Young 1939b); [e] holotype of *Lystrosaurus youngi* (Sun 1964); * approximate measurement

Africa, South West Africa, Tanzania, Zambia, Argentina, India, and Russia (e.g., Bandyopadhyay 1988).

The upper Ermaying Formation yields three dicynodont genera, the Chinese endemics *Parakannemeyeria* and *Sinokannemeyeria* and the cosmopolitan *Shansiodon*. *Parakannemeyeria* (figures 7-8, 7-12) and *Sinokannemeyeria* are large dicynodonts known from many complete skulls and skeletons (Sun 1963). Overall morphology of these genera indicates a close relationship to *Kannemeyeria*. However, the Chinese genera have broader snouts, smaller temporal fenestrae and lower temporal crests. King (1990) suggests this indicates the Chinese genera modified the masticatory system to emphasize orthal chopping less than did *Kannemeyeria*. She speculates that *Parakannemeyeria* and *Sinokannemeyeria* may have rather indiscriminately seized and torn vegetation, in contrast to the more selective cropping of *Kannemeyeria*.

At a time of dicynodont cosmopolitanism (see *Shansiodon* below), *Parakannemeyeria* and *Sinokannemeyeria* are endemic to north China. This localization of large tetrapods cannot now be explained. However, fragmentary large dicynodont bones from strata in the Russian Urals correlative with the upper Ermaying Formation (Efremov 1940; Vyushkov 1969) may extend the geographic range of the Chinese genera.

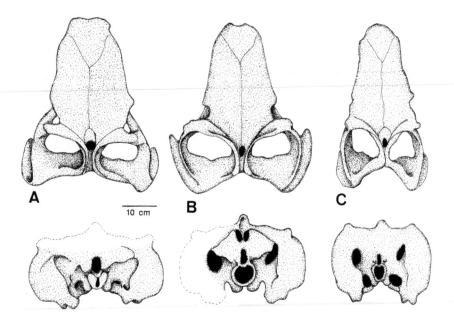

Figure 7-12 *Skulls of Parakannemeyeria, dorsal views above, occipital views below (after Sun 1963): A) P. ningwuensis, B) P. youngi, and C) P. dolicocephala.*

Contemporaneous with these genera is a much different Chinese dicynodont, *Shansiodon* (see figure 7-13). This small- to medium-sized dicynodont has a skull that is triangular in outline with a wide and short face and a downwardly bent and blunt snout. The pre- and postorbital regions of the skull are of equal length, nasal ridges and bosses are present, and the canines are set in moderately developed caniniform processes.

Shansiodon is particularly important for the global correlation of the Ningwuan faunachron. Skulls from the upper Ermaying Formation (Ye 1959; Cheng 1980b) encompass a wide range of cranial variation that identifies *Shansiodon* as a senior subjective synonym of the following non-Chinese genera: *Tetragonias* (Cruickshank 1967), *Rhinodicynodon* (Kalandadze 1970), *Dolichuranus* (Keyser 1973), *Rhopalorhinus* (Keyser 1973), and *Angonisaurus* (Cox & Li 1983). Keyser and Cruickshank (1979), Cooper (1980) and King (1988) already suggested the close relationship of these genera and their possible synonymy. Recognition of this synonymy gives *Shansiodon* a broad distribution across Triassic Pangaea (China, Russia, Tanzania, Zambia, and South Africa) in strata identified as of early Anisian age (Anderson 1980; Cooper 1980; Lucas 1993a).

Shansiodon is the youngest well-known dicynodont from China. The youngest Chinese dicynodont fossils are the fragments that made up the type specimens of the supposed aetosaur *Fukangolepis* discussed earlier. These fragments probably are of Middle Triassic age.

The "Nine-Dragon Wall"

An ancient Chinese myth of the nine realms refers to a perfect world where there are eight cardinal directions with a ninth in the center for the domain of the sun (Schafer 1967). The dragon in ancient China was the rain spirit and the sacred symbol of the East, whose beneficence was essential to a rich harvest. The ancient Chinese thus conceived of nine dragons—usually portrayed as a painting or mosaic wall—as a very powerful image. Indeed, the last Chinese emperors built such a "nine-dragon wall" (or panel or screen), which can still be seen in the old Imperial Palace ("Forbidden City") in Beijing.

It was both propitious and ironic, then, that field workers from the Institute of Vertebrate Paleontology and Paleoanthropology, collected a slab-like block with nine dicynodont skeletons in the Kelamayi Formation near Fukang in Xinjiang. Fully prepared, this slab became known among Chinese paleontologists as the "nine-dragon wall." Sun (1978) described the "dragons" on the wall as a new species of *Parakannemeyeria*, *P. brevirostris*. The name derives from the relatively short preorbital region of the skull (see figure 7-14), a supposedly diagnostic feature. However, the "dragons" are juvenile individuals, and it is likely the difference is allometric and not a valid basis for a distinct species.

Lotosaurus

One of the most unusual reptiles from the Triassic of China is *Lotosaurus*, known from abundant skeletal material found in nonmarine strata intercalated in the marine Middle Triassic Badong Formation at Sangzhi, Hunan (F. Zhang

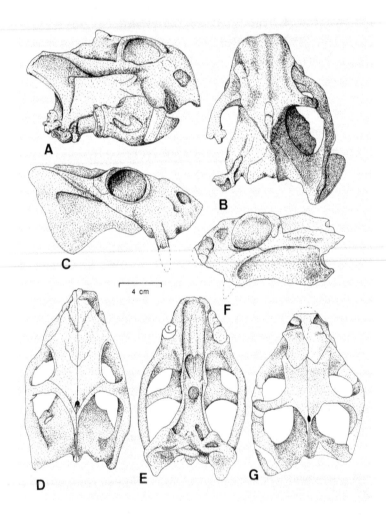

Figure 7-13 *Skulls of Shansiodon from the Ermaying Formation in Shanxi (after Ye 1959 and Cheng 1980b): A), C), and F) are lateral views, B), D), and G) are dorsal views, and E) is a ventral view.*

6 cm

Figure 7-14 *A skull from the "nine-dragon wall," a death assemblage of juvenile Parakannemeyeria from the Middle Triassic of Xinjiang. Note the relatively short snout and large eyes (compare to figure 7-8), characteristic of a juvenile.*

1975). This 3-m-long thecodont has an edentulous skull that terminates in a pointed beak (see figure 7-15). The body lacks armor scutes, and the dorsal neural spines are very tall and plate-like, forming a "sail" reminiscent of the sphenacodontid pelycosaurs of the Early Permian. The large antorbital foramen close to the orbit and the configuration of the pelvis and tarsus support assignment of *Lotosaurus* to the "thecodonts," though its affinities among "thecodonts" are very unclear. Clearly, it is closely related to *Ctenosauriscus* from the Middle Triassic of Germany and "*Hypselorhachis*" from the Middle Triassic of South America. These three genera form an obvious clade of sail-backed thecodonts, the Ctenosauriscidae, but their placement among thecodonts remains *incertae sedis* (Olshevsky 1991).

Triassic Fishes

Triassic fish occurrences are nonmarine records of Early, Middle, and Late Triassic age from the Junggur and Ordos basins, and a few marine records from southern China (M. Chang and Jin 1996).

Early Triassic records from the Ordos basin include lungfish (*Ceratodus heshanggouensis* Cheng, mentioned above), and the Hengshan locality in Shaanxi, which yield palaeoniscids (*Palaeoniscum* sp., *Gyrolepis* sp.), an acrolepid (*Boreosomus* sp.), a perleidid (*Perleidus,* cf. *P. woodwardi* Stensiö), and the saurichthyid *Saurichthys huanshenensis* Chou & Liu. In the Junggur basin, only a palaeonisciform (*Duwaichthys mirabilis* Liu et al.) and the redfieldiid *Sinkiangichthys longipectoralis* Liu have been reported from Lower Triassic strata. Scattered Early Triassic marine records from China include perleidids from Hexian, Anhui (*Perleidus yangtzensis* Su), and Huangshi, Hubei (*Plesioperleidus dayeensis* Su & Li), an actinistan (*Sinocoelocanthus fengshanensis* Liu) from Fengshan,Guangxi, and the edestid *Helicampodus qomolangma* Zhang from Dingri, Tibet.

Middle Triassic fish records from China are even more sparse. *Sinosemionotus urumchii* Yuan & Koh and *Fukangichthys longidorsalis* Su are osteichthyans from Middle Triassic strata in Xinjiang. *Triassodus yanchangensis* Su is from Middle Triassic nonmarine strata at Yaoxian, Shaanxi. The hybodont selachian *Hybodus youngi* Liu is from nonmarine strata at Yanchang, Shaanxi. Fishes associated with the nothosaur *Kueichousaurus* at Xingyi, Guizhou are the peltopleurid *Peltopleurus orientalis* Su, the furid *Sinoeugnathus kueichowensis* Su, and the semionotid *Asialepidotus shingyiensis* Sun.

Late Triassic records are equally sparse. Hybodont selachians have been reported from Anning, Yunnan (*Hybodus houtiensis* Young), and Nielamu, Tibet (genus indeterminate: Dong 1972). The Xujiahe Formation in Sichuan yielded the palaeonisciform *Shuniscus longianalis* Su and the pholidophorid *Jialingichthys serratus* Su. Upper Triassic strata in Shaanxi yield the palaeoniscid *Wayaobulepis zichangensis* Su.

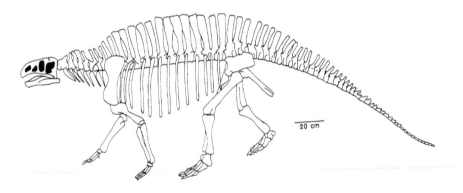

Figure 7-15 *The skeleton of the bizarre, toothless, and sail-backed "thecodont" Lotosaurus from the Middle Triassic of Hunan (after F. Zhang 1975).*

Chinese Triassic fishes include many cosmopolitan genera, such as *Perleidus*, *Saurichthys*, *Gyrolepis*, *Ceratodus*, and *Hybodus*, and thus well represent the broadly distributed fish faunas of Triassic Pangea. They also support the broad distinction between nonmarine deposition in northern China and marine deposition to the south during the Triassic. Beyond that, little can be deduced from Chinese Triassic fishes because of their sparse record.

Triassic Marine Reptiles

Marine Triassic strata are widespread in southern China, especially in a belt extending from Yunnan and Guizhou through the Yangtze Valley region. These rocks have yielded several ichthyosaurs and sauropterygians (Dong 1980a; Rieppel 1999).

The sauropterygians are *Keichousaurus hui* Young, *K. yunnanensis* Young, *Chinchenia sungi* Young, and *Sanchiaosaurus dengi* Young from the Middle Triassic, and *Kwangsisaurus orientalis* Young, *K. lusiensis* Young, and *Hanosaurus hupehensis* Young from the Lower Triassic. These taxa are mostly from marine Triassic strata along and south of the Yangtze River in southern China. Of them, *Keichousaurus* is the best known, being represented by many complete skeletons (see figure 7-16).*Keichousaurus* is a small pachypleurosaur with very large orbits and a pointed rostrum. The most distinctive feature of the genus, the ulna, is very short and broad (Storrs 1991). The genus is known from the Middle Triassic of Xingyi, Guizhou, and Yuanan, Hubei (Young 1958d, 1965c).

Less well known is *Shingyisaurus*, based on a poorly preserved skull and the anterior cervical vertebrae, also from Xingyi, Guizhou (Young 1965c). This nothosauriform is larger than *Keichousaurus* and has a blunt rostrum with small external nares located halfway between the rostrum tip and the orbits.

Chinchenia from Qingzhen, Guizhou is better known, being represented by several partial skeletons. This small nothosaur has a very thick lower jaw with a short symphysis and anisodont teeth. *Sanchiaosaurus* is known also from a partial skeleton from Guiyang, Guizhou. It most resembles *Chinchenia* among the other Chinese nothosaurs.

The oldest Chinese nothosaur, from the Lower Triassic of Guangxi and Yunnan, is *Kwangisaurus*. Partial skeletons of this genus show a very robust femur, but a short and small pes. These are the most distinctive features of *Kwangisaurus* among the nothosauriforms (Storrs 1991).

Hanosaurus hupehensis Young from the Middle Triassic Jialingjiang Formation at Nanzhang, Hubei is known for an incomplete skull and some postcrania. It has an elongate skull with long, posteriorly placed orbits, no lower

2 cm

Figure 7-16 *An exquisitely preserved skeleton of the nothosaur Keichousaurus from the Middle Triassic of Tingxiao, Guizhou.*

temporal opening, and an elongate oval supratemporal opening. It was originally described as a thalattosaur, but has recently been reinterpreted as a pachypleurosaur (Rieppel 1998).

More unusual Middle Triassic marine reptiles from China are *Nanchangosaurus* and *Hupehsuchus*. These animals are known from complete and incomplete skeletons found in the Middle Triassic Daye Formation of Hubei (K. Wang 1959; Young and Dong 1972). *Nanchangosaurus* and *Hupehsuchus* retain many characteristics of terrestrial precursors to the ichthyosaurs and may be a "missing link" between them and their non-aquatic ancestors. Thus, Carroll and Dong (1991) created the new order Hupehsuchia for *Nanchangosaurus* and *Hupehsuchus*. In hupehsuchians, the skull has a long and toothless snout and an upper temporal opening. The limbs are somewhat reduced, and the trunk is fusiform, yet there are dermal armor plates along the vertebral column.

The oldest Chinese ichthyosaur, and one of the oldest ichthyosaurs, is *Chaohusaurus geishanensis* Young & Dong from the Lower Triassic Majianshan Formation at Chaoxian, Anhui. *Chaohusaurus* was named from an incomplete skeleton, but much more complete material is now known. It shares primitive features with other Early Triassic ichthyosaurs such as a wide skull with a short snout, heterodont teeth, vertebrae that are longer than tall and weakly amphicoelous, and a large forelimb only slightly modified toward a paddle with a phalangeal count of 2-3-4-4-2 (see figure 7-17). *Chensaurus chaoxiensis* (Chen) and *C. faciles* (Chen) (formerly assigned to the preoccupied name *Anhuisaurus*) are also from the Lower Triassic of Anhui. Like *Chaohusaurus*, these ichthyo-saurs show a mosaic of primitive and advanced ichthyosaur features, but they are less well known at present.

Worldwide, the most common and characteristic Middle Triassic ichthyo-saur is *Mixosaurus* (Callaway and Massare 1989; Lucas and González-León 1995). In China, *Mixosaurus* is known from a partial skeleton from Jenhui, Guizhou for which Young (1965c) created the distinct species *M. maotiensis*.

Two large ichthyosaurs have been identified from Upper Triassic marine strata in the Himalaya Mountains of Tibet. *Himalayasaurus tibetensis* Dong is based on a partial skull and skeleton from the Mt. Everest (Qomolangma Feng) region of Tibet; the fossils were collected at an altitude of about 4800 m above sea level (Dong 1980b). *Tibetosaurus tingjiensis* Young, Liu & Zhang is based on an incomplete skeleton from Xixabangma Feng in the Dingri district of Tibet (see figure 7-18).

Both of these taxa are not well known or adequately described. When first proposed, no diagnosis of *Himalayasaurus tibetensis* was published by Dong (1972). So, it was a *nomen nudum* until 1992, when Sun et al. (1992: 123) published a diagnosis. A loose translation of the diagnosis of *Tibetosaurus tingjiensis* reads:

> Large ichthyosaur; length about 10 m; tooth root long and large; in cross section, tooth enamel is crenulated; base of tooth crown wide, tapers to tip and some teeth have slightly recurved crowns; a longitudinal ridge (carina) divides the tooth crown into two portions: in occlusal view, one portion is

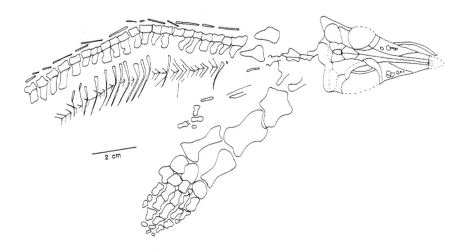

2 cm

Figure 7-17 *The skeleton of the primitive ichthyosaur Chaohusaurus (after Young and Dong 1972). Note that the forelimb is not fully modified to a paddle as in more advanced ichthyosaurs.*

Figure 7-18 *The type specimen of the Himalayan Triassic ichthyosaur Tibetosaurus is mostly vertebrae and ribs (approximately x 1/10).*

slightly curved whereas the other is obtusely curved; tooth located in a groove on the jaw; vertebrae biconcave (amphicoelous) but the anterior side is less concave than the posterior side; ribs with single head; humerus short and wide.

Unfortunately, this diagnosis fails to distinguish *Tibetosaurus tingjiensis* from *Himalayasaurus tibetensis*. Both are very similar morphologically and are from the same stratigraphic unit (Langjiexue Group) in nearby localities. I thus consider *H. tibetensis* to be a junior subjective synonym of *T. tingiensis*.

This still leaves a problem; even combining Young et al.'s (1982) and Sun et al.'s (1992) diagnoses to produce a single diagnosis for *T. tingjiensis* fails to distinguish this taxon from other large, Late Triassic ichthyosaurs. Particularly striking are the similarities of the Chinese taxon to the only other Triassic ichthyosaur in its large size class, *Shonisaurus* from the Late Triassic of Nevada (Camp 1976, 1980). Because no features have been identified that distinguish *Tibetosaurus* from *Shonisaurus*, nor can I identify any, I consider *Tibetosaurus* to be a junior subjective synonym of *Shonisaurus*, as the species *S. tingjiensis*. This indicates a trans-Pacific distribution for *Shonisaurus* fossils. Surely, a 10-m long (or longer) *Shonisaurus* could have swam across the Triassic Panthalassan ocean, so its distribution in China and the United States should not surprise us. Mazin and Sander's (1993) conclusion that *Tibetosaurus* and *Himalayasaurus* were endemic to the Tethys thus lacks a taxonomic basis.

Chinese Triassic Tetrapods, Pangea, and Facies

China has an excellent fossil record of Early and Middle Triassic tetrapods. Most striking is the great similarity of the tetrapod successions in the great Chinese Triassic basins—Junggur and Ordos—and the great South African Triassic basin, the Karoo (see figure 7-19). Many genera are shared between China and South Africa during the Early and Middle Triassic, making this an easy and secure correlation. The close similarity of the South African and northern Chinese Early-Middle Triassic tetrapods certainly argues for a land connection between the two far-flung areas by Triassic time. This evidence of Triassic membership of the Kazakstan-north China blocks in a united Pangea is further supported by the presence in China of the three cosmopolitan dicynodont genera of Early-Middle Triassic Pangea—*Lystrosaurus*, *Kannemeyeria*, and *Shansiodon*. These genera and the South African-Chinese correlations proposed here make clear the integrity and cosmopolitanism of Early and Middle Triassic Pangea.

Correlation of the Chinese Triassic tetrapods with the classic Lower-Middle Triassic tetrapod successions of the Russian Urals (Ochev and Shishkin 1989; Shishkin et al. 1995) also can be accomplished with shared dicynodont taxa. *Lystrosaurus*, *Kannemeyeria* (= *Uralokannemeyeria*) and *Shansiodon* (= *Rhinodicynodon*) are known from the Urals (Kalandadze 1970, 1975; Danilov 1971; Shishkin et al. 1995). However, unlike the correlative Chinese tetrapod faunas, which are dominated by dicynodonts, dicynodonts are rare in the Russian faunas, which are composed mostly of temnospondyl amphibians. The most obvious explanation of this pattern is that it is facies related. The dicynodont-dominated faunas of China (similar faunas are also found in India and South

AGE				CHINA			RUSSIA	SOUTH AFRICA	
TRIASSIC	MIDDLE	Perovkan		lower Kelamayi Formation	upper Ermaying Formation	Ningwuan lvf	Donguz Formation	*(hatched)*	Cynognathus assemblage zone
	EARLY	Nonesian		lower Ermaying Formation		Ordosian lvf	Petropavlovsk Formation (Yarenskiy horizon)	Burgersdorp Formation	
				upper Hesshangou Formation		Fuguan lvf			
		Lootsbergian		lower Jiucaiyuan Formation		Jimsarian lvf	Vetlugan Series (Vokhmian horizon)	Katberg Formation	Lystrosaurus assemblage zone
				upper Guodikeng Formation				Balfour Formation	

Figure 7-19 *Correlation of Triassic tetrapod successions in China with those in Russia and the Karoo basin of South Africa is based largely on the similarity of dicynodonts across Triassic Pangea.*

Africa) are terrestrial faunas found in relatively dry inland/upland depositional settings. The labyrinthodont-dominated faunas of Russia (similar faunas are also found in western Europe, the western United States, and Australia), in contrast, are aquatic faunas from relatively wet lowland/coastal depositional settings. The Chinese Triassic tetrapod faunas thus not only support the integrity of Pangea but also suggest a significant facies difference between coeval Early-Middle Triassic tetrapod faunas of the vast supercontinent.

Chapter 8

Jurassic

During the Jurassic, China was part of eastern Pangea. Marine deposition took place in southwestern China along the northern margin of Tethys, but the rest of the country was vast terrestrial lowland (see figure 8-1). Extensive volcanism took place in eastern China during the Late Jurassic and continued into the Cretaceous (Z. Xu 1990).

Jurassic vertebrate fossils have a much broader geographic distribution in China than do Triassic vertebrate fossils. This is because sedimentary deposition across China was almost totally nonmarine during the Jurassic (see figure 8-1). Thick and extensive accumulations of nonmarine Jurassic strata in southern, northwestern, north-central, and northeastern China contain one of the most significant records of Jurassic vertebrates on earth (Lucas 1996a; P. Chen

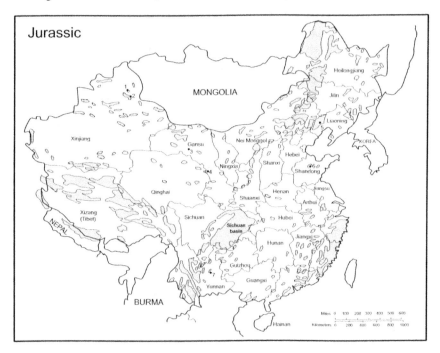

Figure 8-1 *Distribution of Jurassic strata and principal Jurassic vertebrate fossil localities in China (after Yang et al. 1986). Most Jurassic vertebrate fossils come from the Sichuan basin. Other localities are: 1 – Shishigou Formation, Xinjiang; 2 – Wucaiwan Formation, Xinjiang; 3 – Jiayuguan, Gansu; 4 – Lanzhou, Gansu; 5 – Fuxin, Liaoning; 6 – Mengyin, Shandong; 7 – Lufeng, Yunnan.*

1996; K. Li et al. 1996). These vertebrates come from strata in the Sichuan, Junggur, Ordos, and north China basins (see figure 8-1). They record many important mileposts in Jurassic vertebrate evolution, especially in the evolution of dinosaurs.

Sichuan Basin

The Sichuan basin contains Early, Middle, and Late Jurassic vertebrate faunas in a 3000+ meter thick sequence of directly superposed strata (see figure 8-2).

System	Series	Formation		land-vertebrate faunachron
JURASSIC	Upper	Penglaizhen Formation		Ningjiagouan
		Suining Formation		
	Middle	upper Shaximiao		Tuojiangian
		lower Shaximiao		Dashanpuan
		Xintiangou Formation		
	Lower	Ziliujing Group	Daanzhai Formation	
			Maanshan Formation	
			Dongyuemiao Formation	
		Zhenzhuchong Formation		Dawan

Figure 8-2 *The Jurassic strata and vertebrate fossils in the Sichuan basin are the standard Jurassic succession in China (after Dong et al. 1983).*

This is one of the most complete single successions of Jurassic vertebrate faunas in the world. The rocks present here are red-bed fluvial and lacustrine strata, except for the Lower Jurassic strata, which in northwestern Sichuan are coal-bearing lacustrine rocks. Deposition during the Jurassic was mostly lacustrine in a fluvial drainage internal to the Sichuan basin. This lacustrine system has been referred to as the ancient Bashu Lake.

The basal Jurassic unit, the Zhenzhuchong Formation, and overlying strata of the Ziliujing Group, comprise a nearly 500 m thick sequence of variegated mudstones and marlstones with minor quartzose sandstones that yield a rich fossil record of bivalves, conchostracans and fossil plants. *Lufengosaurus* from the Zhenzhuchong Formation is the oldest Jurassic dinosaur (and Jurassic fossil vertebrate) from the Sichuan basin. The Ziliujing Group has a sparse vertebrate record: (1) a lacustrine pliosaur from the Dongyuemiao Formation (Dong 1980b); (2) the crocodilian *Peipehsuchus teleorhinus* Young, a "stegosaurid" and the sauropod? *Sanpasaurus yaoi* Young from the Maanshan Formation; and (3) the semionotid fish *Lepidotes chungkingensis* Liu & Wang, a prosauropod (cf. *Lufengosaurus magnus* of Dong 1984) and the vulcanodontid sauropod *Zizhongosaurus chuanchengensis* Dong, Zhou & Zhang from the Daanzhai Formation.

The oldest Middle Jurassic strata in the Sichuan basin belong to the Xintiangou Formation, 100 to 250 m of interbedded grayish sandstone and mudstone. No fossil vertebrates are known from the Xintiangou Formation. The overlying lower part of the Shaximiao Formation (or Xiashaximiao Formation of some authors; "xia" means "lower") contains the extensive Dashanpuan vertebrate fauna discussed below. It is 100 to 250 m of variegated (mostly purplish) lacustrine mudstone.

The oldest Upper Jurassic strata in the Sichuan basin belong to the extremely thick (767–2200 m) upper part of the Shaximiao Formation (or Shangshaximiao Formation of some authors; "shang" means "upper"). These rocks are interbedded red-bed mudstones and green arkosic sandstones. They yield the Tuojiangian vertebrate fauna discussed below. The onset of deposition of the upper part of the Shaximiao Formation also has been interpreted to mark a significant climate change in the Sichuan basin, from hot and humid during the Early-Middle Jurassic to dry in the Late Jurassic.

Upper Jurassic strata of the Suining Formation, which overlies the Upper Shaximiao Formation, are 200 to 500 m of brownish-red mudstone intercalated with siltstone. This lacustrine unit contains an extensive fossil record of ostracods, bivalves, and conchostracans, but has only yielded a single vertebrate, the lungfish *Ceratodus szechuanensis*.

The youngest Jurassic strata in the Sichuan basin belong to the Penglaizhen Formation, 200 to 1700 m of alternating purplish-red mudstones interbedded with yellowish-green calcareous shales and grayish-white quartzose sandstones. Dinosaur footprints are the vertebrate-fossil record of the Penglaizhen Formation.

Other Jurassic Basins

Fossils from the Sichuan basin dominate China's Jurassic vertebrate-fossil record, but significant Jurassic fossil vertebrates are known elsewhere in the country (see figure 8-3). Perhaps the most famous location is in the Lufeng area of western Yunnan. Here, the Lufeng Formation (also called Lufeng Group) is 700 to 1600 m of red-bed mudstones and sandstones (see figure 8-4) that yield the Dawan fossil vertebrates (the world-famous "Lufeng saurischian fauna").

In the Junggur basin of northwestern China, the Middle Jurassic Wucaiwan Formation contains fossil vertebrates of Dashanpuan age, including dinosaurs and a tritylodontid (Sun and Cui 1989; Dong 1990, 1992; Zhao and Currie 1993). The Wucaiwan Formation is 200 m of mostly red-bed sandstone and mudstone. Its correlative, the Toutunhe Formation to the northwest, has yielded an ankylosaur (Zhao and Currie 1993).

AGE		Shandong	Xinjiang – Junggur	Xinjiang – Turpan	Gansu	Ordos	Nei Monggol		Sichuan	Yunnan	Hunan	Zhejiang	Anhui	Liaoning
EARLY CRET.	Tsagantsabian	Qingshan Frmn.	Tugulu Group	Tugulu Group	Xinminbao Formation	Zhidan Group	Egannur Formation	Dashuigou Formation	Zilong Frmn. / Baizi Frmn. / Yanting Frmn.		Shenwang Shan Frmn. / Dongjing Frmn.	Chaochuan Frmn.	Huizhou Formation	Jehol Group
JURASSIC – LATE	Ningjiagouan	Laiyang Frmn. / Mengyin Group	Shishugou Formation	Qigu Kalaza Frmn.	Chijinbao Formation	Anding Frmn. / Fenfeng-he Frmn.			Penglai zhen Fm.			Guantou Frmn.	Yantiang Formation	Fuxin Group
JURASSIC – LATE	Tuojiangian								Sunning Frmn.		Jiande Formation			
JURASSIC – MIDDLE	Dashanpuan		Wucaiwan Formation	Toufunghe Formation		Zhilou Frmn.			Shaximiao Formation / Xintian gou Frmn.	Lufeng Formation	Tangjianwu Formation		Tucheng-zi Frmn. / Lanchi Frmn.	Haifangou Formation
JURASSIC – EARLY	Dawan		Shuixigou Group / Xiaquangou Group	Shuixigou Group / Xiaquangou Group		Yanan Frmn. / Yanchang Group			Zhujing Formation / Xujiahe Formation	Lufeng Formation / Yonlian Group	Paijincheng Formation			Haifangou Formation / Beipiao Formation

Figure 8-3 *Correlation of the Jurassic formations of China organizes the vertebrate-fossil-producing strata into four Jurassic faunachrons (after Y. Wang and Sun 1983).*

Figure 8-4 *This outcrop of the Lufeng Formation shows the lower dull purplish beds (light-banded strata at base of hill) overlain by the dark red beds.*

In the Junggur basin, Upper Jurassic (Tuojiangian) vertebrates are known from the Shishigou (Shishu of some authors) Formation, which overlies the Wucaiwan Formation. The Shishigou Formation is interbedded sandstones, siltstones and mudstones with minor lenses of limestone deposited in fluvial and lacustrine environments.

The Ordos basin of north central China was the site of nonmarine deposition during the Jurassic, but few fossil vertebrates are known from the nearly 800 m thick Jurassic succession. Only the crocodilian *Sunosuchus* and the sauropod *Mamenchisaurus* are known from the Xiangtang (Hantong) Formation, indicating a Tuojiangian age (Young 1948).

In Shandong, the Mengyin Formation is as much as 714 m of gray-green and purple-red sandstone, siltstone, mudstone and shale (P. Chen 1982; P. Chen et al. 1982a, b). It contains a Late Jurassic vertebrate fauna characteristic of the Ningjiagouan land-vertebrate faunachron.

In northeastern China, latest Jurassic—earliest Cretaceous deposition took place in a series of small, isolated, volcanic basins (see figure 8-5). I consider most of the vertebrate-fossil-bearing strata of the north China basins to be of Early Cretaceous age (see chapter 9). An exception is the Tuchengzi Formation of western Liaoning, which is 320 m of red sandstones and conglomerates that

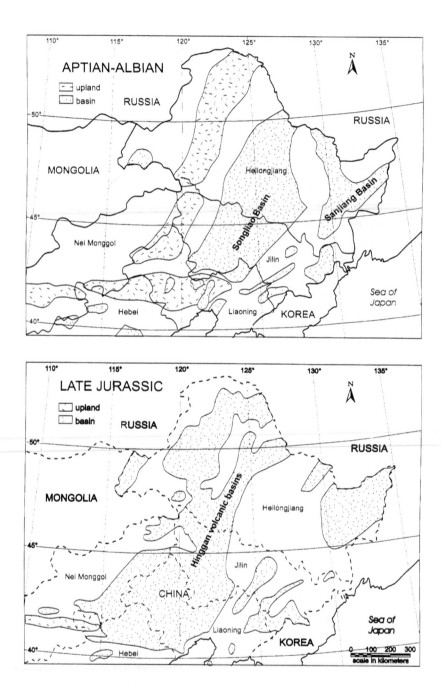

Figure 8-5 *Late Jurassic (below) and Early Cretaceous (above) configuration of the sedimentary basins of northeastern China (after H. Wang 1985).*

in its lower part contains dinosaurs of probable Middle Jurassic age. Overlying Upper Jurassic strata are largely volcaniclastic units that yield a few fossil fishes.

Dawan Vertebrates

The classic Lufeng fauna ("Lufeng saurischian fauna" or *"Lufengosaurus"* fauna of earlier workers) of southwestern Yunnan Province was first collected by C. C. Young and M. N. Bien in the 1930s (Bien 1940). It forms the basis of the Dawan land-vertebrate faunachron of Lucas (1996a and c). Dawa (Tawa) is a village 4 km northeast of Lufeng near the principal fossil vertebrate localities in the Lufeng Formation. The vertebrate fauna of the lower, "dull purplish beds" and the upper, "dark red beds" of the Lufeng Formation (see figure 8-6) are the basis of the Dawan faunachron.

The Lufeng Formation has yielded a diverse vertebrate fauna (see table 8-1) dominated by fossils of the prosauropod dinosaur *Lufengosaurus* (see figure 8-7) and a diversity of tritylodontid synapsids. Labyrinthodont amphibians are known from isolated neorhachitomous centra (Sun 1962), and turtles by indeterminate proganochelyids. Some of the oldest known crocodylomorphs are from the Lufeng Formation—*Platyognathus hsuii* Young, *Dibothrosuchus elaphros* Simmons, and *Strigosuchus licinus* Simmons. Protosuchians are *Microchampsa scutata* Young and *Dianosuchus changchiawanensis* Young. *Clevosaurus petilus* Young and *C. wangi* Wu are sphenodontians. *Fulengia youngi* Carroll & Galton was originally identified as a lepidosaur but is actually a juvenile prosauropod (Evans and Milner 1989). Young (1951a) described *Pachysuchus imperfectus* as a phytosaur, but the type material is indeterminate and certainlynot a phytosaur. A series of small centra from the dark red beds of the Lufeng Formation represent a temnospondyl amphibian, one of China's youngest temnospondyls (Sun 1962). China's oldest fossil turtles are proganochelyid shell fragments from the Lufeng dark red beds.

The crocodylomorph *Platyognathus hsui* is relatively poorly known from only the anterior portion of the skull and lower jaws. This medium-sized crocodylomorph has a short snout, fused symphysis, procumbent anterior teeth, a large, canine-like sixth dentary tooth, and irregular-shaped anterior teeth with polygonal cross sections. Young (1944) originally identified *Platyognathus* as a pseudosuchian, but it is now regarded as a protosuchian crocodylomorph by most workers, following Romer (1956) (e.g., Luo and Wu, 1994; X. Wu and Sues 1995, 1996).

The best-known Lufeng crocodylomorph is *Dibothrosuchus*, recently restudied by X. Wu and Chatterjee (1993). They restored *Dibothrosuchus* as a 1.3 m long, tall, slender quadruped with a pointed muzzle, narrow chest, arched

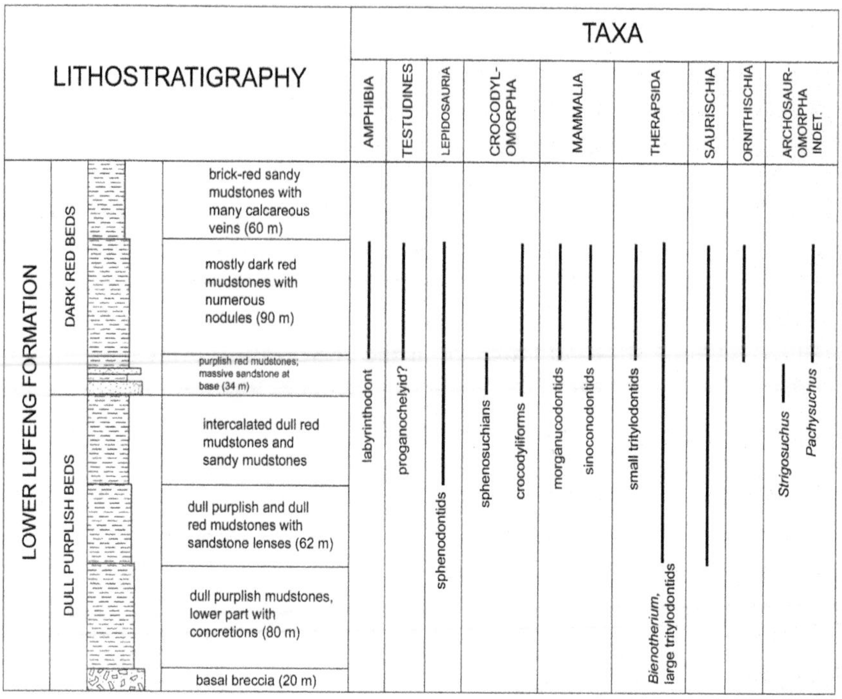

Figure 8-6 *Stratigraphy and distribution of fossil vertebrates in the Lufeng Formation (after Luo and Wu 1994).*

Table 8-1 List of Fossil Vertebrates from the Lufeng Formation (after Sun and Cui 1986; Luo and Wu 1994)

	Lower Red Beds	Upper Purple Beds
Amphibia		
Labryinthodontia indet.	X	
Reptilia		
Chelonia		
Proganochelyidae indet.	X	
Crocodylomorpha		
Platyognathus hsui Young	X	
Dibothrosuchus elaphros Simmons	X	
Strigosuchus licinus Simmons	X	
Protosuchia		

Table 8-1 List of Fossil Vertebrates from the Lufeng Formation (after Sun and Cui 1986; Luo and Wu 1994) (Continued)

	Lower Red Beds	Upper Purple Beds
Microchampsa scutata Young	X	
Dianosuchus changchiawaensis Young	X	
Sphenodontia		
Clevosaurus petilus Young		X
C. wangi Wu		X
C. mcgilli Wu		X
?Thecodontia		
Pachysuchus imperfectus Young	X	
Saurischia		
Lufengosaurus huenei Young	X	
Lufengosaurus magnus Young	X	
Yunnanosaurus huangi Young	X	
Yunnanosaurus robustus Young	X	
Gyposaurus sinensis Young	X	
Sinosaurus triassica Young	X	
Lukousaurus yini Young	X	
Ornithischia		
Tatisaurus oehleri Simmons	X	
Tawasaurus minor Young	X	
Dianchungosaurus lufengensis Young	X	
Therapsida		
Bienotherium yunnanense Young		X
Bienotherium minor Young		X
Bienotherium magnum Chow	X	
Kunminia minima Young	X	
Lufengia delicata Chow and Hu	X	

Table 8-1 List of Fossil Vertebrates from the Lufeng Formation (after Sun and Cui 1986; Luo and Wu 1994) (Continued)

	Lower Red Beds	Upper Purple Beds
Oligokyphus lufengensis Luo and Su	X	
Yunnanodon brevirostre Cui	X	
Dianzhongia longirostrata Cui	X	
Mammalia		
Morganucodon (Eozostrodon) heikoupengensis		
Young	X	
Morganucodon oehleri Rigney		
Sinocondon rigneyi Patterson and Olson	X	

back, long tail, and long, slender legs (see figure 8-8). The skull lacks any of the aquatic specializations seen in more advanced crocodylomorphs, and this, plus the postcranial morphology, identify *Dibothrosuchus* as a fully terrestrial animal just like its closest relatives, the other sphenosuchians. Furthermore, cranial evidence (highly elongated lagena, large tympanum, and elaborate tympanic recesses) suggests acute hearing ability in *Dibothrosuchus*, and probably indicates some sort of social and/or defensive behavior. Cladistic analysis by X. Wu and Chatterjee (1993) identifies *Dibothrosuchus* as the most derived sphenosuchian.

Strigosuchus licinus is known from a fragment of a left mandible. The jaw is slender and has an upturned symphysis. This poorly known taxon has been deemed either a crocodylomorph (Sun and Cui 1986), a pseudosuchian (Simmons 1965; Sun et al. 1992) or a *nomen dubium* (Luo and Wu 1994).

Microchampsa scutata is known only from postcrania-vertebrae, ribs, dermal scutes and some bones of the manus (Young 1951a). The dorsal vertebrae are short and stout, the ribs are double headed and fused to the scutes in the lumbar region, and both dorsal and ventral scutes are rectangular, overlapping, and form three rows dorsally. Sun et al. (1992) assign *Microchampsa* to the Notochampsidae, but there is no real basis for doing so, and it is best regarded as a poorly known protosuchian, ergo a *nomen dubium* (Luo and Wu 1994).

Dianosuchus changchiawaensis is known from a nearly complete skull and lower jaw. This small crocodylomorph has a strongly flattened snout, extremely small antorbital fenestrae, small and widely separated supratemporal fenestrae

Figure 8-7 *The most common Dawan dinosaur, the prosauropod Lufengosaurus (total length of skeleton is about 6 m).*

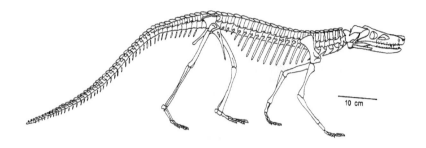

Figure 8-8 *Skeletal reconstruction of the 1.3 m long crocodylomorph Dibothrosuchus (after X. Wu and Chatterjee 1993).*

and isodont, conical teeth that lack striations. The type specimen of *Dianosuchus* probably is a juvenile of a protosuchid (Luo and Wu 1994; X. Wu and Sues 1995).

"*Dianosaurus*" (= *Clevosaurus*: X. W, 1994), long considered a possible protorosaur, is known only from a skull missing everything anterior to the orbits (Young 1982). Recent restudy, however, indicates it is a diapsid (both upper and lower temporal fenestra are present) and has a parietal foramen. It belongs to the Sphenodontia because it possesses an enlarged tooth row on the lateral edge of the palatine, a dentary with a tall coronoid process, a long posterior process that extends back to the articular fossa, and an external mandibular foramen enclosed by the dentary and surangular (Sun et al. 1992). Three Lufeng species of *Clevosaurus* are recognized: *C. petilus, C. wangi,* and *C. mcgilli* (X. Wu 1994).

The supposed lepidosaur *Fulengia youngi* Carroll & Galton is now considered to have been based on a juvenile prosauropod fossil (Evans and Milner 1989; Sereno 1991). Young (1951a) described *Pachysuchus imperfectus* as a parasuchid (phytosaur), but the poorly preserved holotype skull fragment is not diagnostic. Buffetaut (1993) reopened the possibility that *Pachysuchus* is a phytosaur but presented no convincing arguments.

Lufeng dinosaurs, as noted above, are mostly prosauropods, best known from *Lufengosaurus* (see figure 8-7). *Lufengosaurus* displays the characteristic prosauropod features: relatively small skull with spatulate, serrated teeth, jaw joint well below the tooth row, long neck, saurischian pelvis, long tail, massive hind limbs, large claw on digit 1 of the forefoot, and a rudimentary digit 5 on the hind foot. It is very similar to the anchisaurid *Plateosaurus*, although *Plateosaurus* is geologically much older, being of Late Triassic age.

The species-level taxonomy of *Lufengosaurus* is oversplit, and the type species *L. huenei* Young may be the only valid species. Synonyms of *L. huenei* besides the many named species assigned to the genus are *Gyposaurus sinensis* Young and *Tawasaurus minor* Young (Galton 1990). The other saurischian dinosaurs from the Lufeng Formation are *Sinosaurus triassica* Young; *Lukousaurus yini* Young; *Yunnanosaurus huangi* (= *Y. robustus*) Young; and *Dilophosaurus sinensis* Hu. *Sinosaurus* and *Lukosaurus* are poorly known, but *Yunnanosaurus* is a well known and very distinctive prosauropod with unique teeth that are cylindrical and somewhat flattened from side to side. These teeth lack the coarse serrations characteristic of other prosauropods and are more similar to the teeth of later sauropods.

The recent report of the ceratosaurian theropod *Dilophosaurus* from Lufeng is based on a skull (S. Hu 1993). Previously, *Dilophosaurus* was known only from the Lower Jurassic Kayenta Formation of Arizona (Welles 1984). Its discovery in the Lufeng Formation thus bolsters evidence of an Early Jurassic age (see below).

Although saurischian dinosaurs are abundant in the Lufeng Formation (*Lufengosaurus* is the most common vertebrate fossil in the formation), ornithischians are rare and known from very fragmentary remains. Two taxa have been named for jaw fragments: *Scelidosaurus oehleri* (Simmons) and *Dianchungosaurus lufengensis* Young.

Lufeng Formation therapsids are tritylodontids (see figure 8-9) of the genera *Bienotherium*, *Dianzhongia*, *Lufengia*, *Oligokyphus*, and *Yunnanodon* (Young 1947, 1974a; Chow and Hu 1959; Chow 1962; Cui 1976, 1981; Luo and Wu 1994). These relatively small therapsids have one of their most important fossil records in the Lufeng Formation.

Bienotherium is the most common, characteristic, and largest Lufeng tritylodontid (see figure 8-9). Three species have been named: the type species, *B. yunnanense* Young, to which most specimens belong; the much larger and rare *B. magnum* Chow in which the upper cheek teeth are about 1 cm long; and the much smaller and rare *B. minor* Young, which may actually belong in the genus *Lufengia* according to Hopson and Kitching (1972). *Bienotherium* is readily identified by its very large size and robustness, the exposure of the maxillaries on the lateral and palatal surfaces of the skull, relatively long diastemata and slender zygomata, postcanine teeth with obliquely subquadrangular outlines, cusp formula of 2.3.3, and doubled roots on the lower postcanine teeth.

Somewhat smaller is *Dianzhongia*, known from a single species, *D. longirostrata* Cui. Known only from a skull lacking the zygomata, *Dianzhongia* has a slender skull with a long, robust snout, long diastemata, seven relatively small upper postcanines, and a cusp formula of 2.3.2.

Unlike *Bienotherium* and *Dianzhongia*, *Lufengia* (type and only species = *L. delicata* Chow & Hu) is a very small tritylodontid. The skull has a narrow, short and pointed snout, no sagittal crest, a flat frontal region, slender zygomatic arch,

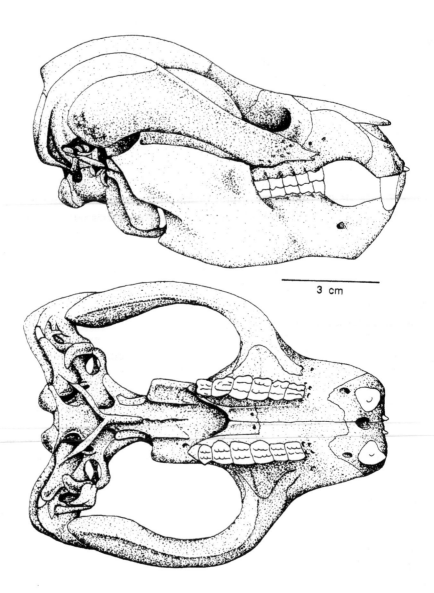

Figure 8-9 *Skull and lower jaws of the characteristic Dawan tritylodontid Bienotherium (after Young 1947).*

postcanine teeth in which the width is greater than the length, and a cusp formula of 2.3.3. It may be based on juvenile specimens of *Bienotherium*.

Young (1947) originally identified the widespread tritylodontid genus *Oligokyphus* from the Lufeng Formation based on a partial dentary with postcanine teeth, the holotype of *O. sinensis*. However, subsequent workers have identified this specimen as a juvenile of *Bienotherium* (Sues 1985; Sun and Cui 1986; Sun et al. 1992). Nevertheless, Luo and Sun (1993) recently assigned a lower jaw fragment from the Lufeng Formation to *Oligokyphus* as the new species *O. lufengensis*. The presence of three principal cusps in each postcanine tooth row is the key dental feature that distinguishes *Oligokyphus* from all other tritylodontids, and the Chinese specimen displays this feature (see figure 8-10). Outside of China, *Oligokyphus* is known from Liassic strata in England, Germany, and United States (Arizona) (e.g., Hennig 1922; Kühne 1956; Sues 1985, 1986). Its presence in the Lufeng Formation thus supports other Liassic age indicators.

Yunnanodon (originally *Yunnania*, but that name was preoccupied by the name of a gastropod) is another small Lufeng tritylodontid about the size of *Lufengia*. However, unlike *Lufengia*, it has a short, wide snout, a vaulted preorbital region, a low sagittal crest, and a cusp formula of 2.3.2 (Cui 1976). One species, *Y. brevirostre* Cui, is known from a single skull.

One last possible therapsid from the Lufeng Formation is *Kunminia minima*, which Young (1947) thought was an ictidosaur. The poorly preserved partial skull and lower jaw are difficult to interpret, and Hopson and Kitching (1972) regard the taxon as a *nomen dubium* and possible synonym of *Morganucodon*.

Mammals also have a very important record in the Lufeng Formation. Species of two of the best known early mammals are present, the genera *Morganucodon* and *Sinoconodon*. *Morganucodon* was first described from Wales and is very well known from-fissure fill deposits there of latest Triassic and/or Early Jurassic age (Kermack et al. 1973, 1981). Chinese *Morganucodon* (see figure 8-11) are very similar to Welsh specimens. But, *Sinoconodon* is endemic to China, and to the Lufeng Formation. It presents a unique and important perspective on mammal origins, discussed below.

Outside of the Lufeng Formation, *Lufengosaurus* is known from the Zhenzhuchong Formation in the Sichuan basin (Y. Wang and Sun 1983; Dong et al. 1983; Dong 1984). Strata correlatives to the Lufeng Formation in western Yunnan are the Fengjiahe Formation, which has yielded fossils of *Lufengosaurus* and has a theropod dinosaur footprint assemblage (Zhen et al. 1989).

As mentioned previously, the Lufeng Formation traditionally has been divided into two units (Bien 1940): (1) a lower, dull purplish mudstone unit as much as 230 m thick overlain by (2) dark red beds as much as 184 m thick (Figs. 8-4, 8-6). (A formal nomenclature identifying these strata as the Shawan Formation overlain by the Zhangjiawa Formation of the lower Lufeng Group also exists [X. Wu and Chatterjee 1993, see figure 1] but is not used here to maintain continuity with previous usage in the vertebrate paleontological literature.) The fossil vertebrate record of the Lufeng Formation begins roughly in

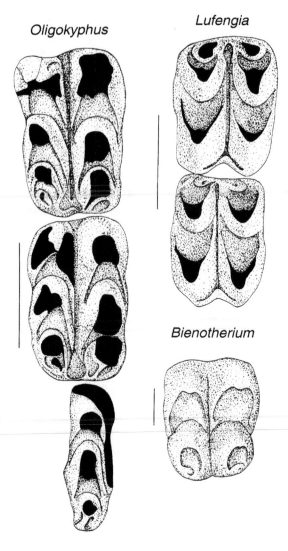

Figure 8-10 *Occlusal views of the postcanine teeth of the tritylodontid Oligokyphus compared to those of the contemporaneous tritylodontids Lufengia and Bienotherium (after Luo and Sun 1993). Scale bars = 2 mm.*

the middle of the dull purplish beds and ends abruptly in the upper part of the dark red beds (see figure 8-6).

Luo and Wu (1994) noted that there is considerable faunal turnover between the vertebrate fauna of the dull purplish beds and the dark red beds of the Lufeng Formation. This turnover occurs low in the dark red beds, and it mostly involves the appearance of many new taxa, including crocodylomorphs, ornithischian dinosaurs, advanced and diverse tritylodontids, and mammals. The fauna of the dull purplish beds is merely a depauperate subset of the fauna of the dark red beds, except for the tritylodontid *Bienotherium yunnanense* Young, which is restricted to the dull purplish beds. This might provide a basis for subdivision of the Dawan faunachron, but an earlier sub-faunachron based on the vertebrate fauna of the dull purplish beds is impossible to characterize except for the presence of *Bienotherium yunnanense*. Because the faunal turnover in the Lufeng Formation corresponds to a major lithofacies change, it is unlikely that it is of real evolutionary or paleobiogeographic significance. Instead, this faunal turnover probably reflects preferential preservation of vertebrate bone in the dark red beds rather than in the underlying dull purplish beds.

Global Correlation of the Dawan

For decades most vertebrate paleontologists considered the fossil vertebrates from the Lufeng Formation to be of Late Triassic age or of Late Triassic (lower vertebrate-producing horizons) and Early Jurassic (upper vertebrate-producing horizons age). Some recent workers (e.g., Colbert 1986, Dong 1992; X. Wu 1994) continue to advocate one of these age assignments, although by 1982 it was clear that the entire Lufeng Formation is of Early Jurassic age. The lower

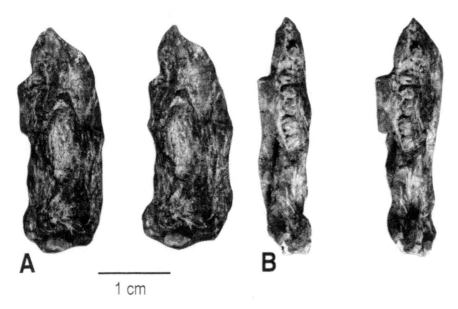

A _____ B

1 cm

Figure 8-11 A skull of the early mammal Morganucodon oehleri from the Lufeng Formation seen as two stereophotos: A, dorsal view; B, lateral view showing teeth.

Lufeng Formation and its correlatives, the Fengjiahe Formation of central Yunnan and the Zhengzhuchong Formation in Sichuan contain ostracods (*Gomphocythere-Darwinula* assemblage), conchostracans (*Palaeolimnadia*), and bivalves (*Qiyangia, Apseudocardina*) that indicate a Liassic age (P. Chen et al. 1982a, b). Three key fossil vertebrate genera of the Lufeng Formation—*Dilophosaurus, Scelidosaurus,* and *Oligokyphus*—are only found in Liassic vertebrate faunas outside of China. The overall composition of the Dawan vertebrate fauna, especially its dominance by crocodylomorphs, prosauropod dinosaurs, tritylodontids, and mammals, is characteristic of a cosmopolitan Liassic vertebrate fauna across Pangea (see figure 8-12). This fauna is readily identified as Liassic, probably late Liassic (Sinemurian) in age (Luo and Wu 1995).

Dashanpuan Vertebrates

The term Dashanpuan faunachron of Lucas (1996a, c) refers to the interval of time represented by the vertebrate fossil assemblage from Dashanpu just east of Zigong in Sichuan Province (see figure 8-1). This assemblage is from the lower Shaximiao (= Xiashaximiao) Formation. The following taxa are present: scales and teeth of hybodontid and ceratodontid fishes; the labyrinthodont amphibian *Sinobrachyops placenticephalus* Dong; the chengyuchelyid turtles *Chengyuchelys zigongensis* Ye, *C. baenoides* Young & Chow, and *C. dashanpuensis* Fang; the pterosaur *Angustinaripterus longicephalus* He, Yan, & Su; the "cetiosaurid" sauropod *Shunosaurus lii* Dong, Zhou, & Zhang; the camarasaurid sauropods *Abrosaurus dongpoensis* Ouyang and *Datousaurus bashanensis* Dong & Tang; the megalosaurid theropod *Xuanhanosaurus gilixianensis* Dong; the "fabrosaur" *Agilisaurus louderbacki* Peng; the hypsilophodontids *Yandusaurus hongheensis* He (= *Y. multidens* He & Cai: Sues and Norman 1990), and *Xiaosaurus dashanpensis* Dong & Tang (probably a *nomen dubium*); the most primitive stegosaur, *Huayangosaurus taibaii* Dong, Tang & Zhou; and the tritylodontids *Bienotheroides zigongensis* Sun and *Polistodon chuannanensis* He & Cai (He and Cai 1984; Sun 1984, 1986; Sun and Li 1985; Cui and Sun 1987).

The lungfish and hybodont fishes from the lower Shaximiao Formation have not been described. The turtle *Chengyuchelys* (see figure 8-13) is known from nearly complete carapaces and plastra that represent three named species (Young and Chow 1953; Ye 1982; Fang 1987). Ye (1990, 1994) united *Chengyuchelys* with the genus *Xinjiangchelys* to form the family Chengyuchelyidae, a group of Jurassic cryptodirans with oval carapaces that lack ornamentation, have eight hexagonal neural plates and have a wide bridge, round posterior margin, mesoplastron, intergulars, and inframarginals, among other features.

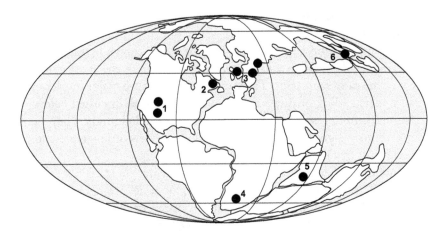

Figure 8-12 *Distribution of principal late Liassic vertebrate faunas across Pangea: 1 - Kayenta and La Boca formations, western USA and northern Mexico; 2 - McCoy Brook Formation, Canada; 3 - Liassic fissure fills and marginal marine deposits of western Europe; 4 - middle-upper Elliot Formation and Clarens Formation, South Africa; 5 - Kota Formation, India; 6 – Lufeng Formation, China.*

Most of the Dashanpuan vertebrate fauna comes from a single quarry at Dashanpu. Taxa in the fauna are mostly endemic to this locality, though *Bienotheroides zigongensis* is also known from the Wucaiwan Formation in the Junggur basin of Xinjiang (Sun and Cui 1989). The Wucaiwan Formation also has yielded fossils of the amphibian *Superstogyrhinus ultimus* Zhang; the lizard *Archovaranus klameliensis* Zhang; the sauropod dinosaur *Bellusarus sui* Chao; the theropod *Monolophosaurus jiangi* Zhao & Currie; and ornithischian dinosaurs (Dong 1992).

Middle Jurassic vertebrate faunas are generally not well known. The Dashanpu quarry assemblage is outstanding for its time interval because most of the taxa present are known from complete skulls and skeletons. *Sinobrachyops* (see figure 8-14) is the youngest Chinese labyrinthodont and contributes to recent discoveries that confirm that this great diversification of early amphibians, long thought to have suffered extinction at the end of the Triassic, continued into the Early Cretaceous (though post-Triassic diversity was very low).

The fishes and turtle from Dashanpu confirm the aquatic (fluvial) nature of deposition at the site. *Angustinaripterus* is a rhamphorhynchid pterosaur known from a skull missing its rear end (see figure 8-15). Its elongated skull (estimated total length about 165 mm) suggests a pterosaur with a wingspan of about 1.6 meters (Wellnhofer 1991). The procumbent, alternating tooth rows were those of a fish catcher. The very narrow external narial opening, from which *Angustinaripterus* takes its name (which is Latin for "wing with narrow nostril"), is the most distinctive feature of the genus and readily distinguishes it from other rhamphorhynchids.

The Dashanpu dinosaur fauna includes the oldest known hypsilophodontid and stegosaur. *Shunosaurus* and *Datousaurus* are among the most completely known of the early sauropods. These dinosaurs thus have major impact on our overall understanding of Jurassic dinosaur evolution, as discussed below. The tritylodontids from Dashanpu are among the last tritylodontids. They are part of an extremely significant record of tritylodontid evolution represented by Chinese fossils (see figure 8-13).

Because of its endemism, the Dashanpuan vertebrate fauna is not easily correlated to the standard global chronostratigraphic scale. Stratigraphic position and bivalves (*Eolamprotula—Psilunio* fauna), ostracods, conchostracans (*Euestheria zilinjinensis*), and charophytes (*Euaclistochara*) suggest a Middle Jurassic (probably Bajocian) age (Chen et al. 1982a, b). The Dashanpuan vertebrates are consistent with this age—especially the primitive sauropods, megalosaurid, hypsilophodontids, and stegosaur—but do not provide stage-age resolution.

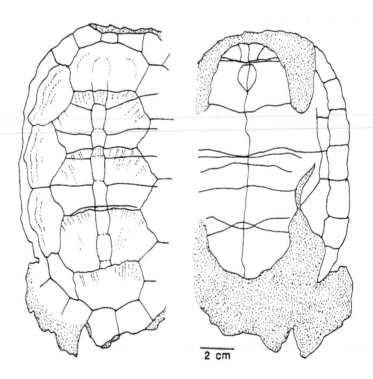

2 cm

Figure 8-13 *The Dashanpuan turtle Chengyuchelys, carapace on left, plastron on right (after Ye 1982).*

Tuojiangian Vertebrates

The Tuo Jiang is a river in Sichuan Province with headwaters northeast of Chengdu that joins the Chang Jiang (Yangtze River) at Luzhou. The Tuojiangian land-vertebrate faunachron is the time represented by the vertebrate fauna of the upper Shaximiao (= Shangshaxiamiao) Formation in the Sichuan basin. This fauna is the "sauropod—*Mamenchisaurus* fauna" or faunal complex of Dong et al. (1983) and Zhen et al. (1985), and the "*Mamenchisaurus* fauna" of Dong (1992).

The vertebrate fauna of the upper Shaximiao Formation (Camp 1935; Young 1939a; Dong et al. 1983; He et al. 1984) includes: the ptycholepid fishes *Yuchoulepis szechuanensis* Su and *Chungkingichthys tachuensis* Su; the semionotid fish *Tianfuichthys spinodorsalis* Su; the turtles *Tienfuchelys tzuyangensis* Young & Chow, *Plesiochelys radiplicatus* Young & Chow, *Plesiochelys tatsuensis* Ye, *Chengyuchelys baenoides* Young and Chow, and *Sinaspideretes wimani* Young & Chow; the protosuchian crocodilian *Sichuanosuchus huidongenesis* Peng (see G. Peng 1995, 1996); the sebecosuchian crocodilian *Hsisosuchus chungkingensis* Young & Chow; the theropods *Szechuanosaurus campi* Young, *Yangchuanosaurus shangyouensis* Dong, Zhang, Li & Zhou, and *Y. magnus* Dong, Zhou & Zhang; the sauropods *Omeisaurus junghsiensis* Young, *O. changshouensis* Young, *O. tianfuensis* He, Li, Cai & Gao, and *O. fuxiensis* Dong,Zhou & Zhang; the diplodocids *Mamenchisaurus constructus* Young, *M. jing-yanensis* Zhang, Li & Zeng, and *M. hochuanensis* Young & Chao; the ornithopod *Gongbusaurus shiyii* Dong, Zhou & Zhang; the stegosaurids *Tuojiangosaurus multispinus* Dong, Li, Zhou & Zhang, *Chialingosaurus kuani* Young, and *Chungkingosaurus jiangbeiensis* Dong, Zhou & Zhang; the youngest known tritylodontid, *Bienotheroides wanshienensis* Young; and the mammal *Shuotherium dongi* Chow & Rich.

The ptycholepid fishes are primitive Mesozoic neopterygians. Tuojiangian turtles are primitive cryptodires, the plesiochelyids, and chengyuchelyids, and a very early soft-shelled turtle (trionychid), *Sinaspideretes*.

The only Tuojiangian crocodilian, *Hsisosuchus*, is the sole genus of Young and Chow's (1953) family Hsisosuchidae. It has a deep, elongate snout with an antorbital fenestra and relatively posteriorly located external nares. The infratemporal fenestrae are slit-like openings, and the secondary palate is poorly developed. The few teeth are widely separated, and those in the maxillary have serrated anterior and posterior edges. This unusual crocodilian has either been included in the Sebecosuchia, Mesosuchia, or suggested to be a representative of a new suborder.

Tuojiangian theropods are *Szechuanosaurus* and *Yangchuanosaurus*, both known from nearly complete skeletons. *Szechuanosaurus* is a medium-sized

Figure 8-14 *Skull of the Dashanpuan brachyopid amphibian Sinobrachyops, dorsal view on left, ventral on right.*

allosaurid, whereas *Yangchuanosaurus* (see figure 8-16) is an 8 m long form of less certain affinities among the carnosaurs (Molnar et al. 1990).

The largest Chinese sauropods, *Omeisaurus* and *Mamenchisaurus*, are of Tuojiangian age (see figure 8-17). The distribution of these sauropods in China provides a key means of identifying and correlating Tuojiangian strata. The highest diversity of Chinese stegosaurs was during the Tuojiangian. Best known is *Tuojiangosaurus* (see figure 8-17). This animal was about the size of North American *Stegosaurus* but differs notably in its spike-shaped, not plate-like, dorsal armor.

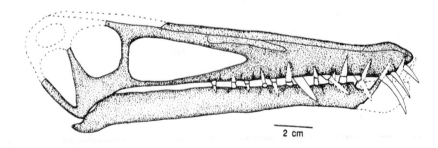

Figure 8-15 *Lateral view of the skull and lower jaw of the Dashanpuan pterosaur Angustinaripterus (after Wellnhofer, 1991). The procumbent anterior teeth were probably used to grab fish.*

In China, Tuojiangian age strata outside the Sichuan basin include the Keilozo (= Kalazha, = Karaza, = Hongshan, = Huoyanshan) Formation of Xinjiang, which yielded the theropods *Shanshanosaurus huoyanshanensis* Dong and *Szechuanosaurus*, and the sauropod *Hudiesaurus sinojapanorum* Dong (Dong 1977, 1992; 1997f and g); and the Xiangtang (Hantong) Formation of Gansu, which contains the goniopholid crocodilian *Sunosuchus*, also known from the Jurassic of Thailand (Buffetaut 1982), and the sauropod *Mamenchisaurus hochuanensis* (Young 1958c; Young and Chao 1972). Another probable correlative, the vertebrate fauna of the Shishigou Formation (or Qigu Formation of the Shishigou Group: J. Peng and Brinkman 1993) in the Junggur basin, yielded the turtle *Xinjiangchelys junggarensis* Ye; the crocodilian *Sunosuchus junggarensis* Wu, Brinkman & Russell; the sauropods *Tienshanosaurus chitaiensis* Young, and *Mamenchisaurus sinocanadorum* Russell & Zheng; the theropods *Sinraptor hepingensis* (Gao) and *S. dongi* Currie & Zhao; the ornithopod *Gongbusaurus wucaiwanensis* Dong; and the amphilestid mammal *Klamelia zhaopengi* Chow & Rich.

Most workers have correlated the Tuojiangian dinosaur fauna with the dinosaur faunas of the Morrison Formation in the western United States and middle and upper dinosaur beds of the Tendaguru "series" in Tanzania, east Africa (e.g., Young 1951b; Dong et al. 1983; Russell 1993). No Chinese dinosaur taxa are found in these non-Chinese faunas, so the argument for their correlation identifies taxonomic counterparts and concludes they are at the same stage of evolution and therefore the same age (see table 8-2). This analysis suggests the Tuojiangian dinosaurs are of Late Jurassic (Kimmeridgian—Tithonian) age, as do nonmarine invertebrates (conchostracans, ostracods, and bivalves) from Tuojiangian strata. However, it seems likely the Tuojiangian is on the older end of this age range (Kimmeridgian) or late Middle Jurassic, and thus may be slightly older than the Morrison and Tendaguru dinosaur faunas. This would explain the endemism of the Tuojiangian dinosaurs, rather than arguing for the paleobiogeographic isolation of eastern Asia during the Late Jurassic (Russell 1993). Ningjiagouan vertebrates are younger than Tuojiangian dinosaurs and are Late Jurassic (Tithonian?) in age (see table 8-2).

Ningjiagouan Vertebrates

The vertebrate fauna of the Mengyin Formation in Shandong found near the village of Ningjiagou is the basis of the Ningjiagouan faunachron (Lucas 1996a). This vertebrate fauna consists of the fishes *Sinamia zdanskyi* Stensiö (see figure 8-18) and *Lycoptera* sp.; the turtles *Sinemys lens* Wiman, *Sinochelys appalanata* Wiman, and *Scutemys tecta* Wiman; an indeterminate theropod?;

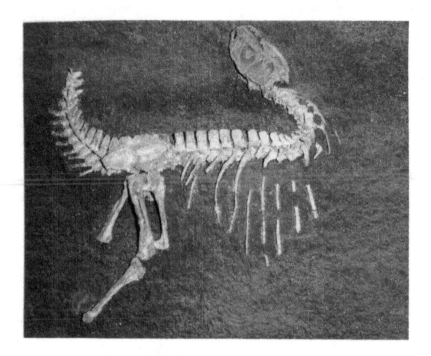

Figure 8-16 *An incomplete skeleton of the 8 m long carnosaur Yangchuanosaurus shows the characteristic death position with the head and neck curled up, indicating dessication of the carcass before burial.*

the sauropod *Euhelopus zdanskyi* (Wiman); an indeterminate sauropod; and an indeterminate stegosaur (Wiman 1929 1930; Stensiö 1935).

Sinamia (see figure 8-18) is an extinct bowfin well known from numerous Ningjiagouan localities in northern and eastern China (H. Liu and Su 1983). At some of these sites it co-occurs with the closely related genus *Ikechaoamia*.

The Ningjiagouan turtles are cryptodires of uncertain affinities. The only well-known Ningjiagouan dinosaur is the sauropod *Euhelopus* (see figure 8-19). Wiman (1929) erred in his reconstruction of the skull of *Euhelopus*. A revised reconstruction by Mateer and McIntosh (1985) identifies *Euhelopus* as a *Camarasaurus*-like sauropod with a delicate skull and uniquely shaped pterygoids and palatines. However, unique features of *Euhelopus* also include its high-humerus:femur ratio (0.99), higher than any sauropod except *Brachiosaurus;* its extremely high number of vertebrae, approaching that of *Mamenchisaurus;* and its diplodocid-like bifurcation of the neural spines.

Tan and J. G. Andersson collected some of these vertebrates in the fall of 1922, and O. Zdansky collected the remainder in the spring of 1923 (Mateer and Lucas 1985). The initial impetus for this work came from the discovery of dinosaur bones in the Mengyin area in 1916 by W. Behagel, a German mining engineer (Tan 1923: 95; Wiman 1929: 5). The Ningjiagouan vertebrates were long considered to be of Early Cretaceous age (e.g., Tan 1923; Wiman 1929;

Figure 8-17 *Two characteristic Tuojiangian dinosaurs are the sauropod Mamenchisaurus (above) and the stegosaur Tuojiangosaurus (below).*

Morris 1936; Rozhdestvensky 1977) based in part on Grabau's (1923) assess-

Table 8-2 Comparison of Tuojiangian Dinosaur Genera with the Dinosaur Faunas of the Morrison and Tendaguru (after Haubold 1989)

	Morrison Formation	Tendaguru Series	Tuojiangian
Stegosauria	Stegosaurus	Kentrosaurus	Chialingosaurus
			Tuojiangosaurus
			Chungkingosaurus
Fabrosauridae	Echinodon	—	Gongbusaurus
Ankylosauria	—	—	—
Ornithopoda	Nanosaurus	Dryosaurus	—
	Othnielia		
	Dryosaurus		
	Camptosaurus		
Sauropoda	Camarasaurus	Brachiosaurus	Omeisaurus
	Uintasaurus	Barosaurus	(Zigongosaurus)
	Brachiosaurus	Dicraeosaurus	
	Ultrasaurus	Tornieria	Mamenchisaurus
	Supersaurus		
	Dystylosaurus		
	Diplodocus		
	Apatosaurus		
	Haplocanthosaurus		
	Barosaurus		
Ceratosauria	Ceratosaurus	Ceratosaurus	—
	Marshosaurus		
Carnosauria	Torvosaurus	—	Szechuanosaurus
			Yangchuanosaurus
	Allosaurus	Allosaurus	—
Coelurosauria	Elaphrosaurus	Elaphrosaurus	—
	Coelurus		
	Ornitholestes		
	Stokesosaurus		

ment of associated invertebrate fossils. Microfossils from Ningjiagouan strata-suggest it is of Late Jurassic age, though it is possible some Ningjiagouan vertebrate localities are of Early Cretaceous (Neocomian) age.

Recent work reported by P. Chen (1982) and P. Chen et al. (1980, 1982a and b) has altered the stratigraphic nomenclature and correlation of the "Mengyin Series" of earlier workers. P. Chen et al. (1980) divided the "Mengyin Series" into two formations—the Mengyin Formation and overlying Xiwa Formation. The Xiwa Formation is a volcaniclastic sequence as much as 1600 m thick and contains the fossil fish *Lycoptera* and conchostracans (especially *Eosestheria*) indicative of a Late Jurassic (Tithonian) age (P. Chen 1982; P. Chen et al. 1982a, b). The Mengyin Formation is as much as 714 m thick and consists of gray-green and purple-red sandstone, siltstone, mudstone, and shale. The fossil vertebrates described by earlier workers from the "Mengyin Series" are from the Mengyin Formation as restricted by P. Chen et al. (1980). Conchostracans from the Mengyin Formation indicate it is of Tithonian age.

Grabau's (1928) term "Jehol fauna" referred to invertebrate and vertebrate fossils from what is now termed the Jehol Group in western Liaoning (e.g., Gu 1992). Fossil localities of the "Jehol fauna" occur in several stratigraphic unitst-hat may encompass portions of Late Jurassic and Early Cretaceous time. Indeed, there is clear disagreement among Chinese biostratigraphers about the placement of the Jurassic-Cretaceous boundary in the units that encompass the Jehol fauna (P. Chen 1988; Gu 1992; Y. Wang et al. 1995). Nevertheless, most of the vertebrate fauna of the Jehol Group is clearly of Ningjiagouan age. It includes the fishes *Sinamia zdanskyi* Stensiö and *Lycoptera* sp., the turtle *Mandchurochelys manchouensis* Endo & Shikama, the lizard *Yabeinosaurus tenuis* Endo & Shikama, the crocodilian? *Rhynchosaurus orientalis* Endo & Shikama, the sauropod dinosaur *Euhelopus zdanskyi* (Wiman), and the mammals *Manchurodon simplicidens* Yabe & Shikama and *Endotherium niinomi* Shikama. This suggests a Late Jurassic age for some of the vertebrates of the "Jehol fauna," although earlier workers generally considered them to be of Early Cretaceous

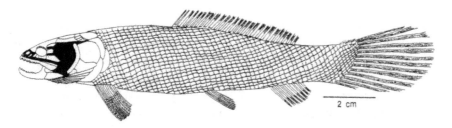

Figure 8-18 *The bowfin fish Sinamia zdanskyi is a characteristic fish taxon of the Ningjiagouan (after H. Liu and Su 1983).*

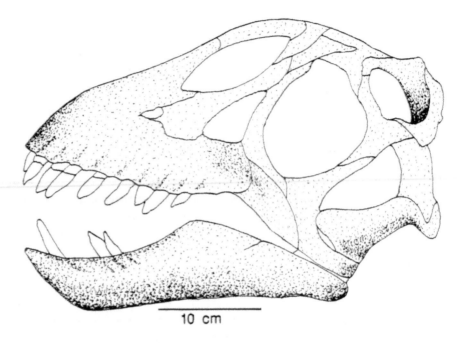

Figure 8-19   *Skull of the Ningjiagouan sauropod dinosaur Euhelopus (after Mateer and McIntosh 1985).*

age (e.g., Grabau 1923; Endo 1934, 1939; Morita 1939). Some Jehol Group vertebrates are of Early Cretaceous age—*Psittacosaurus* and the recently described early birds, including *Sinornis, Confuciusornis,* and *Cathayornis* (Sereno and Rao 1992; Zhou 1992) (see chapter 9). Clearly, the Ningjiagouan encompasses parts of Late Jurassic and Early Cretaceous time.

Jurassic Fishes

There are a few scattered records from China of Early Jurassic fishes. They are of ceratodontid lungfishes (*Ceratodus szechuanensis* Young, *C. shenmuensis* Liu & Yeo, and *C. youngi* Liu & Yeo) from Shaanxi, Hunan, and Sichuan; the palaeonisciform *Xingshikous xishanensis* Liu from near Beijing; the ptycholepid *Yuchoulepis gansuensis* Su and the furid *Plesiofuro mingshuica* Su from Gansu; and a small assemblage of semionotids, amiiforms, the coccolepid *Plesiococcolepis hunanensis* Wang, the pholidophorid *Hengnania gracilis* Wang, an archaeomanid, and a ceratodontid from Hunan (N. Wang 1977a, b; X. Liu and Wang 1985). These are all nonmarine records.

Middle Jurassic fish records are more widely distributed in China, coming from Xinjiang, Gansu, Nei Mongol, Shaanxi, Hunan, and Sichuan (M. Chang and Jin 1996). They range from marine records of hybodont selachians (*Hybodus antingensis* Liu from Shaanxi, *Hybodus huangnidanensis* Wang from Hunan) to nonmarine records dominated by semionotids and pholidophorids. The most significant nonmarine record is from the lower Shaximiao Formation in the Sichuan basin, where the ptycholepid *Yuchoulepis szechuanensis* Su and the chungkingichthyid *Chunkingichthys tacheunsis* Su are common.

The Late Jurassic fishes of China can be included in a *Lycoptera* "fish fauna" (M. Chang and Jin 1996) that includes assemblages of fossil fishes that are not only Late Jurassic but also some that are Early Cretaceous in age. This is the first teleost-dominated fish fauna from China, and is discussed with Cretaceous fishes in the next chapter.

Jurassic Dinosaur Footprints

Most dinosaur footprints known from China are of Jurassic age, and come from the Sichuan basin (see figure 8-20), though a few Cretaceous footprint sites are known (e.g., Young 1943, 1960; Zhen et al. 1989, 1994, 1996; Matsukawa et al. 1995). At least 28 ichnogenera have been named, but many are synonyms of already used names. Lower Jurassic footprint sites are located principally in Liaoning and Yunnan and are theropod dominated, especially by the ichnogenera *Grallator* and *Eubrontes*. Only a few Middle Jurassic sites are known, in Sichuan, Shaanxi, Shanxi, and Jilin, and they contain tracks of theropod dinosaurs (see figure 8-21) and of crocodilians (*Batrachopus*) (e.g. Young 1943). Late Jurassic footprints also are not common, but do include those of both theropod (see figure 8-21) and ornithopod dinosaurs ("*Yangtzepus*"). Cretaceous dinosaur footprints are common at some sites, but remain theropod dominated. Perhaps the most that can be said about the Chinese dinosaur footprint record is that it is taxonomically oversplit and poorly reflects the dinosaur body-fossil record.

Chinese Tritylodontids

Tritylodontids are a group of mostly small, herbivorous synapsids. These quadrupeds have large, chisel-shaped incisor teeth in the front of the mouth, followed by a diastema (gap) brought up by cheek teeth with broad, flat, multicusped crowns (see figures 8-9–8-10). Looked at in modern terms, these are very rodent-like features that suggest tritylodontids independently evolved

Figure 8-20 *Principal Jurassic and Cretaceous dinosaur footprint localities in China (after Zhen et al. 1989): 1 – Jinning, Yunnan; 2 – Chaoyang, Liaoning; 3 – Laiyang, Shandong; 4 – Tungchuan, Shaanxi; 5 – Xiangxi, Hunan; 6 – Shenmu, Shaanxi; 7 – Guangyuan, Sichuan; 8 – Yuechi, Sochuan; 9 – Shanxi; 10 – Chabu, Nei Monggol; 11 – Emei Shan, Sichuan; 12 – Lianyungang City, Jiangsu; 13 – Xigaze, Tibet; 14 – Nanxiong, Guangdong; 15 - Xiaguan, Henan; 16 – Yiping, Sichuan; 17, 18, 19 – Sichuan; 20 – Hebei; 21 – Chengde, Hebei; 22 – Huinan, Jilin; 23 – Yanji, Jilin.*

a rodent-like body form during the Jurassic, more than 100 years before the evolution of rodents. Indeed, some paleontologists argue that tritylodontids are the closest relatives of mammals, largely because they underestimate the degree of morphological convergence in these closely related groups (Lucas and Luo 1993).

Chinese tritylodontids are strictly of Jurassic age, and their fossils are known from Europe, North America, South Africa, and China. The Chinese trityl-odontid record is particularly significant because it includes a well-preserved and diverse array of Early Jurassic tritylodontids as well as some of the youngest known tritylodontids. The Lufeng Formation of Yunnan has yielded the most-diverse collection of Early Jurassic tritylodontids known—various species of the genera *Bienotherium, Lufengia, Oligokyphus, Yunnanodon* and *Dianzhongia*

(Young 1947, 1974a; Chow and Hu 1959; Chow 1962; Cui 1976, 1981). The fossils of these tritylodontids are from the upper part of the Lufeng Formation (dark red beds), except for some *Bienotherium* from the underlying dull purplish beds (Sun and Cui 1986).

Chinese Middle Jurassic tritylodontids are known from the Shaximiao Formation of Sichuan Province: *Bienotheroides zigongensis* and *Polistodon chuannanensis* (Sun 1984, 1986; Sun and Li 1985; Cui and Sun 1987). *Bienotheroides zigongensis* is also known from the Middle Jurassic Wucaiwan Formation in the Junggur basin of Xinjiang (Sun and Cui 1989). The youngest Chinese tritylodontid is *Bienotheroides wanshienensis* from the upper Shaximiao Formation in Sichuan, a unit of Late Jurassic age (P. Chen et al. 1982a, b; Lucas 1993c).

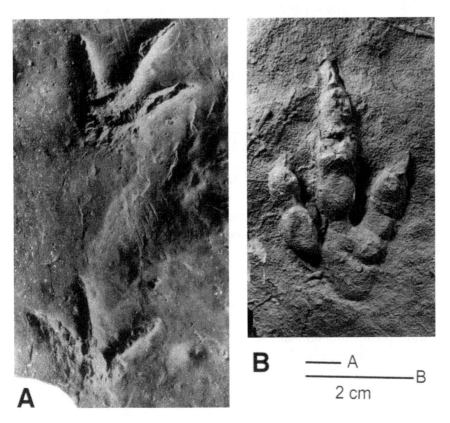

Figure 8-21 *Characteristic dinosaur tracks from the Jurassic of China: A) tracks of a theropod dinosaur that Young (1966) termed Shensipus, from the Middle Jurassic of Shaanxi; B) track of a theropod dinosaur, Jialingpus from the Penglaizhan Formation in Sichuan.*

Chinese Jurassic Mammals

Most Chinese Jurassic mammals are of Early Jurassic age, and their fossils come from the dark red beds of the Lufeng Formation. Two genera have been identified—*Morganucodon* and *Sinoconodon*.

Morganucodon oehleri Rigney and *Morganucodon* (= *Eozostrodon*) *heikoupengensis* Young are the Chinese species of this genus (see figure 8-11). They closely resemble the best known and type species, *Morganucodon watsoni* Kühne, from Wales. The Welsh *Morganucodon* have been described extensively (Kermack et al. 1973, 1981), and the Chinese specimens add little new information, though they do confirm many of the anatomical details of the Welsh specimens (Luo et al. 1995).

Although much older and more primitive mammals are known (Lucas and Luo 1993), *Morganucodon* can still be regarded as the best known of the earliest mammals. The skull, lower jaw, and postcrania of this small, mouse-sized mammal are well preserved and have provided the basis for many ideas of mammal origins. However, *Sinoconodon* from China presents morphological data that forces a modification of ideas of mammal origins based primarily on *Morganucodon*.

Sinoconodon is known only from the skull and lower jaws (see figure 8-22). It is younger geologically than some *Morganucodon*, but *Sinoconodon* has many features which are more primitive than *Morganucodon*. Particularly significant is the dentition of *Sinoconodon*, in which the postcanine tooth row consists of five multicuspid trenchant teeth with only the vestiges of cingula. These teeth do not precisely occlude with one another (Crompton and Luo 1993). They thus do not look like typical mammalian teeth, which do occlude precisely and have distinct cingula and cusps offset from a single longitudinal row. Indeed, precise occlusion is thought to have evolved very early in the evolution of mammals, in *Morganucodon* itself (e.g., Kermack and Kermack 1984).

Sinoconodon shows how much more complex mammalian origins were than previously thought. Although it has many features of the skull (especially the petrosal promontorium) and lower jaws characteristic of mammals, *Sinoconodon* lacks a "mammalian dentition." Precise occlusion of the postcanine teeth thus was not a feature that appeared in all the earliest mammals. It probably evolved several times independently in the early evolution of mammals.

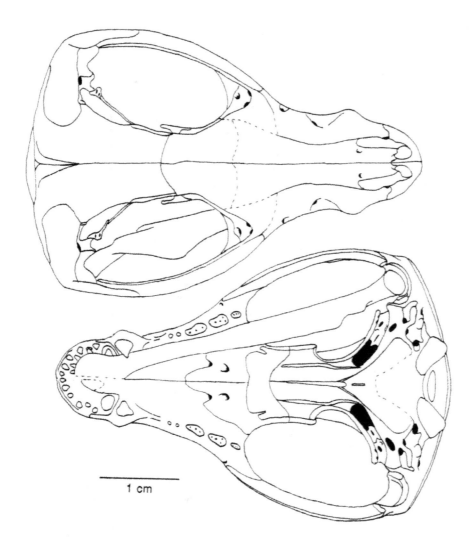

Figure 8-22 *Skull and lower jaw of Sinoconodon, dorsal (above) and ventral (below) views (after Crompton and Luo 1993).*

Chinese Jurassic Dinosaurs

Besides the tritylodontids, China's Jurassic vertebrate fossil record contains a large number of dinosaurs, many of which are of critical importance to deciphering dinosaur phylogeny and evolutionary history.

The prosauropod fossils, especially those of *Lufengosaurus*, from the Lufeng Formation, are part of a Pangea-wide distribution of prosauropods during the

Early Jurassic. This was the zenith of diversity and distribution of the prosauropods, who disappeared at the end of the Early Jurassic, the first major dinosaur clade to suffer extinction.

The oldest and most primitive stegosaur is *Huayangosaurus* (see figure 8-23) from the Middle Jurassic of Sichuan. At 4.3 m long, *Huayangosaurus* has spike-shaped armor along the body midline flanked by rows of small armor plates. *Huayangosaurus* is much more primitive than other stegosaurs, all of which are assigned to a single, derived family Stegosauridae. Primitive features include a deep skull with a short snout, orbits positioned above the posterior cheek teeth, smaller and less massive skeleton, and fore- and hind limbs of more nearly equal length. Although these features (and others) readily distinguish *Huayangosaurus* from more derived stegosaurs, there is still a considerable morphological gap between *Huayangosaurus* and a primitive thyreophoran such as *Scelidosaurus*. This means that we really understand little about the origin of stegosaurs, despite the valuable information *Huayangosaurus* provides. That origin, or at least the next link in the morphological chain, probably is to be found in Lower Jurassic strata. An improved knowledge (new discoveries) of Lufeng Formation ornithischian dinosaurs may clarify stegosaur origins.

China's sauropod dinosaur record has generally been thought to begin in the Middle Jurassic, though a maxillary fragment from the Lufeng Formation may be that of an Early Jurassic sauropod (Barrett 1999). Two of the best known and most primitive sauropods are *Shunosaurus* and *Datousaurus* from the Middle Jurassic of Sichuan. Both are relatively small sauropods up to 12 m long. Their skulls are deep with nostrils in front of the orbits, as in the camarasaurid sauropods, but the muzzles of *Datousaurus* and *Shunosaurus* are relatively long. Tooth structure is intermediate between those of diplodocids and camarasaurids: slender but with small, spatulate crowns. The neck vertebrae are short and essentially lacked pleurocoels. There are 12 cervical vertebrae and 13 dorsals. The neural spines are not divided. The humerus-to-femur ratio is 0.66, about the same as in diplodocids, and the wrist has three bones, whereas there are two in the ankle. Forked chevrons are present in both Chinese genera, and *Shunosaurus* has a club at the end of its tail. These two sauropods were identified as cetiosaurids, a grade of primitive sauropods thought to have given rise to all later sauropods (McIntosh 1990). However, more recent cladistic analysis groups them with *Euhelopus, Mamenchisaurus,* and *Omeisaurus* as a family Euhelopodidae (Upchurch 1998) or as primitive sister taxa of other sauropods (Wilson and Sereno 1998).

Figure 8-23 *Skeleton of the most primitive stegosaur, Huayangosaurus (about 5 m long). Note the spiked armor in two rows on the back.*

Chapter 9

Cretaceous

China was almost totally emerged during the Cretaceous, unlike most other vast land areas, which were periodically submerged under epicontinental seas. Nonmarine deposition was focused mostly on the same depositional centers that existed during the Jurassic, the Junggur, Ordos, Sichuan, and northeastern China basins, as well as the Nanxiong basin of Guangdong (see figure 9-1). Most Chinese terrestrial Cretaceous strata are red beds that have an abundant fossil flora and fauna (e.g., P. Chen 1983). Early Cretaceous vertebrates are well

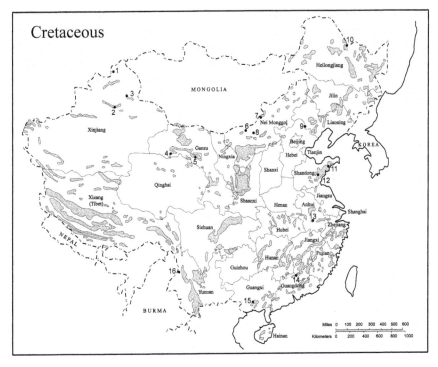

Figure 9-1 *Distribution of Cretaceous strata (after Yang et al. 1986) and principal vertebrate fossil localities in China. Localities are: 1 – Urho (Wuerho), Xinjiang; 2 – Turpan, Xinjiang; 3 – Wucaiwan Formation, Xinjiang; 4 – Tebsch, Gansu; 5 – Minhe Formation, Gansu; 6 – Bayan Mandahu, Nei Monggol; 7 – Iren Dabasu, Nei Monggol; 8 – Dashigou Formation, Nei Monggol; 9 – Jehol Group, Liaoning; 10 – Amur River, Helongjiang; 11 – Laiyang, Shandong; 12 – Mengyin, Shandong; 13 – Xiaoyuan Formation, Anhui; 14 – Nanxiong basin, Guangdong; 15 – Napan Formation, Guangxi; 16 – Jingxiwu Formation, Yunnan.*

represented in China, especially in the Junggur basin of Xinjiang (Shen and Mateer 1992). The "middle" Cretaceous (Albian-Turonian) record however, is very poor, a significant gap in the Chinese Mesozoic vertebrate record. Late Cretaceous vertebrates are rather well known, especially in Nei Monggol, Shandong and Guangdong.

As is true elsewhere, dinosaur fossils dominate the Chinese Cretaceous vertebrate-fossil record. Important groups are the ornithopods, theropods, ankylosaurs and ceratopsians, paralleling the dinosaur faunas of the Cretaceous of western North America.

Vertebrate-Bearing Strata

Cretaceous strata that contain fossil vertebrates—especially dinosaur eggs—are very widespread in China (see figure 9-1). Therefore, this review focuses on the principal vertebrate-producing strata, which are representative of the whole.

The Tugulu Group (150–1640 m thick) of the Junggur basin in Xinjiang yields the largest assemblage of Early Cretaceous vertebrates known from China. It consists of four formations (see figure 9-2): (1) Qingshuihe Formation, about 150 m of thin-bedded, yellow-green, and gray-green sandstones intercalated with siltstones and mudstones, from which no vertebrates are known; (2) Hutubihe Formation, mostly purple mudstones and gray-green siltstones with a few sandstones that contain a few fossil fishes; (3) the thin (50–60 m thick) Shengjinkou Formation, gray-green mudstones and shales that have produced a diverse and endemic ichthyofauna (Su 1985); and (4) Lianmuqin Formation, interbedded red green and yellow variegated mudstones and siltstones, 213–360 m thick. Near Urho, the Lianmuqin Formation yields a dinosaur-dominated tetrapod fossil assemblage. Upper Cretaceous red-bed fluvial clastics—the Donggou and Honglishan formations—overlie the Tugulu Group in the Junggur basin. A homotaxial and rather similar Cretaceous sequence is exposed in the Turpan basin south of the Tien Shan (see figure 9-2). The Subashi Formation in the Turpan basin is fluvio-lacustrine, gray-green and yellow sandstones, siltstones, and mudstones that contain Late Cretaceous dinosaurs.

Tugulu Group deposition took place in and around a large inland lake that formed in the Junggur basin under a subtropical to tropical climate. This type of depositional regime was characteristic of northern China, Mongolia and parts of Siberia during the Early Cretaceous (Ponomarenko and Popov 1980). Chinese Early Cretaceous vertebrate localities in Xinjiang, Gansu, and Nei Monggol are in strata deposited in and around such lakes.

These deposits contrast with Early Cretaceous vertebrate localities in eastern China, which are variegated, volcaniclastic fluvial strata interbedded with small

Age	Junggur Basin		Turpan Basin
LATE CRETACEOUS	Donggou Formation		**Subashi Formation**
			Kumutake Formation
EARLY CRETACEOUS	**Tugulu Group**	**Lianmuqin Frmn.**	**Lianmuqin Formation**
		Shengjinkou Frmn.	**Shengjinkou Formation**
		Hutubihe Frmn.	Shalidatun Formation
		Qingshuihe Frmn.	

boldface indicates vertebrate-fossil-bearing units

Figure 9-2 *Stratigraphic subdivisions of the Tugulu Group and adjacent Cretaceous strata in the Junggur and Turpan basins in northwestern China (after Shen and Mateer 1992).*

lacustrine deposits and estuarine coal-bearing beds. The Qingshan Formation of Shandong has yielded a *Psittacosaurus*-dominated tetrapod fauna and is representative of these eastern Chinese Lower Cretaceous strata. The Qingshan Formation is about 2000 m thick and consists of green-purple and red volcaniclastic sands, gravels and mudstones. Correlative rocks of the Jehol Group in northeastern China (see figure 9-3) are volcaniclastic strata interbedded with coal-bearing strata. Here, the Jehol Group strata yield *Psittacosaurus* and a significant avifauna discussed below.

Upper Cretaceous vertebrate-bearing strata in China contrast dramatically with those of the Lower Cretaceous. This reflects fundamental changes in tectonics and climate that took place during the Cretaceous in China. The Early Cretaceous was a time of vast, impounded drainage basins in western and northern China and volcanic eruptions along the Pacific plate margin of eastern China (see figure 9-4). Deposition took place under a subtropical to tropical climate.

Major tectonism took place in China during the early Late Cretaceous (Cenomanian-Santonian). Climates remained subtropical, and fluvio-lacustrine

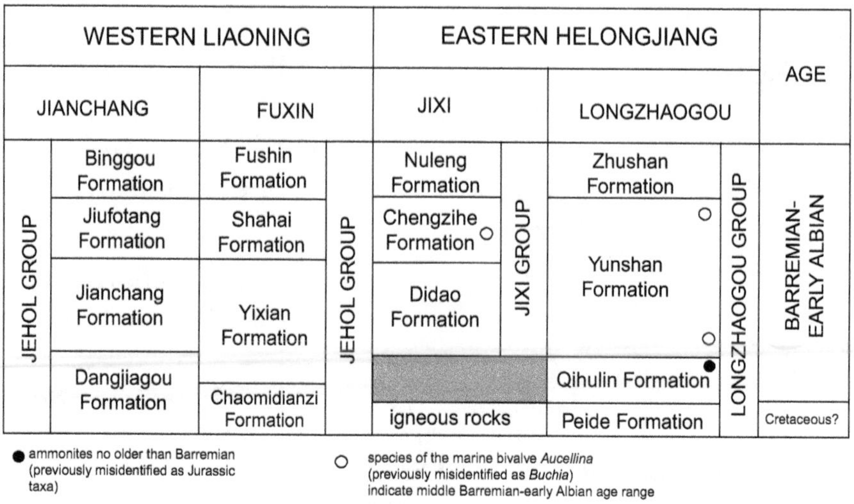

WESTERN LIAONING		EASTERN HELONGJIANG			AGE
JIANCHANG	FUXIN	JIXI	LONGZHAOGOU		
Binggou Formation	Fushin Formation	Nuleng Formation	Zhushan Formation		BARREMIAN–EARLY ALBIAN
Jiufotang Formation	Shahai Formation	Chengzihe Formation ○	Yunshan Formation ○	LONGZHAOGOU GROUP	
Jianchang Formation	Yixian Formation	Didao Formation			
Dangjiagou Formation	Chaomidianzi Formation	igneous rocks	Qihulin Formation ●		Cretaceous?
			Peide Formation		

● ammonites no older than Barremian (previously misidentified as Jurassic taxa)

○ species of the marine bivalve *Aucellina* (previously misidentified as *Buchia*) indicate middle Barremian-early Albian age range

Figure 9-3 *Stratigraphy and correlation of the Jehol Group in northeastern China (after Gu 1992) indicates the Early Cretaceous age of the Jehol Group in Liaonong by correlation to marine Cretaceous strata in Helongjiang.*

deposits are largely confined to northeastern China (see figure 9-4), where the few fossil vertebrates from this time interval have been collected.

Much warmer and arid climates characterize the Campanian-Maastrichtian interval, with limited deposition taking place in inland river and lake basins scattered across China (see figure 9-4). Deposits in these intermontane basins are mostly red-bed clastics that are dominated by coarse-grained sandstones and conglomerates. The Nanxiong Formation in southeastern China is representative of these deposits. It is as much as 3000 m of red-bed mudstones, sandstones and conglomerates that yield a dinosaur bone and egg assemblage of Maastrichtian age (see figure 9-5). The Wangshi series of Shandong is similar and contains China's most extensive Late Cretaceous vertebrate fauna (see figure 9-6). Thinner, but broadly similar Upper Cretaceous nonmarine strata, some of which are of eolian origin, yields vertebrates across northern China from Nei Monggol to Xinjiang.

Land-Vertebrate Faunachrons

Jerzykiewicz and Russell (1991) proposed a succession of "Mongolian land-vertebrate ages" (MOLVAs) for Late Jurassic-Late Cretaceous time based on formations and vertebrate-fossil assemblages from Mongolia. They took the names of the MOLVAs from formation names; a practice I eschew because it leads to easy confusion: is the age coextensive with the formation or coextensive with the vertebrate-fossil assemblage? Also, three of Jerzykiewicz and Russell's (1991) MOLVAs lack fossil vertebrate characterization—Late Jurassic Khamarkhuburian and Sharilinian and Cretaceous Saynshandian. These MOLVAs are neither biostratigraphic nor biochronological units and are rejected here. Furthermore, the Shinkhudukian MOLVA contains only four precisely identified taxa: the long-ranging lycopterid fish *Lycoptera* Müller, the turtle *Hangaiemys* Shuvalov & Chkhikvadze, the primitive ceratopsian dinosaur *Psittacosaurus* Osborn, and the endemic early bird *Ambiortus* Kurochkin. It cannot be distinguished from the younger Khukhtekian, so I reject the Shinkudukian as a valid vertebrate faunachron. Despite these caveats, most of the MOLVAs named by Jerzykiewicz and Russell (1991) can be recognized as vertebrate faunachrons. Five of these faunachrons can be identified by Chinese Cretaceous vertebrate-fossil assemblages and are employed here (see figure 9-7).

Tsagantsabian Vertebrates

The vertebrate fossil assemblages of the Gurvan Eran, Tevsh, Undurukhin and Tsagantsab formations of Mongolia form the basis of the Tsangantsabian vertebrate

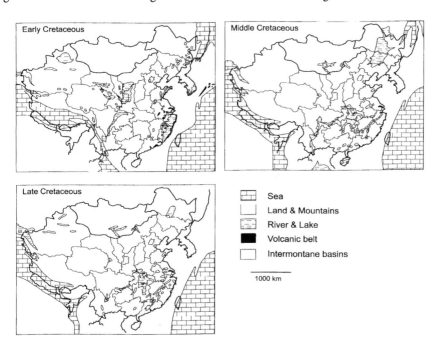

Figure 9-4 *Three paleogeographic reconstructions of China during the Cretaceous (after P. Chen 1987).*

Figure 9-5 *Outcrops of the Nanxiong Formation in Guangdong are amidst cultivated fields.*

faunachron (Jerzykiewicz and Russell, 1991: 363). Key taxa of biochronological value are the theropod *Prodeinodon* Osborn, the sauropod *Asiatosaurus* Osborn (both form genera based on teeth), the ceratopsian *Psittacosaurus* Osborn, and the pterosaur *Dsungaripterus* Young (see figure 9-8). The lycopterid fish *Lycoptera* also is present and obviously has a long temporal range that spans the Jurassic-Cretaceous boundary (Jerzykiewicz and Russell 1991).

Prodeinodon is a form genus for relatively large (carnosaur-size) theropod teeth (Osborn 1924; Bohlin 1953; Hou et al. 1975). *Asiatosaurus* likewise is a tooth-based form genus for sauropod dinosaurs with spatulate (camarasaurid-like) teeth (Osborn 1924; Hou et al. 1975). The ceratopsian *Psittacosaurus* (see figure 9-9) is critical to recognizing the Tsagantsabian, and its distribution defines a *Psittacosaurus* biochron discussed below.

Dsungaripterus (see figure 9-8) is a large pterosaur with a 3.5 m wingspan and a 50 cm long skull. Its jaw tips lack teeth, and the teeth are knob-like. It probably was a durophage, eating bivalves, crabs, and other shelled invertebrates. Other distinctive features of *Dsungaripterus* include its cranial crests—an elongate crest from above the orbits forward along the snout and a short crest projecting backward from the head (Young 1964c, 1973b).

Tsagantsanbian vertebrates can be recognized across much of China. *Prodeinodon* and *Asiatosaurus* have been identified in the Napai Formation of

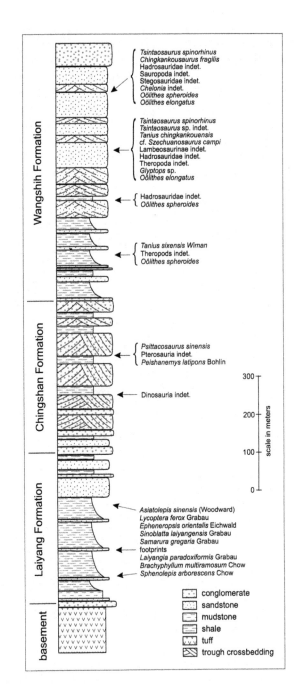

Figure 9-6 *Stratigraphic distribution of fossil vertebrates in the Laiyang area of Shandong (from Young 1958b).*

Epoch	land-vertebrate faunachron
Late Cretaceous	Nemegtian
	Djadokhtan
	Baynshirenian
	▓▓▓▓▓▓▓▓▓▓▓▓
Early Cretaceous	Khukhtekian
	Tsagantsabian
	Ningjiagouan

Figure 9-7 *Mongolian land-vertebrate "ages" of Jerzykiewicz and Russell (1991) that can be recognized in China.*

Guangxi (Hou et al. 1975; Dong 1980b). *Psittacosaurus* has a broad distribution in northern China and is also known from the younger Khukhtekian faunachron. *Asiatosaurus, Psittacosaurus,* and *Dsungaripterus* are present in the upper Tugulu Group of the Junggur basin of Xinjiang (Shen and Mateer 1992).

The Upper Tugulu Group (Shenjinkou and Lianmuqin formations) is the most extensive Chinese vertebrate fauna of Tsagantsabian age and includes the following taxa: the endemic fishes *Dsungarichthys bilineatus, Neobaleiichthyus chikuensis, Siyuichthys ornatus, S. pulcher, S. pulchellus, Bogdaichthys fukangensis, B. serratus, Manasichthys elongatus, M. tuguluensis, Uighuroniscus sinkiangensis,* and *Wukangia houyanshanensis* (Su 1985); the turtles *Sinemys wuerhoensis* Ye, and *Dracochelys bicuspis* Gaffney & Ye; the protosuchian crocodilian *Edentosuchus tianshanensis* Young; the pterosaurs *Noripterus complicidens* Young and

Figure 9-8 *Wall mount of the skeleton of Dsungaripterus, the characteristic Tsagantsabian pterosaur, which had a wingspan of 3.5 m.*

Dsungaripterus weii Young; the coelurosaurs *Tugulusaurus faciles* Dong and *Phaedrolosaurus ilikensis* Dong (both *nomina dubia*); the carnosaur *Kelmayisaurus petrolicus* Dong; the sauropod *Asiatosaurus mongoliensis* Osborn; the stegosaurid *Wuerhosaurus homheni* Dong; and the ceratopsian *Psittacosaurus* sp.

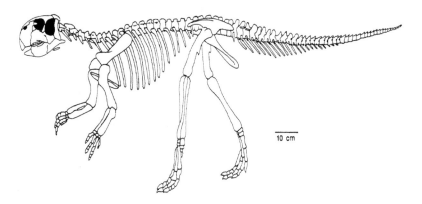

Figure 9-9 *Skeleton of the archetypal ceratopsian Psittacosaurus, a facultatively bipedal herbivore.*

The Tugulu vertebrates give us the best picture we have of Tsagantsabian vertebrates that lived in and around the giant lake basins of central and eastern Asia during the Early Cretaceous. Diverse and endemic actinopterygian fish faunas inhabited these lakes. Turtles seem little changed from the Late Jurassic and include both trionychids and more terrestrial forms. Crocodilians are little known. For example, *Edentosuchus* (see figure 9-10) is a small protosuchian? with a heterodont dentition, known from only two specimens, the holotype and paratype (Young 1973a; J. Li 1985; X. Wu and Sues 1995).

The Tsagantsabian pterosaurs are highly distinctive and endemic to eastern Asia. In addition to *Dsungaripterus*, discussed above, *Noripterus* is a smaller, closely related form. Both genera are placed in the family Dsungaripteridae, also known from the Lower Cretaceous of Mongolia, the Upper Jurassic of East Africa, and the Lower Jurassic of South America (Galton 1980; Bennet 1989).

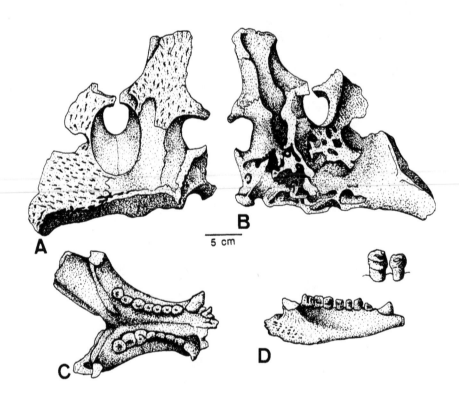

Figure 9-10 *Partial skull and lower jaw of Edentosuchus (after J. Li 1985). Dorsal (A) and ventral (B) views of posterior part of skull and occlusal (C) and lateral (D) views of lower jaw.*

Tsagantsabian theropod dinosaurs are not well known but include carnosaurs (*Prodeinodon, Kelmayisaurus*) and coelurosaurs (*Tugulusaurus* and *Phaedrolosaurus*). *Prodeinodon* is a form genus for teeth; *Kelmayisaurus* is based on jaw fragments similar to those of *Ceratosaurus* (Dong 1973); *Tugulusaurus* is based on four caudal vertebrate and part of a hind limb (Dong 1973); and *Phaedrolosaurus* is known from a *Deinonychus*-like tooth and a few hind limb bones (Dong 1973).

The Tsagantsabian sauropod *Asiatosaurus* also is a tooth form genus. The Tsagantsabian stegosaur, *Wuerhosaurus,* however, is relatively well known from partial skeletons of large size (length approximately 7–8 m). Distinctive features include osteoderm plates that are long, large, and low, a solid dorsal plate to the sacrum, and very elongated neural spines on the proximal caudals (Dong 1973).

The best known Tsagantsabian dinosaur is *Psittacosaurus* (see figure 9-11), the archetypal ceratopsian. As many as seven species of *Psittacosaurus* are known from the Lower Cretaceous of eastern Asia, and their remains include many complete skulls and skeletons. For many years, psittacosaurs had been allied with the ornithopods, but recent analysis of their excellent fossil record supports the identification of *Psittacosaurus* as the earliest ceratopsian.

Psittacosaurus possesses the key evolutionary novelties of the Ceratopsia even though it has only the most rudimentary of frills. The posterior end of the skull roof just barely overhangs the back end of the skull. The short snout, the high position of the nostrils, the tall rostrum that superficially resembled a parrot's beak, and the reduction of the functional toes of the hand to three, are diagnostic features of *Psittacosaurus* among ceratopsians.

The cheek teeth of *Psittacosaurus* have broad, flat wear surfaces with self-sharpening edges, but they did not occlude precisely. Their placement in the jaws is inset from the sides of the skull, suggesting the presence of cheek pouches. *Psittacosaurus* does not exceed 2 m in length, and its skeletal structure is much more like that of a primitive ornithischian rather than other ceratopsians. In particular, the hind limb is longer than the forelimb, and the forefoot has only three functional toes, whereas the hind foot has four slender toes. The neck is short, and the tail is moderately long. Ossified tendons are present in some species of *Psittacosaurus* along the spine in the back and hip region.

The teeth of psittacosaurs are characteristic of plant eaters. They are usually well worn, and polished stones (gastroliths) associated with some psittacosaur skeletons suggest that significant amounts of vegetation were milled in the stomach. The forelimbs of *Psittacosaurus* are about 58 percent of hind limb length, indicating this dinosaur was a facultative biped. Psittacosaurs were probably able to grasp with their hands; their first finger diverges from the

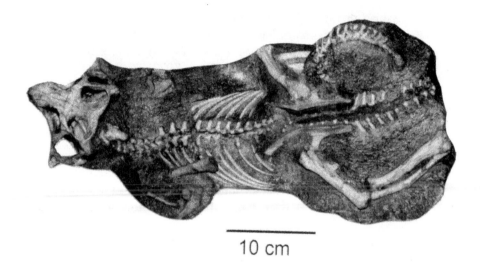

10 cm

Figure 9-11 *Dorsal view of a skeleton of Psittacosaurus sinensis from Shandong.*

other two. If this was the case, psittacosaurs were primarily bipeds, using the hands to grasp vegetation while eating.

The psittacosaurs were widespread and reasonably common dinosaurs in Asia during the Early Cretaceous. They represent well the ancestry of a subsequent, much more diverse and impressive group of dinosaurs, the Neoceratopsia. Their temporal distribution also is of significance to vertebrate biochronology.

The *Psittacosaurus* Biochron

The "parrot dinosaur" *Psittacosaurus* (figures 9-10 and 9-11) well represents the primitive, ancestral structure of the Ceratopsia, though an older, Late Jurassic ceratopsian, *Chaoyangosaurus*, is known from Liaoning (Zhao et al. 1999). Its biochronologic value merits some mention and the crucial role it plays in understanding ceratopsian phylogeny will also be discussed.

In Mongolia, fossils of *Psittacosaurus* occur in strata Jerzykiewicz and Russell (1991) assigned Tsagantsabian and Khukhtekian ages (see above). These are specimens of *P. mongoliensis* from the Tsagantsab and Khukhtek formations of Mongolia. In China, *P. mongoliensis* occurs in the Binggou Formation, Liaoning, and unnamed strata in Nei Monggol. In Russia, *P. mongoliensis* is present in the Shestakov Formation in the Gorno-Altay Autonomous Region. Other records of *Psittacosaurus* include:

1. *P. guyangensis* Cheng is from the Lisangou Formation of Nei Monggol (Young 1931), and it is probably the same species as *P. mongoliensis*.

2. *P. osborni* Young (= *P. tingi* Young) is from the Lisangou and Xinpong-naobao formations of Nei Monggol.

3. *P. sinensis* Young (= *P. youngi* Chao) is from the Qingshan Formation of Shandong.

4. *P. sinensis* is also reported from the Laohuondong Formation in Nei Monggol, which also produced the fish "*Sinamia,*" the turtle *Ordosemys leios*, the champsosaur *Ikechosaurus*, the crocodilians *Eotomistoma* and *Shantungosuchus*, stegosaurs, sauropods, ornithopods, a pterosaur, and a mammal (Sigogneau-Russell 1981; Brinkman and Dong 1993; Brinkman and Peng 1993).

5. *P. neimongoliensis* Russell & Zhao is from the Ejinhoro Formation of Nei Monggol, where it co-occurs with *Wuerhosaurus* and cf. *Chiayusaurus* (Russell and Zhao 1996).

6. *P. meileyingensis* Sereno, Chao, Cheng & Rao is from the Binggou Formation of Liaoning.

7. *Psittacosaurus* sp. is from the Yixian Formation in Liaoning (X. Xu and Wang 1998).

8. *P. xinjiangensis* Sereno & Chao is from the Lianmuqin Formation of Xinjiang.

9. *Psittacosaurus* sp. is from red beds at Muhaxiao in Nei Monggol that also yielded the troodontid dinosaur *Saurornithoides youngi* Russell and Dong.

10. In the Alxa Desert of Nei Monggol, the type locality of the theropod *Alxasaurus elesitaiensis* Russell & Dong yields *Psittacosaurus* sp.

11. Fossils of *Psittacosaurus* sp. are reported from the Ulan-tsonch/Tebsch area, Gansu (Bohlin 1953).

12. *Psittacosaurus mazongshanensis* is from the Lower Cretaceous Xinminbao Group in the Mazongshan area ("Gongposhuan basin") of western Gansu (X. Xu 1997; Dong 1997a).

13. *P. sattayarki* Buffetaut & Suteethorn is from the Khok Kruat Formation of northeastern Thailand (Buffetaut and Suteethorn 1992).

These occurrences define a *Psittacosaurus* biochron of Early Cretaceous age across much of eastern Asia (also see Matsukawa and Obata 1994). An Ar^{40}/Ar^{39} age from the Tebch basalts at Tebch, Nei Monggol is 110 ± 0.52 Ma. This is above the *P. mongoliensis* occurrence there (Eberth et al. 1993). Tsagantsabian basalts in Mongolia produce K/Ar ages of about 130 Ma (Samoilov et al., 1988), which are consistent with ostracods, molluscs, conchostracans, and fossil plants that suggest a late Neocomian age (Jerzykiewicz and Russell, 1991). This means that the *Psittacosaurus* biochron ranges in age from about Barremian to Aptian or possibly too as young as early Albian. This is very consistent with age assignments previously made to the *Psittacosaurus*-bearing formations with one exception, the Jiufotang Formation of Liaoning.

The lacustrine strata of the Jiufotang Formation contain invertebrate and plant fossils that some Chinese paleontologists assign a Late Jurassic age (P. Chen 1988; Gu 1992; Jin 1996). This seems unlikely because it would be the only Late Jurassic occurrence of *Psittacosaurus* and thus one considerably older than all its other occurrences. The *Psittacosaurus* biochron thus is a recognizable interval of the Early Cretaceous across eastern Asia.

Khukhtekian Vertebrates

The vertebrate fossil assemblages of the Dzun Bayan, Dushilin, and the Khulsyngol formations of Mongolia are the basis of the Khukhtekian vertebrate faunachron (Jerzykiewicz and Russell 1991: 364–365). Key taxa of biochronological value are the turtle *Peishanemys* Bohlin, the ceratopsian *Psittacosaurus* Osborn, and the mammal *Gobiconodon* Trofimov.

There are only two definite Khukhtekian age vertebrate faunas known from China. *Peishanemys* and *Psittacosaurus* co-occur in the Qingshan Formation in Shandong (Chow 1954; Young 1958b; Chao 1962). Other Chinese *Psittacosaurus* localities may be of Khukhtekian (or Tsagantsabian) age, as discussed above. The Khukhtekian may prove to be indistinguishable from the older Tsagantsabian because of its poor characterization but it is used here provisionally.

The recently described vertebrate fossil assemblages from the Xinminbao Group of western Gansu are also of probable Khukhtekian age. They include a mesosuchian crocodile, the troodontid *Sinornithoides* sp., a dromaeosaur, the sauropod *Chiayusaurus* sp., a "nemegtosaurid," the hypsilophodontid *Siluosaurus zhangqiani* Dong, the iguanodontid *Probactrosaurus mazongshanensis* Lu, the ceratopsians *Psittacosaurus mazongshanensis* Xu and *Archaeoceratops oshimai* Dong & Azuma, the segnosaur *Nanshiungosaurus bohlini* Dong & Yu, and a triconodont mammal (Dong,1997a, b, c, d, e, and f; Lu 1997; X. Xu 1997; Dong and Azuma 1997; Dong and Yu 1997).

The Dashigou (Tashikou) Formation of Nei Monggol may be of Khukhtekian age. It has yielded the turtle *Aspideretes alashanensis* Ye, indeterminate sauropods, the theropod *Chilantaisaurus maortuensis* Hu and the iguanodontids *Probactrosaurus gobiensis* Rozhdestvensky and *P. alashanicus* Rozhdestvensky. *Chilantaisaurus* is also known from the Chaochuan Formation in eastern Zhejiang (Dong 1979), and this occurrence may also be of Khukhtekian age.

Peishanemys (see figure 9-12) is a rather large dermatemydid turtle with a broad, circular shell. The skull is not known (Bohlin 1953; Chow 1954). The triconodontid? mammal *Gobiconodon* was originally described by Russian paleontologist Trofimov (1978) from the Early Cretaceous of Mongolia, and is also known from the Early Cretaceous (Aptian-Albian Cloverly Formation) of Montana, USA (Jenkins and Schaff 1988). Though not known from China, *Gobiconodon* provides key evidence the Khukhtekian is no younger than early Albian, which is the minimum age of the Cloverly Formation (Lucas 1993c).

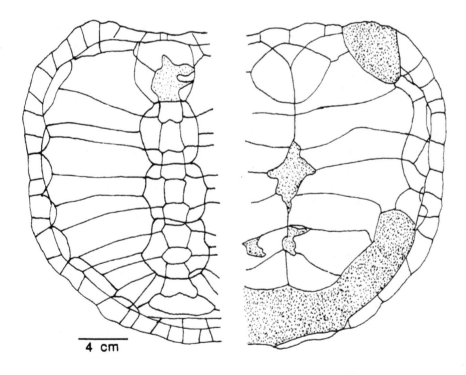

4 cm

Figure 9-12   *The shell of Peishanemys: carapace (left) and plastron (right) (after Chow 1954).*

Age of the Liaoning Birds

The last decade has witnessed an explosion of knowledge of the early evolution of birds (Padian and Chiappe 1998). Much of this has been due to the new discoveries of Mesozoic birds from Liaoning in northeastern China discussed below. The age of these Chinese birds has been asserted as Late Jurassic or Early Cretaceous, so there is a substantial difference of opinion about their place in time relative to the oldest known bird, *Archaeopteryx*, which is without question of Late Jurassic (Tithonian) age. This difference of opinion is important because the age of the Chinese Mesozoic birds is critical to interpretation of the origin and earliest diversification of the Aves (e.g., Luo 1999; Barrett 2000). However, any reasonable reading of the data relevant to the age of the Mesozoic birds from northeastern China indicates they are of Early Cretaceous (Barremian-Aptian) age.

The Mesozoic birds from northeastern China come from the Chaomidianzi, Yixian and Jiufotang formations in western Liaoning (figures 9-3 and 9-13). These strata were long included in the Jehol Group, a stratigraphic concept that dates back to Grabau (1923, 1928). The age of the Jehol Group was originally, and for many years, considered to be Cretaceous, but in recent years some have concluded it is Jurassic.

Fossils from the Chaomidianzi, Yixian, and Jiufotang formations are of palynomorphs, megafossil plants, conchostracans, insects, ostracods, gastropods, bivalves, and vertebrates. Each one of these groups is of potential biochronological significance, but the overwhelming majority of the taxa known from these Chinese units are endemics and thus of no real significance to correlation. One of the few exceptions is the ceratopsian dinosaur *Psittacosaurus*, which is known from numerous Early Cretaceous localities (see earlier discussion), so it indicates an Early Cretaceous age for its records in Liaoning. Another is the palynomorph assemblage, which also indicates an Early Cretaceous age (W. Li and Liu 1994).

The principal reason some Chinese geologists and paleontologists consider the Jehol Group to be Jurassic is its correlation with marine strata to the east, in Helongjiang, that supposedly yield Jurassic bivalves and ammonites (see Gu 1992, for an excellent summary of this point of view). However, recent restudy of these marine fossils, which are from the Longzhaogou and Jixi groups, indicates that they were misidentified and actually are Early Cretaceous (Barremian-early Albian) in age (e.g., Sha et al. 1994; Futakami et al. 1995) (see figure 9-3). Furthermore, radioisotopic ages from the Jehol Group volcanic ashes are in the 120-125 Ma range, and thus indicate an Early Cretaceous age (Smith et al. 1995; Swisher et al. 1999).

Against this seemingly incontrovertible evidence that the bird-bearing strata are Early Cretaceous, are arguments based on the stage-of-evolution of the birds (and of some other organisms) to indicate that the beds are Jurassic. Just how tricky these arguments are is revealed by the lack of agreement on just what the stage-of-evolution is of the Liaoning birds. For example, Zhou et al. (1992) argued that *Cathayornis* (see figure 9-14) from the Jiufotang Formation is more evolutionarily advanced than *Archaeopteryx* and closest in evolutionary grade to Early Cretaceous birds from Spain. Therefore, they regarded the Jiufotang Formation as Early Cretaceous. In contrast, Hou et al. (1995) judged *Confuciusornis* and the other birds from Liaoning to be more primitive than Early Cretaceous birds, so they assigned them a Late Jurassic age. To add to the uncertainty, Martin et al. (1998) identified *Confuciusornis* as an evolutionary mosaic of features more primitive than, equivalent to, and more advanced than the features of *Archaeopteryx*.

The conclusion I draw from this is that age assignments based on stage-of-evolution are not the best way to correlate strata. Index fossils such as *Psittacosaurus*,

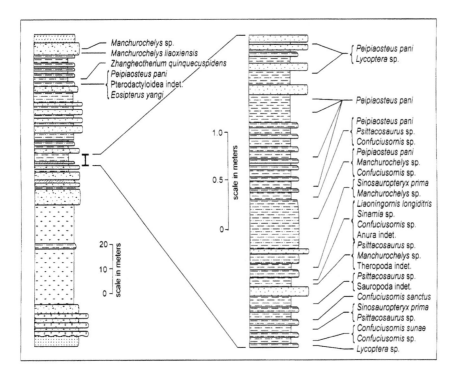

Figure 9-13 *Composite section (left) and detail (right) of vertebrate-fossil-bearing interval of Yixian Formation at Sihetun, Liaoning (after X. Wang et al. 1998).*

Figure 9-14 *Superbly preserved holotype skeleton of the sparrow-sized bird Cathayornis from the Lower Cretaceous of Liaoning.*

radioisotopic ages, and correlation to Lower Cretaceous marine strata establish without question the Early Cretaceous (Barremian-Aptian) age of the bird-bearing strata in Liaoning.

Chinese Early Cretaceous Birds and Avian Origins

Lake beds of the Lower Cretaceous Jiufotang, Chaomidianzi and Yixian formations in Liaoning have produced myriad avian fossils. Several taxa of birds have been named, including *Sinornis santensis* Sereno & Rao (see figure 9-15), *Cathayornis yandica* Zhou, Jin & Zhang (see figure 9-14), *Balouchia zhengi* Zhou, *Longchengornis sanyansis* Hou, *Gansus yumenensis* Hou & Liu, *Liaoningornis longidiris* Hou, *Chaoyangia beishanensis* Hou & Zhang and *Confuciusornis sanctus* Hou, Zhou, Gu & Zhang (see Sereno and Rao 1992; Hou 1995, 1998; Hou et al. 1993, 1995, 1996, 1999; Chiappe et al. 1999). These birds are included in the enantiornithines, a major Cretaceous radiation of flying birds (e.g., Chiappe 1995). They are mostly sparrow-to-pigeon-sized forms that show remarkably advanced structures for flying and perching. Thus, their trunks and tails are short, their forelimb structures indicate an advanced wing-folding mechanism, and the pes has an opposable hallux for perching. These advanced features, nevertheless, contrast markedly with some features in *Sinornis* and *Cathayornis* that indicate they are very primitive birds: short snout, teeth present, pelvis with pubic foot, gastralia, flexible clawed manus and limited skeletal co-ossification.

These birds co-occur in some strata with a remarkable assemblage of non-avian vertebrates, including lizards, the theropod dinosaurs *Sinosauropteryx* and *Caudipteryx* (both of which have preserved integumentary structures) and the

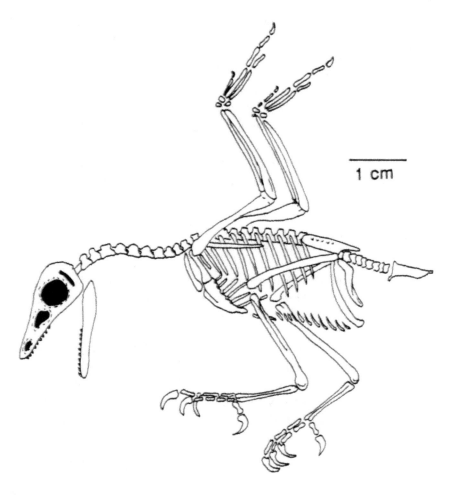

Figure 9-15 *Skeleton of Sinornis from the Lower Cretaceous of Liaoning (after Sereno and Rao 1992).*

symmetrodont mammal *Zhangheotherium* (Young 1958a; Hou et al. 1995, 1996; Ji and Ji 1996, 1997; Ji et al. 1998; Hu et al. 1998). As already discussed, the age of these Chinese birds is no older than Barremian, about 125 Ma. They are at least 25 million years younger than *Archaeopteryx*, widely acknowledged as the oldest known bird. This age difference, and the burgeoning diversity and evolutionary advancement of Early Cretaceous birds, has led Kurochkin (1985, 1995) to suggest that *Archaeopteryx* is too late in geological time to be the ancestor of birds. Kurochkin simply argues that the Chinese Early Cretaceous birds are so advanced in their flight and perching mechanisms that 25 million years was not enough time for them to have evolved from *Archaeopteryx*, the "feathered dinosaur."

Nevertheless, 25 million years is 25 million years, an immense amount of time. We can just as easily argue that the initial phase of avian evolution was rapid and explosive. The Chinese Early Cretaceous birds thus demonstrate how quickly modern levels of organization evolved among birds, not an avian ancestry more ancient than *Archaeopteryx*.

As Sereno and Rao (1992) aptly point out, a Mesozoic avian fossil record derived almost exclusively from nearshore marine or estuarine (lagoonal) sediments has heavily colored our previous views of the early evolution of birds. The Early Cretaceous Chinese birds are from sediments of an inland lake and suggest rapid modernization of the early avifauna took place in inland, wooded environments.

Baynshirenian Vertebrates

The Baynshirenian vertebrate faunachron is based on the vertebrate fossil assemblage of the Bayn Shiren Formation of Mongolia (Jerzykiewicz and Russell 1991: 366). The protoceratopsid dinosaur *Microceratops* Bohlin and the theropod *Alectrosaurus* are key taxa of biochronological value in China. An abundant and diverse turtle fauna and the presence (first Asian appearance) of hadrosaurid dinosaurs are also characteristic.

In China, *Microceratops* is present in the Minhe Formation of Gansu and Nei Monggol (Bohlin 1953; Maryanska 1977) and in the Zhumabao Formation of Shanxi (Weishampel 1990). *Protoceratops* also is present at the Gansu-Nei Monggol *Microceratops* localities, and in the Bayn Shiren Formation of Mongolia, and may also be characteristic of the Baynshirenian in China, though it also is present in the younger Barungoyotian strata of Mongolia (Jerzykiewicz and Russell 1991: 367). *Microceratops* is known from a partial skull and skeleton as well as many fragmentary specimens (Bohlin, 1953). It is the oldest known protoceratopsid, with a short and fenestrated frill and highly cursorial hind limbs.

Protoceratops (see figure 9-16) is much better known, particularly from its fossil record in Mongolia where more than 80 skulls, various skeletons representing a wide range of growth stages, and nests of eggs have been collected (e.g., Granger and Gregory, 1923; Maryanska and Osmólska,1975; Thulborn 1992; Mikhailov et al. 1994). *Protoceratops* and allied genera, the Protoceratopsidae, represent a stage of ceratopsian evolution intermediate between the psittacosaurs and the latest Cretaceous, North American ceratopsids. Thus, *Protoceratops* has a relatively larger skull and a much longer frill than *Psittacosaurus*. The fore and hind limbs of *Protoceratops* are of more nearly equal lengths than in *Psittacosaurus*, and the limbs are more massive with broader feet. Yet, unlike ceratopsids, the frill of *Protoceratops* is still rather short, the dinosaur has no horns, and its nostrils are small.

The classic Iren Dabasu (=Iren Nor, =Erlien Dabasu) Formation of Nei Monggol also contains vertebrates of Baynshirenian age. Its dinosaur fauna (Weishampel and Horner 1986) includes the theropods *Alectrosaurus olseni* Gilmore, *Archaeornithomimus asiaticus* Gilmore and *?Saurornithoides* sp and the primitive hadrosaurids *Gilmoreosaurus mongoliensis* Gilmore and *Bactrosaurus johnsoni* Gilmore. The trionychid turtle *Amyda gregaria* Gilmore and the crocodilian *Shamosuchus* sp. also are present, as are dinosaur eggs (Dong 1992).

Alectrosaurus is known from a skull and partial limbs that represent a large carnosaur closely related to *Tyrannosaurus* (Mader and Bradley 1989). *Archaeornithomimus* is known from vertebrae and limb elements and is the

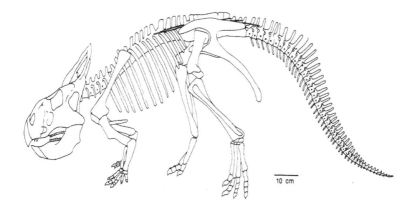

Figure 9-16 *Skeleton of the ceratopsian dinosaur Protoceratops (after Granger and Gregory 1923).*

oldest known ornithomimid coelurosaur; its metatarsus is more primitive than those of other ornithomimids (Russell 1972; Nicholls and Russell 1981). *Saurornithoides* is a troodontid better known from the Djadokhtan in Mongolia (Osborn 1924; Barsbold 1974).

Gilmoreosaurus is one of the most primitive hadrosaurids, known from isolated skull elements and assorted postcrania (Brett-Surman 1979). *Bactrosaurus* is known from similar material and is one of the most primitive lambeosaurine hadrosaurids.

The Iren Dabasu fauna is the best known assemblage of Baynshirenian vertebrates from China. Its global correlation is unclear. Baynshirenian could range in age from Turonian to early Campanian (Jerzykiewicz and Russell 1991). It does, however, encompass a characteristic dinosaur fauna of the Late Cretaceous, dominated by hadrosaurids with a lesser representation of tyrannosaurid carnosaurs and coelurosaurs.

Djadokhtan Vertebrates

Only two fossil assemblages from China, still not completely documented, can be assigned to the Djadokhtan (= Barungoyotian) of Jerzykiewicz and Russell (1991). An unnamed unit in Ningxia yielded the characteristic Djadokhtan ankylosaurid *Pinacosaurus grangeri* Gilmore (= *P. ningshiensis* Young) (Young 1935a; Maryanska 1977).

The Djadokhta Formation crops out at Bayan Mandahu in Nei Monggol, where it contains a diverse tetrapod fauna of Djadokhtan age including: the turtles *Basilemys* and *Zangleria testudinomorpha* Brinkman & Peng; the lizards *Sineoamphisbaena hexatabularis* Wu, Brinkman & Russell, *Anchaosaurus gilmorei* Gao & Hou, *Xihaina aquilonia* Gao & Hou, *Mimeosaurus crassus* Gilmore, *Priscagama gobiensis* Borsuck-Bialynicka & Moody, *Pleurodontagama aenigmatodes* Borsuk-Bialynicka & Moody, *Conicodontosaurus djadochtaensis* Gilmore, *Adamisaurus magnidentatus* Sulimski, *Carusia intermedia* Borsuk-Bialynicka, *Bainguis* sp., a nerosaurid, a varanid?, and *Isodontosaurus gracilis* Gilmore; the crocodilian *Shamosuchus*; the dinosaurs *Protoceratops, Udanoceratops*, possible *Bagaceratops, Pinacosaurus, Velociraptor, Oviraptor, Saurornithoides,* and *Tarbosaurus* sp.; dinosaur eggs; and mammals, including *Kennalestes* (Jerzykiewicz et al. 1989, 1993; Gao and Hou 1996; Dong and Currie 1996).

Nemegtian Vertebrates

The vertebrate fossil assemblage of the Nemegt Formation in Mongolia is the basis of the Nemegtian land-vertebrate faunachron (Jerzykiewicz and Russell 1991: 370). Characteristic taxa are the theropod *Tarbosaurus*, the sauropods *Nemegtosaurus* (see figure 9-17) and *Opisthocoelicaudia* and the hadrosaurid *Saurolophus*. In northeastern China, Riabinin (1930) described *Tarbosaurus?*, *Tanius*, and *Saurolophus* from strata in Heilongjiang of Nemegtian age. In Xinjiang, the Subashi Formation yields *Tarbosarus* and *Nemegtosaurus* and thus is of Nemegtian age (Dong 1977, 1997f and h).

I consider the vertebrate fauna of the upper Wangshi Formation of Shandong (see figure 9-6) (Z. Cheng et al. 1995; X. Wang 1996) to be of Nemegtian age, although this correlation is not certain. It has yielded the tyrannosaurid *Chingkankousaurus fragilis* Young, the hadrosaurids *Tanius sinensis* Wiman (= *T. chingkankouensis* Young, = *T. laiyangensis* Zhen), *Shantungosaurus giganteus* Hu, and *Tsintaosaurus spinorhinus* Young (see figure 9-18); the ankylosaur *Pinacosaurus* cf. *P. grangeri* Gilmore (see Buffetaut 1995); the pachycephalosaurid *Micropachycephalosaurus hongtuyanensis* Dong; and dinosaur eggs (Wiman 1929; Chow 1951; Young 1958b; Dong 1978, 1992*)*. However, note that *Pinacosaurus* suggests that part of the Wangshi Series may be of Djadokhtan age (Buffetaut 1995; Buffetaut and Tong 1995), as does the suggestion that *Bactrosaurus*

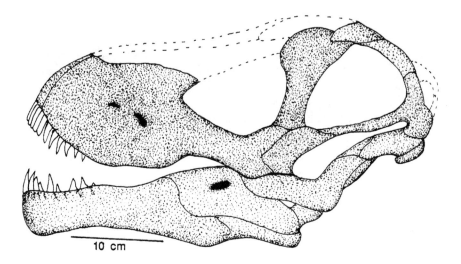

10 cm

Figure 9-17 *Skull and lower jaw of the sauropod Nemegtosaurus (after Nowinski 1971).*

Figure 9-18 *Skeleton of the bizarre hadrosaur Tsintaosaurus. The strange, tubular spike on top of the head was originally thought to be real, then considered an artifact of incorrect reconstruction, but is now again considered to be part of the skull.*

johnsoni from Iren Dabasu is a synonym of *Tanius sinensis* (Z. Cheng et al. 1995).

What may be even younger dinosaur dominated assemblages from China must be assigned a Nemegtian age because they cannot at present be distinguished from the classic Nemegtian assemblage. Nemegtian thus represents the last interval of Cretaceous time that can be recognized from fossil vertebrates in China. The Nemegtian hadrosaurid *Saurolophus* Brown is an early Maastrichtian genus in North America (Horshoe Canyon Formation, Alberta). This suggests an early Maastrichtian age for at least part of the Nemegtian.

The "younger" Nemegtian vertebrate fossil assemblage is known only from the Nanxiong basin of Guangdong, where it occurs in the upper part of the Nanxiong Formation: the turtle *Nanshiungchelys wuchingensis* Ye, the theropod *Tarbosaurus bataar* Maleev, the segnosaur *Nanhsiungosarus brevispinus* Dong (a *nomen dubium*) and a vast number of dinosaur eggshells assigned to the ichnogenera *Ovaloolithus, Nanhsiungoolithus, Macroolithus, Shixingoolithus, Stromatoolithus* and *Elongatoolithus* (Zhao 1975, 1994; Dong 1979).

Tarbosaurus is the classic large tyrannosaurid dinosaur of eastern Asia. Originally described by Russian paleontologist Maleev (1955a, b, 1974), it is so similar to *Tyrannosaurus* that some authors have synonymized the two genera (Carpenter 1992). Chinese fossils identified as *Tarbosaurus* are almost exclusively isolated teeth. *Chingkankousaurus* is a probable tyrannosaurid known only from a scapula (Young 1958b).

Nemegtosaurus is a diplodocid sauropod known principally from the skull (see figure 9-17). The skull shows many of the classic diplodocid features, including its long, low profile and pencil-like teeth restricted to the front of the mouth (Kurzanov and Bannikov 1983).

The Wangshi Group hadrosaurids are the best known Nemegtian dinosaurs from China. *Tanius* is known from skull and postcranial material (Wiman 1930) that may represent a primitive hadrosaurid comparable to *Gilmoreosaurus* in evolutionary grade. *Shantungosaurus* (see figure 9-19) is known from nearly complete skull and postcranial material (Hu 1973). It was the largest hadrosaurid, with an estimated weight of about 16 metric tons (Weishampel and Horner 1990).

Tsintaosaurus (see figure 9-18) has long been reconstructed as one of the most unusual hadrosaurids, with a nasal crest resembling a unicorn's horn. Varied opinions have been expressed about its affinities among hadrosaurs (e.g., Young 1958b; Hopson 1975; Brett-Surman 1979) and doubt has been cast on the reality of its nasal crest (Taquet 1991). Weishampel and Horner (1990) argued the genus is a chimera based on a combination of lambeosaurine and hadrosaurine cranial material. If so, one of the most bizarre and distinctive Chinese dinosaurs needs to be erased from the popular mind. However, Buffetaut and Tong-Buffetaut (1993) recently reviewed the cranial anatomy of *Tsintaosaurus*, reaffirming Young's original restoration of the bizarre skull.

Remaining Nemegtian dinosaurs from China are very poorly known. The hadrosaurid *Microhadrosaurus* is based on a jaw fragment and is a *nomen*

Figure 9-19 *The largest known hadrosaurid Shantungosaurus has a skeleton that is 15 m long.*

dubium. The pachycephalasaurid *Micropachycephalosaurus* is known from skeletal fragments. The segnosaur *Nanshiungosaurus* is known from a vertebral column and pelvis that show a few segnosaurian features and little else.

Most of China's record of dinosaur eggs comes from Nemegtian strata, especially in the Nanxiong basin of Guangdong. The significance of this record merits a separate discussion below.

The last Nemegtian vertebrate from southeastern China worthy of special mention is the unusual turtle *Nanshiungchelys* (see figure 9-20). This turtle has a long tubular snout, probably for rooting out food in analogy to the modern pig-nosed turtle, *Carettochelys* of Australia and New Guinea.

There are several other Late Cretaceous vertebrate occurrences in China that cannot yet be assigned to a specific faunachron. They include: (1) Ulungurhe Formation, Junggur basin, Xinjiang, which yielded a tyrannosaurid (Dong 1992) and the hadrosaurid *Jaxartosaurus fuyunensis* Wu; (2) a hadrosaurid from the Alikehu (Ilike) Formation in the Junggur basin of Xinjiang (Dong,1992); (3) the Honglishan Formation, which underlies the Ulungurhe Formation in the Junggur basin, has yielded a turtle, tyrannosaurid and hadrosaurid (Dong 1992); (4) Xiaoyan Formation in Anhui, which yielded the pachycephalosaurid *Wannanosaurus yansiensis* Hou; and (5) Huiquanpu Formation in Shanxi, which yields the ankylosaur *Shanxia tianzhenensis* Barrett, You, Upchurch & Burton (Barrett et al. 1998).

Cretaceous Fishes

M. Chang and Jin (1996) organized the nearly 40 genera and 60 species of named Cretaceous fishes from China into three "fish faunas." The oldest is their "*Lycoptera* fauna," which includes fish records of Late Jurassic and Early Cretaceous age, mostly from northern China. Lycopterids dominate this fauna,

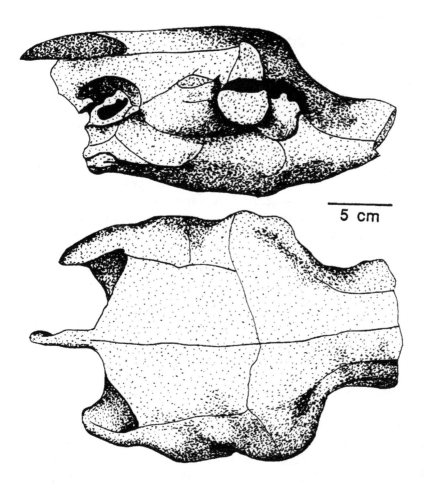

Figure 9-20 *Skull of the tubular-snouted turtle Nanshiunchelys, lateral (above) and dorsal (below) views (after Ye 1994).*

and important subordinates are peipiaosteids (e.g., Ma 1980; Jin 1991, 1995). This fish fauna is essentially endemic to northeastern China and adjacent areas.

Overlapping in age is the *Siyuichthys* fauna from the Tugulu Group in Xinjiang (see above). This Early Cretaceous assemblage is dominated by advanced pholidophoriforms and has a few leftover paleonisciforms. It looks more like a Middle Jurassic fish assemblage than the teleost-dominated *Lycoptera* fauna.

The *Mesoclupea* fish fauna of southeastern China (Zhejiang, Anhui, Jiangxi and Fujian) is also teleost dominated (e.g., M. Chang and Chou 1977; M. Chang and Chow 1986). This fish fauna was apparently isolated from the *Lycoptera* fauna to the north by highlands that prevented nearly all exchange, except for the amiids (*Sinamia*).

Ceratopsian Evolution

The early evolution of the Ceratopsia has long been based wholly on their fossil record from China and Mongolia. Well-corroborated cladistic analysis of the Ceratopsia (see figure 9-21) has supported the following evolutionary and paleobiogeographical scenario. Ceratopsians apparently first evolved in eastern Asia during the Late Jurassic, about 150 million years ago. *Psittacosaurus* is the archetypal ceratopsian, and cannot be excluded from the ancestry of later protoceratopsids. Likewise, protoceratopsids are the likely ancestors of the more derived ceratopsids. During the Late Cretaceous, protoceratopsids reached North America, where they gave rise to an endemic ceratopsid radiation.

However, a recent discovery of Early Cretaceous ceratopsian teeth in eastern North America (Chinnery et al. 1998) may raise doubt about the Asian origin of ceratopsians. Furthermore, *Zuniceratops*, a recently described horned ceratopsian from the Turonian of western North America (Wolfe and Kirkland 1998), also raises questions about the geographic origin of the ceratopsids. At present, ceratopsian and ceratopsid origins are unclear, though the most extensive fossil record of early ceratopsians still comes from China and Mongolia.

Chinese Cretaceous Dinosaur Eggs

Dinosaur eggs are known from at least 41 separate Cretaceous locations in China (see table 9-1) and have received extensive study, especially by Zhao (1979a, b, 1994; Zhao and Ding 1976; Zhao and Li 1988; Zhao et al. 1991). The abundance of dinosaur eggs in the Chinese Cretaceous directly contrasts with other contemporaneous records, especially in western North America, where dinosaur eggs are rare. The Chinese dinosaur egg record thus has heavily

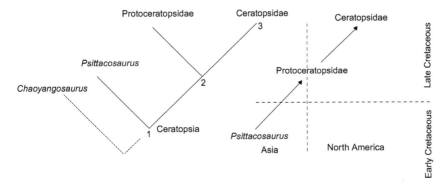

Figure 9-21 *Traditional cladogram of ceratopsians (left) and evolutionary tree (right) based on the cladogram. Selected characters corresponding to the numbered node points are: 1—rostral bone present, skull with narrow beak and flaring jugals, deep jugal, frill present, vaulted palate; 2—head extremely large relative to body, broad and prominent frill, sharply keeled rostrum, limb structures indicative of obligate quadrupedalism; 3—very large skulls (1–2.4 m long), large nostrils, horns.*

Table 9-1 Principal Cretaceous Dinosaur Egg Localities of China (after Carpenter and Alf 1994)

Locality	Formation	Age	Publication
1. Ningxia	?	Early K	Young (1979a)
2. Xixia	?	Early K	Zhao (1979a)
3. Erenhot	Iren Dabasu	Campanian	Dong et al. (1989)
4. Bayan Manduhu	Djadokhota	Campanian	Dong et al. (1989)
5. Chiangchungting	Wangshi	Campanian?	Chao & Chiang (1974)
6. Laiyang	Wangshi	Campanian?	Chao & Chiang (1974)
7. Chaochun	Wangshi	Campanian?	Chow (1951)
8. Chingkangkou	Wangshi	Campanian?	Young (1965b)
9. Jiaozhou	Wangshi	Campanian?	Carpenter & Alf (1994)
10. Zhucheng	Wangshi	Campanian?	Carpenter & Alf (1994)
11. Nanxiong	Nanxiong	Maastrichtian	Zhao et al. (1991)

Table 9-1 Principal Cretaceous Dinosaur Egg Localities of China (after Carpenter and Alf 1994) (Continued)

Locality	Formation	Age	Publication
12. Ulungar	Donggou?	Maastrichtian?	Carpenter & Alf (1994)
13. Shenjikou	Subash	Maastrichtian	Hao and Guan (1984)
14. Jiangjunmiao	Hongshaguan	Maastrichtian	Young (1965b)
15. Shanyang	Shanyang	Maastrichtian	Carpenter & Alf (1994)
16. Taoyuan	Fenshuiao	Maastrichtian	Zeng & Zhong (1979)
17. Chaling	Daijiaping	Maastrichtian	Young (1965b)
18. Taihe-Ganzhou	Yuanpu	Maastrichtian	Young (1965b)
19. Qitai	Subashi	Late K	Young (1965b)
20. Alxa	?	Late K	Zhao & Ding (1976)
21. Lingbao	Nanzhao	Late K	Carpenter & Alf (1994)
22. Liguanqiao	Hugang	Late K	Carpenter & Alf (1994)
23. Anlu	Gonganzhai	Late K	Zhao & Li (1988)
24. Xuanzhou	Xuannan	Late K	Carpenter & Alf (1994)
25. Tiantai	Laijia B	Late K	Mateer (1989)
26. Gaotangshi	Quxian	Late K	Mateer (1989)
27. Quzhou	Qujiang	Late K	Carpenter & Alf (1994)
28. Anwen	Xiaoyan	Late K	Dong (1980b)
29. Gaoan	Qingfengqiao	Late K	Carpenter & Alf (1994)
30. Xinyu	Qingfengqiao	Late K	Carpenter & Alf (1994)
31. Heyuan	Nanxiaong	Late K	Carpenter & Alf (1994)
32. Hizhou	Nanxiaong	Late K	Carpenter & Alf (1994)
33. Shiguguan	Hugang	Late K	Zhao (1979b)

Table 9-1 Principal Cretaceous Dinosaur Egg Localities of China (after Carpenter and Alf 1994) (Continued)

Locality	Formation	Age	Publication
34. Tsatzeyuanshu	?	Late K	Young (1965b)
35. Xinyang	?	Late K	Carpenter & Alf (1994)
36. Changchun	Quantou	Late K	Carpenter & Alf (1994)
37. Quantou	Quantou	Late K	Carpenter & Alf (1994)
38. Yixing	?	K	Carpenter & Alf (1994)
39. Anging	?	K	Carpenter & Alf (1994)
40. Guangzhou	Sanshui	K	Carpenter & Alf (1994)
41. Xiaguan	?	K	Zhao (1979b)
42. Yunxian	?	K	Carpenter & Alf (1994)
43. Taohe	Majiacun	K	Carpenter & Alf (1994)

influenced ideas about dinosaur egg taphonomy, stratigraphic utility, taxonomy, and the process of dinosaur extinction.

Carpenter (1982) suggested that dinosaur eggs were preferentially preserved in environments characterized by well-drained soils and high pH. Chinese Upper Cretaceous sediments are mostly relatively coarse-grained and oxidized; many are red beds. These sediments are indicative of well-drained environments with high pH. The Chinese dinosaur egg record thus strongly supports the idea that well-drained environments with high pH favored the preservation of dinosaur eggs.

Young (1965b) attempted to correlate Chinese Cretaceous localities based on their fossil eggs, which is one of the few attempts at fossil-egg-based biostratigraphy. Dinosaur eggs from the Nanxiong basin (see figure 9-22) are stratigraphically important because they document the stratigraphically highest occurrence of dinosaurs in that basin (see figure 9-23).

A rather extensive parataxonomy of Chinese fossil eggs has been created by Zhao (1975, 1979a, b, 1994). This parataxonomy is based on a variety of macromorphological features (shape and size of egg, sculpturing of outer shell surface, thickness of eggshell), histomorphological features (basic structural units and pore canal including size, shape and arrangement of mammilla, cone, columna and type of pore canal system; texture of eggshell, especially the composition and sequence of horizontal ultrastructural zones) and inferred ethological features of the nest. Using these features Zhao has named "species," "genera" (all with the suffix "–oolithus") and "families" of dinosaur eggs. Zhao's parataxonomy of Cretaceous dinosaur eggs from China has become the standard for most parataxonomic nomenclature of dinosaur eggs (Carpenter et al. 1994).

The remarkable record of dinosaur eggs from the Upper Cretaceous Nanxiong Formation in Guangdong has become the basis for inferences about the cause of dinosaur extinction (see figure 9-23). More than 20,000 eggshell fragments, about 300 complete eggs and some 24 complete or nearly complete nests (e.g., see figure 9-22) are present in the upper part of the Nanxiong Formation. Analyzing this record, Zhao et al. (1991) noted that: (1) eggshell thickness decreases upsection, (2) there is an increase in pathological histomorphological structures upsection, (3) carbon and oxygen isotopes ($C^{12/13}$ and $O^{16/18}$) in the eggshells become heavier upsection, and (4) trace element (Mn,

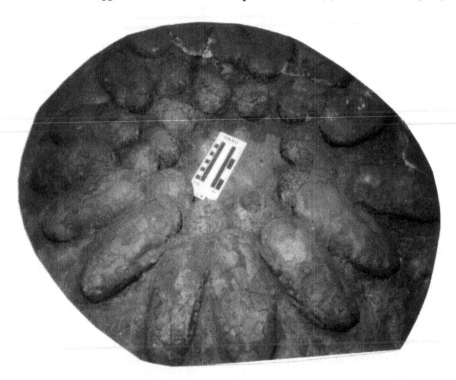

Figure 9-22 *Nest of eggs from the Nanxiong basin; scale in cm (left) and in inches (right).*

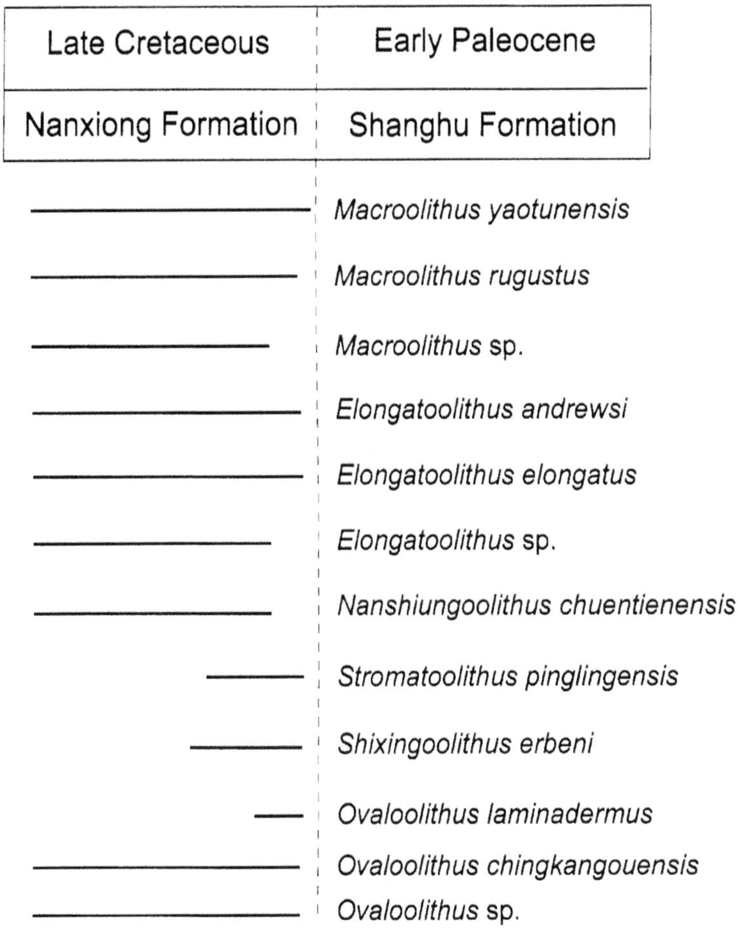

Late Cretaceous	Early Paleocene
Nanxiong Formation	Shanghu Formation

—————————————— *Macroolithus yaotunensis*

———————————— *Macroolithus rugustus*

———————— *Macroolithus* sp.

————————— *Elongatoolithus andrewsi*

————————— *Elongatoolithus elongatus*

———————— *Elongatoolithus* sp.

———————— *Nanshiungoolithus chuentienensis*

————— *Stromatoolithus pinglingensis*

————— *Shixingoolithus erbeni*

——— *Ovaloolithus laminadermus*

———————— *Ovaloolithus chingkangouensis*

———————— *Ovaloolithus* sp.

Figure 9-23 *Distribution of dinosaur egg parataxa in the Nanxiong basin of Guangdong (after Zhao et al. 1991).*

Zn, Sr, etc.) quantities in the eggshells increase upsection. They concluded that excessively dry climates at the end of the Cretaceous caused these changes in the eggshells, producing abnormal embryonic development that led to a collapse of dinosaur populations and their extinction. Zhao et al. (1991) further concluded that dinosaur extinction in the Nanxiong basin took place 200–300,000 years before dinosaur extinction elsewhere because the youngest dinosaur eggshells in the Nanxiong Formation occur in the lower part of chron 29r on the magnetic polarity timescale.

This last claim is easily refuted because Zhao et al. (1991) failed to consider local sedimentation rates affecting the thickness of chron 29r. The Cretaceous-Tertiary boundary, the time of dinosaur extinction, corresponds to some datum in chron 29r, but where it falls in this interval of reversed polarity depends on the thickness of the stratigraphic interval of reversed polarity; thickness is determined by local sedimentation rates.

Zhao et al.'s (1991) conclusion that thinning eggshells and changes in heavy metal and isotopic composition of the eggshells caused dinosaur extinction also is questionable. Their analysis does not establish that these changes caused a reduction of egg viability. In the absence of dinosaur embryos from the Nanxiong basin, their conclusion that these changes produced abnormal dinosaur embryos is sheer speculation. Furthermore, Zhao et al. (1991) failed to study the background geochemistry of the sediments that contain the eggs. They thus cannot demonstrate that the increase in heavy metals and shift in carbon and oxygen isotopes is simply not a result of changes in the entombing sediments, and therefore something that took place during fossilization and diagenesis of the eggshells. I conclude that Zhao et al.'s (1991) analysis does not demonstrate anything conclusive about the timing or the cause of dinosaur extinction.

Cretaceous-Tertiary Boundary and Extinctions

Mateer and Chen (1992) identified 10 places in China where relatively continuous, nonmarine stratigraphic sections encompass the Cretaceous-Tertiary transition. Of these, only the section in the Nanxiong basin of Guangdong contains a significant record of both Late Cretaceous and early Paleocene vertebrates.

The Nanxiong basin is a small graben formed during late Mesozoic rifting. The Upper Cretaceous strata here belong to the Nanxiong Formation, a lacustrine red-bed sequence. The Nanxiong Formation contains extensive numbers of dinosaur eggs and lesser numbers of turtle and dinosaur bones (see figure 9-24). The turtle and dinosaur bones represent the following taxa: *Nanshiungchelys wuchingensis*, *Tarbosaurus bataar*, *Nanshiungosaurus brevispinus* and *Microhadrosaurus nanhsiungensis* (a *nomen dubium*). All of these taxa, except *Tarbosaurus*, are endemic to the Nanxiong basin. In Mongolia, *Tarbosaurus* (or cf. *Tarbosaurus*) is known from the Nemegt Formation of Maastrichtian age and possibly in the underlying Barun Goyot Formation of late Campanian age (Jerzykiewicz and Russell 1991). This is consistent with charophyte evidence that suggests a Maastrichtian age for the Nanxiong Formation (R. Huang 1988), though neither the charophytes nor the vertebrates provide a more precise correlation within the

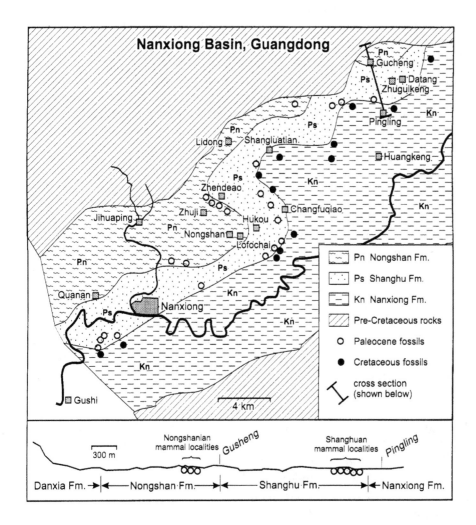

Figure 9-24 *Vertebrate fossil localities across the Cretaceous-Paleogene boundary in the Nanxiong basin (modified from Russell and Zhai 1987).*

stage. The dinosaur and turtle body fossils from the Nanxiong basin thus are of Nemegtian age, but do not provide better age resolution.

The stratigraphically highest dinosaur bones are well below the highest occurrence of dinosaur eggs (see figure 9-25), which are found to the top of the Nanxiong Formation (Zhao et al. 1991). These represent the highest dinosaur occurrence in the Nanxiong basin, so the Cretaceous-Tertiary boundary is conventionally placed at the boundary between the Nanxiong and Shanghu formations.

The Shanghu Formation conformably overlies the Nanxiong Formation (P. Chen 1983; P. Chen and Wang 1984; Russell and Zhai 1987; Huang 1988). (Note that Mateer and Chen [1992] described this contact as an unconformity, contrary to the conclusions of other workers, including Chen himself in earlier articles.) The fossil mammals from the lower part of the Shanghu Formation (as low as 10 m above the last dinosaur egg) are the type fauna of the Shanghuan land-mammal "age" of Russell and Zhai (1987). These mammals are several very primitive anagalids, the most primitive pantodonts, the most primitive tillodont, the most primitive mesonychids, and a very primitive carnivore (see chapter 10). Except for the anagalids, which were Asian endemics, these mammals are more primitive than are North American Torrejonian close relatives. This provides a strong argument that the Shanghu Formation mammals are earliest Paleocene (Puercan correlative) in age (Lucas and Williamson 1995). It does not support correlation of the Shanghuan mammals with the North American Torrejonian or Tiffanian (contra Chow et al. 1977; Sloan 1987; Mateer and Chen 1992; Russell et al. 1993; Rigby et al. 1993; Y. Wang et al. 1998). The Nongshan Formation overlies the Shanghu Formation and contains fossil mammals (notably *Archaeolambda*) typical of the Nongshanian land-mammal "age," a late Paleocene correlative of the North American Tiffanian. Recent attempts to correlate the Shanghuan with the Tiffanian by adjusting published magnetostratigraphy in the Nanxiong basin so that the Shanghuan interval corresponds to chron 26N (Russell et al. 1993) simply ignore the mammal-based correlations.

Palynomorphs and charophytes from the Shanghu Formation have been assigned an early Paleocene (Danian) age (R. Huang 1988; M. Li 1989). P. Chen's (1986) claim that conchostracans from the Shanghu Formation (his "*Fushunograpta changzhouensis* fauna") are late Paleocene contradicts the correlations just summarized. Indeed, it seems likely that Chen's (1986) Paleogene conchostracan correlations are about one stage off because he considers the Nongshan Formation (of well-established late Paleocene age) to be early Eocene. I thus conclude that there is no evidence the Shanghuan mammals are younger than an early Paleocene correlative of the North American Puercan (see chapter 10).

The picture that emerges of vertebrate fossil distribution across the Cretaceous-Tertiary transition in the Nanxiong basin thus resembles what we see in western North America. The highest dinosaur occurrence (in Nanxiong based on eggs) is immediately overlain by fossils of very primitive, Paleocene mammals. No convincing evidence for Paleocene dinosaurs is available from the Nanxiong basin (but see Rigby et al. 1993 and Erben et al. 1995). The eggshell record here tells us little about dinosaur extinction, as argued above. The dinosaur bone record in the Nanxiong basin Upper Cretaceous is very limited, so it

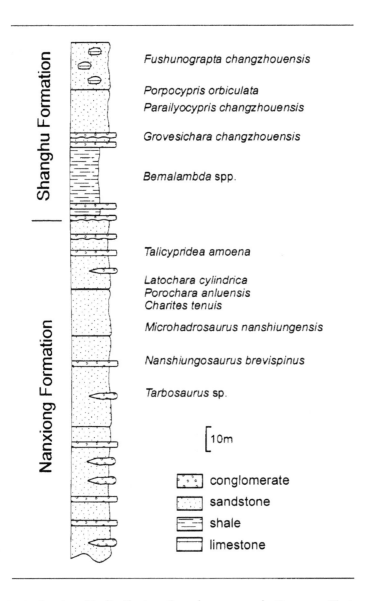

Figure 9-25 *Stratigraphic distribution of vertebrates across the Cretaceous-Tertiary boundary in the Nanxiong basin (after Mateer and Chen 1992).*

yields no information on dinosaur diversity changes and extinction patterns. The great importance to understanding dinosaur extinction offered by the record in the Nanxiong basin seems to be twofold: (1) it demonstrates dinosaurs were laying great quantities of eggs right until the end of the Cretaceous, and (2) it reveals no overlap of dinosaurs with Paleocene fossils and thus shows the same abrupt dinosaur disappearance followed by primitive eutherian mammal dominance seen in western North America. A more extensive bone record of Late Cretaceous dinosaurs and other vertebrates is needed from the Nanxiong basin or elsewhere before more can be said about dinosaur extinction based on the Chinese record.

Two Vertebrate Faunas

Perhaps the most surprising conclusion we can reach from this review of the Cretaceous fossil vertebrates of China is that there are few adequately known vertebrate fossil assemblages from the Chinese Cretaceous. The Early Cretaceous (Tsagantsabian) vertebrates of the Tugulu Group in Xinjiang and the Late Cretaceous (Baynshirenian and Djadokhtan) vertebrates of Nei Monggol represent most of what we know about Chinese Cretaceous vertebrates (also see Dong 1995).

The Tugulu vertebrates come from lake and lake margin facies, so they include a large array of bony fishes. The Tugulu dinosaurs are coelurosaurs, a carnosaur, a sauropod, a stegosaur (*Wuerhosaurus*), and the primitive ceratopsian *Psittacosaurus*. A few turtles, a crocodilian, and pterosaurs complete the Tugulu vertebrate assemblage, which is typical of the Early Cretaceous vertebrate faunas known across Asia.

There is a long temporal gap between the Tugulu vertebrates, which are of Tsagantsabian age, and a comparably well-known younger Cretaceous vertebrate fauna from China. This younger fauna either is that of the Iren Dabasu Formation or the somewhat younger Djadokhta Formation, both in Nei Monggol. These faunas essentially lack fishes because they come from fluvial or eolian units, but have some turtles, crocodilians, and lizards. Their dinosaurs are mostly protoceratopsid ceratopsians and hadrosaurs, dinosaurs that dominate all Asian Late Cretaceous dinosaur assemblages. Small and large theropods, dinosaur eggs, and early mammals complete the Late Cretaceous assemblage.

A large amount of evolutionary change took place between China's well-known Early and Late Cretaceous vertebrate assemblages just outlined. Unfortunately, China's Middle Cretaceous vertebrate record is poor. Furthermore, the record of Cretaceous vertebrates from southern China is much less than that from the northern part of the country. Clearly, vast temporal and geographic gaps exist in China's Cretaceous vertebrate record. By filling them, we stand to gain a much deeper understanding of Cretaceous vertebrate evolution between the now well-known Early and Late Cretaceous records.

Chapter 10

Paleogene

Paleogene rocks of China are widespread (see figure 10-1) and contain numerous mammal-dominated fossil assemblages. Nonmarine red beds and other siliciclastic deposits accumulated as the result of fluvial and lacustrine deposition in numerous basins across China. As in the Cretaceous, volcanism was confined to eastern China. No marine deposition took place in China during the Paleogene, and much of the overall tectonism was a continuation of Cretaceous movements.

The Indo-Pakistani subcontinent collided with southern Asia during the Paleogene, commencing the uplift of the Himalayas and Tibetan Plateau (Himalayan orogeny). However, during the Paleogene this uplifting had just begun, so warm and humid air currents from the Indian Ocean basin still flowed into southern China. A climatic zonation developed (see figure 10-2) much different than that of the later Neogene and Quaternary. Nevertheless, no evidence of any Paleogene provinciality among vertebrates (especially mammals) can be detected, despite some claims to the contrary (Gu and Chen 1987; Du et al. 1992).

The Cenozoic (or Tertiary) is often called the "age of mammals." After dinosaur extinction at the end of the Cretaceous, mammals became the dominant land-vertebrates. China has one of the most extensive fossil records of Paleogene land-mammals, almost all of which were eutherian (placental) mammals. This record has been fundamental to paleontological understanding of the great evolutionary diversification of eutherians during the Paleogene.

Paleogene Vertebrate-Bearing Deposits

Reviewing in detail the wealth of Paleogene deposits in China that contain fossil vertebrates (see table 10-1) is beyond the scope of this book (see Russell and Zhai 1987, for such a review). Most of these deposits are red beds of fluvial and lacustrine origin deposited in half-graben basins.

A good example of this type of deposit is the Yuhuangding Formation in the Liguanqiao basin, an intermontane half graben (see figure 10-3) located in the eastern Qinling Mountains on the border of Hubei and Henan. Strata of the Yuhuangding Formation are mostly red-bed mudstones and sandstones of lacustrine origin (see figure 10-4). Fossil mammals of early and middle Eocene age are found superposed in the Yuhuangding Formation. The overlying, lacustrine Hetaoyuan Formation also produces middle Eocene mammals. Only in Nei Monggol are significant fluvial deposits containing Paleogene vertebrates found.

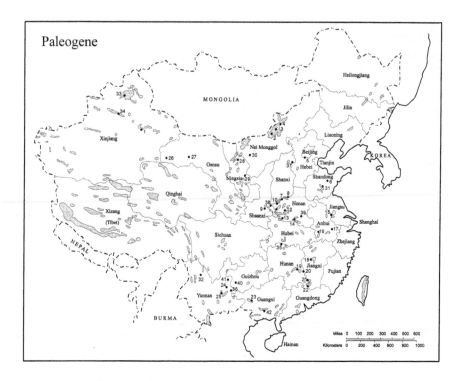

Figure 10-1 *Paleogene rocks of China and fossil vertebrate localities of China (after H. Wang 1985; and Russell and Zhai 1987). Localities are: 1 – Turpan basin; 2 – Shara Murun region; 3 – Nomogen region; 4 – Arshanto; 5 – Changxindian; 6 – Changle and Linqu Districts; 7 – Yuanqu basin; 8 – Jiyuan basin; 9 – Lantian District; 10 – Lingbao basin; 11 – Lushi basin; 12 – Tantou basin; 13 – Xichuan basin; 14 – Wucheng basin; 15 – Laian District; 16 – Qianshan basin; 17 – Xuancheng basin; 18 – Yuanshui basin; 19 – Hengyang basin; 20 – Chaling basin; 21 – Chijiang basin; 22 – Nanxiong basin; 23 – Bose and Yongle basins; 24 – Yuezhou basin; 25 – Lunan basin; 26 – Taben Buluk and Yindirte; 27 – Hui-hui-pu area; 28 – Haoisbuerdu basin; 29 – Lingwi District; 30 – Qianlishan District; 31 – Xintai District; 32 – Lijiang basin; 33 – northern Junggur basin; 34 – southern Junggur basin; 35 – Yidu District; 36 – Luoping basin; 37 – Quyang District; 38 – Shimen basin; 39 – Pingchanguan basin; 40 – Shinao basin; 41 – Xuanwei basin; 42 – Nanning basin.*

Paleogene Land-Mammal "Ages"

The Paleogene vertebrate-fossil record of China can be placed into a concise temporal framework by using the scheme of land-mammal "ages" (LMA) elaborated by Russell and Zhai (1987) (see figure 10-5). Their scheme is used here with only two modifications. In light of relocation of the Eocene-Oligocene

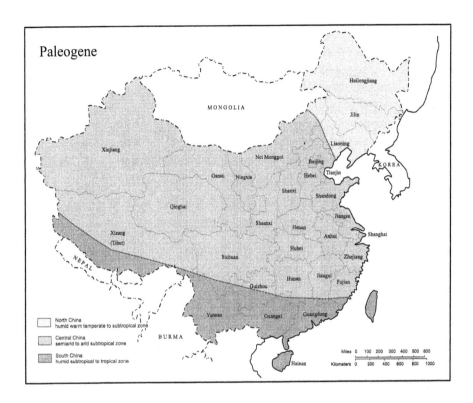

Figure 10-2 *Paleogene climatic zonation of China does not show the aridification of the later Neogene (after H. Wang 1985).*

Table 10-1 Principal Paleogene Fossil Vertebrate Localities of China

Locality	Age	Geologic Formation	Thickness	Dominant Lithology	Principal Reference
1. Nanxiong basin, Guang-dong	early Paleocene (Shanghuan)	Shanghu	470–600 m	purplish red mudstones, sandstones and con-glomerates	Chow et al. (1977); Zhang & Tong (1981)
2. Qianshan basin, Anhui	early Paleocene (Shanghuan)	Wanghudun	1800 m	purplish red sandstones (lower member)	Xu (1976)
3. Chijiang basin, Jiangxi	early Paleocene (Shanghuan)	Shizikou	120 m	brick red sandy mud-stones	B. Wang & Ding (1979)

Table 10-1 Principal Paleogene Fossil Vertebrate Localities of China (Continued)

Locality	Age	Geologic Formation	Thickness	Dominant Lithology	Principal Reference
4. Shimen basin, Shaanxi	early Paleocene (Shanghuan)	Fanghou	165 m	brown-red sandy mud-stones	McKenna et al. (1984)
5. Tantou basin, Henan	early Paleocene (Shanghuan)	Gaoyugou	366 m	red mud-stone	Tong & Wang (1980); Xue et al. (1996)
6. Chaling basin, Hunan	early Paleocene (Shanghuan)	Zaoshi	50 m	purplish red sandy clay-stones	B. Wang (1975)
7. Qianshan basin, Anhui	late Paleocene (Nong-shanian)	Wanghudun	1800 m	purplish red and brown-red sand-stones (upper member)	Qiu (1977)
8. Nanxiong basin, Guangdong	late Paleocene (Nong-shanian)	Nongshan	460 m	purplish red marls (Dtang and hugikeng members)	Ding & Tong (1979); Tong (1982)
9. Qianshan basin, Anhui	late Paleocene (Nong-shanian)	Doumu	600 m	purplish red conglomer-ates, sand-stones and mudstones	Li (1977a); Huang (1977)
10. Xuancheng basin, Anhui	late Paleocene (Nong-shanian)	Shuangtasi	?	coarse-grained red beds	Tang & Yan (1976); Yan & Tang (1980)
11. Tantou basin, Henan	late Paleocene (Nong-shanian)	Tantou, Dashang	136–458 m 104–375 m	mudstone and oil shales, green and red sand-stones and marls	Tong & Wang (1980)

Table 10-1 Principal Paleogene Fossil Vertebrate Localities of China (Continued)

Locality	Age	Geologic Formation	Thickness	Dominant Lithology	Principal Reference
12.Chijiang basin, Jiangxi	late Paleocene (Nong-shanian)	Chijiang Pinghu	500 m 700–300 m	purplish red marls, varie-gated mud-stones and conglomer-ates	Ding & Zhang (1979); Tong (1979)
13.Turpan basin, Xin-jiang	late Paleocene (Nong-shanian)	Taizicun	35–65 m	white sand-stones	Tong (1978)
14.Bayn Ulan, Nei	late Paleocene (Nong-shanian)	Nomogen?	7–30 m	red snady clays and silts	Matthew & Granger (1925); Chow & Qi (1978)
15.Turpan basin, Sinjiang	latest Paleocene (Nong-shanian)	Dabu Shisanjian-fang	22 m 272 m	gray-white sandstones, red sandy clay	Zhai (1978a,b)
16.Liguan-qiao basin, Henan	early Eocene (Liguan-qiaoan)	Yuhangding	360–960 m	pink and gray marl	Xu (1976a); Ma & Cheng (1991)
17.Hengyang basin, Hunan	early Eocene (Liguan-qiaoan)	Limuping	~400 m	red-bed mudstones, and sand-stones	C. Li et al. (1979); Ting (1993)
18.Yuanshui basin, Jiangxi	early Eocene (Liguan-qiaoan)	Xinyu	600–900 m	variegated sandstones and con-glomerates	Chow & Tung (1962); Zheng et al. (1975)
19.Changle, Shandong	early Eocene (Liguan-qiaoan)	Wutu (in part)	?	oil shales, coals, and variegated mudrocks	Chow & Li (1965)
20.Irdin Manha, Nei Monggol	middle Eocene (Irdinmanhan)	Irdan Manha	10 m	white sandy clays, sands, and gravels	Radinsky (1964)

Table 10-1 Principal Paleogene Fossil Vertebrate Localities of China (Continued)

Locality	Age	Geologic Formation	Thickness	Dominant Lithology	Principal Reference
21.Bose basin, Guangxi	middle Eocene (Irdinmanhan)	Dongjun	50 m	gray and white lime-stone	Ding et al. (1977)
22.Lushi basin, Henan	middle Eocene (Irdinmanhan)	Lushi	450 m	reddish and greenish rarls	Chow (1965); Chow et al. (1973)
23.Pingchang-guan basin, Henan	middle Eocene (Irdinmanhan)	Lizhuang	2202 m	gray and brownish red sand-stone and conglomer-ate	B. Wang & Zhou (1982)
24.Wucheng basin, Henan	middle Eocene (Irdinmanhan)	Lishigou	370 m	brown and yellow sandstone	Gao (1976)
25.Xichuan basin, Henan	middle Eocene (Irdinmanhan)	Dacangfang Hetaoyuan	600 m 500 m	sandstones and con-glomerates, green and gray clays	Xu et al. (1979); Tong & Lei (1984)
26.Yidu, Hubei	middle Eocene (Irdinmanhan)	Pailoukou	447 m	sandstone and siltsone	Xu (1980)
27.Arshanto, Nei Monggol	middle Eocene (Irdinmanhan)	Arshanto	15 m?	dark red sandy clays and silts	Qi (1979, 1987)
28.North Mesa, Nei Monggol	middle Eocene (Irdinmanhan)	Ulan Shireh	50 m +	variegated clays	Matthew & Granger (1923)
29.Xintai, Shnadong	middle Eocene (Irdinmanhan)	Guan-zhuang	600–1100 m	variegated clays and sandstones	Zdansky (1930); Chow & Qi (1982)
30.Lijiang basin, Yunnan	middle Eocene (Irdinmanhan)	Xianshan	150–200 m	brick red sandstones, gray-white clay	Zhang et al. (1978)

Table 10-1 Principal Paleogene Fossil Vertebrate Localities of China (Continued)

Locality	Age	Geologic Formation	Thickness	Dominant Lithology	Principal Reference
31.Lunan basin, Yunnan	middle Eocene (Irdinmanhan)	Lumeiyi (in part)	752 m	sandy clays and red clayey sandstones	Zhang et al. (1978); Huang & Qi (1982)
32.Turpan basin, Xinjiang	middle Eocene (Irdinmanhan)	Liankan	82 m	variegated sandstones and claystones	Zheng (1978)
33.Changxindian, Beijing area	middle Eocene (Irdinmanhan)	Changxindian	100 m	fanglomerates	Zhai (1977); Lucas (1996b)
34. Ula Usu, Nei Monggol	middle Eocene (Sharamurunian)	Shara Murun	70–100 m	gray clays	Xu (1966)
35. Jiyuan basin, Henan	middle Eocene (Sharamurunian)	Jiyuan	400–500 m	brownish red sandstones, mudstones, and conglomerates	Chow & Xu (1965)
36.Wucheng basin, Henan	middle Eocene (Sharamurunian)	Wulidui	550 m	brown and green shales	B. Wang (1976)
37.Yuanqu basin, Shanxi-Henan	middle Eocene (Sharamurunian)	Heti	1000 m	red mudstones	Zdansky (1930)
38.Bose-Yongle basins, Guangxi	middle Eocene (Sharamurunian)	Naduo	563–620 m	mudstones and coal beds	Zheng & Chi (1978); Tang & Qiu (1979)
39.Bose-Yongle basins, Guangxi	middle Eocene (Sharamurunian)	Gongkang	1300–1450 m	variegated mudstones and sandstones	Qiu (1979); Tang (1978)

Table 10-1 Principal Paleogene Fossil Vertebrate Localities of China (Continued)

Locality	Age	Geologic Formation	Thickness	Dominant Lithology	Principal Reference
40.Ulan Gochu, Nei Monggol	late Eocene (Ergilian)	Ulan Gochu	17 m	red clay-stone	Burke (1941); Granger & Gregory (1943); Wall (1980)
41.Urtyn Obo, Nei Monggol	late Eocene (Ergilian)	Urtyn Obo	~55 m	red and gray claystones	Granger & Gregory (1943); Chow & Chiu (1963)
42.Changan-bulage, Nei Monggol	late Eocene (Ergilian)	Chagan-bulage	140 m +	variegated mudstones and sand-stones	Matthew & Granger (1924); Qi (1975)
43.Shinao basin, Guizhou	late Eocene (Ergilian)	Shinao	500–600 m	sanstones, conglomer-ates, and coals	Miao (1982)
44.Erenhot, Nei Monggol	late Eocene (Ergilian)	Houldjin	12 m	conglomer-ates and sandstones	Granger & Gregory (1943); Radinsky (1964)
45.Lantian, Shaanxi	late Eocene (Ergilian)	Bailuyuan	400 m	white sand-stones and red-bed mudstones	Xu (1965, 1966)
46.Lunan basin, Yunnan	late Eocene (Ergilian)	Xiaotun	40–50 m	red-bed sandstones	Chow (1958)
47.Yuezhow basin, Yunnan	late Eocene (Ergilian)	Cajiachong	218 m	variegated marls	Zhang et al. (1978); B. Wang & Zhang (1983)

Locality	Age	Geologic Formation	Thickness	Dominant Lithology	Principal Reference
48.Ulantatal, Nei Monggol	early Oligocene (Shandgolian)	unnamed	20 m	variegated mudstones and siltstones	Matthew & Granger (1924); Huang (1982, 1985)
49.Qianlishan, Nei Monggol	early Oligocene (Shandgolian)	Wulanbulage	72 m	red-bed mudstones	Matthew & Granger (1923); B. Wang et al. (1981)
50.Saint-Jacques, Nei Monggol	early Oligocene (Shandgolian)	Wulanbulage?	30 m +	red-bed mudstones and sandstones	Teilhard de Chardin (1926)
51.Lingwu, Ningxia	early Oligocene (Shandgolian)	unnamed-Qingshuiying	2–3 m ?	sandstones green marls	Teilhard de Chardin (1926); Hu (1962)
52.Taben Buluk, Gansu	late Oligocene (Tabenbulukian)	unnamed	40 m	red-bed mudstones, sandstones, and conglomerates	Bohlin (1937, 1942, 1946)
53.Shargaltein Gol, Gansu	late Oligocene (Tabenbulukian)	unnamed	—	red-bed mudstones, sandstones, and conglomerates	Bohlin (1937)
54.Qianlishan, Nei Monggol	late Oligocene (Tabenbulukian)	Yikebulage	58 m	red-bed mudstones, sandstones, and conglomerates	B. Wang et al. (1981)
55.Turpan basin,	late Oligocene (Tabenbulukian)	Taoshuyuanzi (upper part)	800 m +	sandy claystone, and conglomerate lenses	Zhai (1978c)

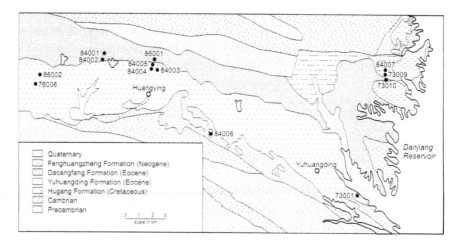

Figure 10-3 *Geologic map of the Liguanqiao basin showing principal Eocene fossil mammal localities (numbers), after Ma and Cheng (1991).*

boundary to 34 Ma, which places the Chadronian LMA of western North America in the latest Eocene, the Eocene-Oligocene boundary is between the Ergilian and Shandgolian. Recent articles that still identify Ergilian mammals as early Oligocene (e.g., B. Wang 1991; Dashzeveg 1993) simply failed to understand how changing the boundary in the West affects its position in Asia. Tong et al. (1995) with good reason subdivided the Irdinmanhan LMA of Russell and Zhai (1987) into the Arshantan and Irdinmanhan. There are thus nine Paleogene LMAs (see figure 10-5) that can be recognized in China using fossil mammals.

Figure 10-4 *Tilted red beds of the Eocene Yuhuangding Formation in the Liguangiao basin are lacustrine strata that contain fossil vertebrates.*

Shanghuan Mammals

The oldest Cenozoic fossil mammals from China are of early Paleocene (Shanghuan) age. Shanghuan mammals come from the Shanghu Formation of the Nanxiong basin, Guangdong (the type fauna of the LMA); the upper part of the Wanghudun Formation of the Qianshan basin in Anhui; the Shizikou Formation in the Chijiang basin of Jiangxi; and the Fangou Formation of the Shimen basin, Shaanxi (see table 10-1; see figure 10-1). Anagalids, bemalambdid pantodonts and mesonychids (Y. Wang et al. 1998) dominate the Shanguan mammal faunas.

Anagalida (see figure 10-6) is an order of eutherian mammals endemic to Asia during the Paleogene. Anagalidans are the most common fossils in Shanghuan and some other Paleogene mammal assemblages from China. They thus

Period	Epoch		land-mammal "age"
PALEOGENE	Oligocene	Late	Tabenbulukian
		Early	Shandgolian
	Eocene	Late	Ergilian
		Middle	Sharamurunian
			Irdinmanhan
			Arshantan
		Early	Bumbanian
	Paleocene	Late	Nongshanian
		Middle	
		Early	Shanghuan

Figure 10-5 *Paleogene land-mammal "ages" of China are nine divisions of Paleocene, Eocene, and Oligocene time.*

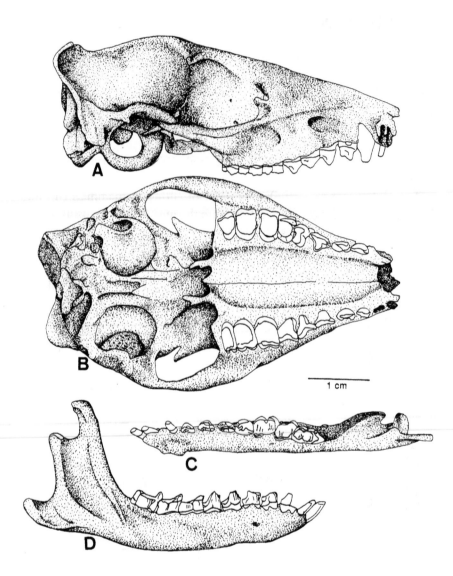

Figure 10-6 *The skull and lower jaw of Anagale, a characteristic anagalid. Skull in lateral (A) and ventral (B) views; dentary in occlusal (C) and lateral (D) views.*

parallel the dominance of North American Paleocene mammal assemblages by "condylarths." The overall habitus of anagalids is that of small herbivores and omnivores, many with rabbit-like hopping modifications of the hind limbs.

Anagalidan relationships have been unclear for many years, but cladistic analysis by Y. Hu (1993) unites the order by synapomorphies that include prismatic and hypsodont cheek teeth. Shanghuan anagalidans are diverse, and include the genera *Linnania, Huaiyangale, Diacronus, Wanogale, Chians-haniaand Anaptogale* (Anagalidae); *Anictops, Cartictops,* and *Paranictops* (Pseud-ictopidae); and *Astigale* and *Zhujegale* (Astigalidae). *Yuodon* and *Palasiodon* from the Shanghu Formation in the Naxiong basin were originally described as hyop-sodontid "condylarths" (Chow et al. 1973b, 1977), but their type lower dentitions clearly belong to anagalidans (Lucas and Williamson 1995). *Decore-don* from the Wanghudun Formation in Anhui also appears to be an anagal-idan, though it too was originally described as a hyopsodontid (Q. Xu 1977).

The Shanghuan mixodonts *Heomys* and *Mimotona* belong to the families Eurymylidae and Mimotonidae. They approximate the ancestry of rodents and of lagomorphs (see later discussion) and thus suggest this phylogenetic split had taken place by early Paleocene time.

The Shanghuan mesonychids are known from isolated teeth and jaw frag-ments. *Dissacusium* is only known from a right M^1 or M^2, and *Hukoutheriumis* known only from a lower jaw; both are probably one genus of mesonychid more primitive dentally than North American Torrejonian *Dissacus* and *Ankalagon* (Lucas and Williamson 1995). The third Shanghuan mesony-chid, *Yantangalestes*, is a small primitive mesonychid about the size of North American *Hapalodectes*.

Pappictidops is the only Shanghuan carnivore, a viverravid. It resembles the North American Torrejonian carnivore *Ictidopappus*, but is more primitive.

Bemalambda (see figure 10-7) and *Hypsilolambda* are the Shanghuan bemalambdid pantodonts. They are the most primitive pantodonts, lacking the w-shaped first and second upper molar ectolophs characteristic of the more derived eupantodonts, to which most subsequent Chinese and all North Amer-ican pantodonts can be assigned. *Hypsilolambda* is known only from the skull and dentition, but *Bemalambda* is the best known Shanghuan mammal, being represented by complete cranial and postcranial material (see figure 10-7). *Bemalambda* was a large dog-sized terrestrial quadruped whose dentition sug-gests omnivory. Van Valen (1988) excluded *Bemalambda* and *Hypsilolambda* from the Pantodonta, arguing that their dentition identifies them as large didelphodontine derivatives. However, a didelphodontine derivation of Pant-odonta seems reasonable (McKenna 1975; Lucas 1982, 1993d), with *Bemalambda* and *Hypsilolambda* identified as pantodonts because of their zalambdodont upper premolars (paracone lingual with large pre- and post-paracristae), a synapomorphy of Pantodonta (Lucas 1993d).

Remaining Shanghuan mammals are the zalambdalestid *Anchilestes*, the old-est tillodont, *Lofochaius*, the didymoconid *Zeuctitherium*, the micropternodon-tid *Prosarcodon* and the problematic *Obtususdon*. These mammals were small insectivores. *Anchilestes* may lie close to the ancestry of tillodonts, first seen in *Lofochaius*.

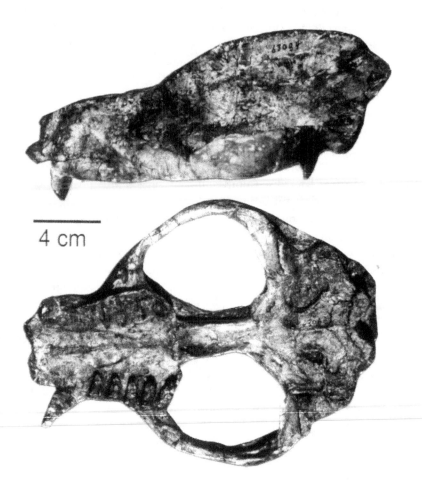

Figure 10-7 *Skull of the Shanghuan pantodont Bemalambda; lateral (above) and ventral (below) views. Note the tiny braincase and very high sagittal crest.*

The correlation of the Shanghuan with North American Paleocene land-mammal assemblages has been a subject of some uncertainty and disagreement. These mammals have been termed Torrejonian ("middle Paleocene") correlatives by Chow et al. (1977), C. Li and Ting (1983), Sloan (1987), Mateer and Chen (1992), Russell et al. (1993) and Y. Wang et al. (1998). Savage and Russell (1983) suggested the Shanghuan might be much younger, a correlative of the North American Tiffanian (late Paleocene). Lucas and Williamson (1995), however, argued that all Shanghuan mammals are more primitive than their closest Torrejonian relatives are (where such relatives exist), so they correlated the Shanghuan with the North American Puercan (early Paleocene).

Correlation of the Chinese Shanghuan and North American Puercan solves a longstanding evolutionary problem: the sudden appearance in North American during the Torrejonian of Carnivora, Mesonychia, Pantodonta, and, possibly, Tillodontia (in the guise of a *Deltatherium*-like form). Lucas and Williamson (1995) suggest these taxa emigrated from China to North America just before the Torrejonian. Their sudden appearance, lack of older North American close relatives, and the presence of older and more primitive Shanghuan relatives in China support an Asian origin and emigration from Asia to North America of the Carnivora, Mesonychia, Pantodonta, and Tillodontia at about the beginning of the Torrejonian.

Nongshanian Mammals

Nongshanian mammals are of late Paleocene age and are much more broadly distributed in China than are Shanghuan mammals. Nongshanian assemblages come from the Nongshan Formation in the Nanxiong basin of Guangdong (the characteristic mammal assemblage), the Doumu formation in the Qianshan basin of Anhui, the Shuangtasi Formation in the Xuancheng basin of Anhui, the Tantou and Dazhang formations in the Tantou basin of Henan, the Chijiang and Pinghu formations in the Chijiang basin of Jiangxi, the Nomogen Formation in Nei Monggol; and the Taizicun Formation in the Turpan basin of Xinjiang (see figure 10-1; table 10-1).

These faunas have abundant anagalids and pantodonts, like the Shanghuan faunas. However, the Nongshanian marks the first appearance of uintatheres (Dinocerata) and arctostylopids (Notoungulata), two new orders of Paleocene-Eocene placental mammals. The association of pantolambdodontid (= archaeolambid) pantodonts with the primitive dinoceratan *Prodinoceras*, phenacolophids (possible relatives of perissodactyls), and the arctostylopids is characteristic of Nongshanian mammal assemblages.

Multituberculates are well known from the Cretaceous of Mongolia, but they do not appear in the Chinese fossil record until the Nongshanian. These multituberculates are from the Nomogen Formation in Nei Monggol, the taeniolabidids *Prionessus* and *Sphenopsalis,* and the lambdopsalid *Lambdopsalis.* Excellent skulls of *Lambdopsalis* make it one of the best known multituberculates cranially (Miao 1988).

Nongshanian anagalidans are less diverse than those of the Shanghuan and include anagalids (*Hsiuannania, Huaiyangale*), pseudictopids (*Allictops, Haltictops,* and *Pseudictops*) and a form of uncertain family position (*Interogale*). *Petrolemur* from the Nongshan Formation was originally described as an adapid? primate (Tong 1979), which would make it the oldest Asian primate.

Szalay (1982) suggested it is a dichobunid artiodactyl, which would make it the oldest artiodactyl anywhere. It appears, however, to be an anagalidan.

Mixodonts are better known from the Nongshanian than from the Shanghuan. Two families still are present, the eurymylids (*Heomys*) and the mimotonids (*Mimotona*). The micropternodontids are still present in the form of *Sarcodon*, as is the problematic *Obtusodon*. Two other Nongshanian mammals of problematic affinities are *Hyracolestes* and *Wanotherium*, two small insectivorous eutherians.

An equally problematic mammal is *Ernanodon*, known from a complete skeleton from the Nongshan Formation. Ding (1979, 1987) described this mammal and tentatively termed it an edentate, which would make it a very early edentate in a very strange place (all other Paleocene edentates are from the New World). It now is best regarded as Eutheria, *incertae sedis*.

The Nongshanian mesonychids are rare as fossils (as are mesonychid fossils everywhere) but more diverse than those of the Shanghuan. *Yantangalestes* still is present, but the other mesonychid genera are new: *Dissacus* (also known from North America), *Jiangxia*, *Sinonyx,* and *Plagiocristodon* (Zhou et al. 1995). Surprisingly, no carnivores have been described from the Nongshangian of China.

Nongshangian pantodonts belong to a single family, Pantolambdodontidae (includes Archaeolambdidae, Harpyodidae, and Pastoralodontidae of previous authors). The most common form is *Archaeolambda*, particularly well known from a nearly complete skeleton (see figure 10-8) collected in the Doumu Formation in the Qianshan basin of Anhui (X. Huang 1977). *Altilambda* and *Pastoralodon* are larger forms known only from cranial and dental material. *Harpyodus* is a very small pantodont that is known from a skull and teeth. It is either the most primitive pantolambdodontid or a more primitive pantodont that is a sister taxon to the pantolambdodontids. *Harpyodus* is very similar to the South American Paleocene pantodont *Alcidedorbignya*, suggesting a broad (Asian-North American-South American) distribution of pantodonts during the late Paleocene (Muizon and Marshall 1992). Pantolambdodontid teeth are those of folivores. The skeleton of *Archaeolambda* indicates they were small, lightly built, possibly arboreal mammals with clawed digits on the manus.

Nongshanian *Prodinoceras* (see figure 10-9) is the first Asian uintathere. The genus is also known from North America and provides solid evidence that the Nongshanian is of late Paleocene age. In Mongolia, complete skeletal material of *Prodinoceras* (= *Mongolotherium*) reveals a large-dog-sized terrestrial quadruped that was omnivorous (see figure 10-9). Unlike most of the later uintatheres, *Prodinoceras* lacked horns and was not a large mammal (though it is the largest Nongshanian mammal from China).

Figure 10-8 *The remarkably complete skeleton of Archaeolambda tabiensis, an arboreal folivore of Nongshanian time.*

Phenacolophids are almost exclusively of Nongshanian age, which is when they have their greatest diversity (an exception is the poorly known Heptaconodon of Irdinmanhan age, discussed below). Known only dentally, with the possible exception of some anklebones, phenacolophids (see figure 10-10) have dilambdodont molars rather similar to those of the earliest perissodactyls.

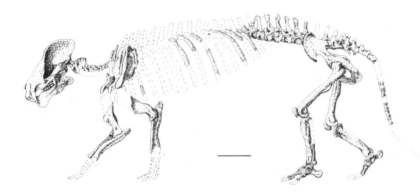

Figure 10-9 *Restored skeleton of Prodinoceras (= Mongolotherium) (after Flerov 1957). Scale bar = 40 cm.*

Nongshanian phenacolophids are *Radinskya* (the most primitive and most probable link to perissodactyls: McKenna et al. 1989), *Yuelophus, Tienshanilophus, Ganolophus* and the large form *Minchenella*, a possible link to tethytheres.

When first discovered in Wyoming (Matthew 1915a) and in Asia (Matthew and Granger 1925; Matthew et al. 1929) by the Central Asiatic Expeditions of the American Museum of Natural History, arctostylopids were thought to belong to the South American order of extinct ungulates, Notoungulata. This suggests a notoungulate distribution during the late Paleocene parallel to that of pantodonts and uintatheriamorphs (Gingerich 1985; Lucas 1986). Cifelli et al. (1989), however, argued (unconvincingly, in my opinion) that the arctostylopids represent a new, distinct order of mammals (Arctostylopida) not closely related to the Notoungulata, thus removing the South American affinities of arctostylopids. Arctostylopids are most diverse in the Chinese Nongshanian (*Sinostylops, Palaeostylops, Gashatostylops, Bothriostylops, Asiostylops, Allostylops,* and *Arctostylops*).

Shanghuan mammals are restricted to eastern China, but Nongshanian mammals are more widespread (see figure 10-1). The uniformity of the Nongshanian mammal fauna across this area suggests China was one zoogeographic region during the late Paleocene.

Bumbanian Mammals

The type fauna of the Bumbanian designated by Russell and Zhai (1987) is that of the Bumban Member of the Naran-Bulak Svita at Tsagan-Khushu, Mongolia. In China, the Bumbanian has a very limited distribution. Bumbanian mammals are known from the Shisanjianfang Formation in the Turpan basin of Xinjiang, the Wutu Formation of Shandong, the lower part of the Yuhuanding Formation in Henan, the Lingcha Formation in Hunan, and the Xinyu Formation in Jiangxi (Ting 1998).

The hallmark of the Bumbanian is the co-occurrence of *Asiocoryphordon, Heterocoryphodon, Hyracotherium,* and *Hyopsodus*. These taxa indicate the Bumbanian correlates to part of the Wasatchian of western North America, and thus is of latest Paleocene–earliest Eocene age (Lucas 1998c). At that time, there was free migration of mammals throughout the Holarctic continents via boreal land connections across Beringia and the North Atlantic (see figure 10-11). The relative endemism of Chinese Paleocene mammal faunas thus ended during the Bumbanian when North America, Europe, and Asia became one zoogeographic region.

Chinese Bumbanian endemics are few in number—the eurymylid *Rhombomylus* and the arctostylopid *Anatolostylops*. Cosmopolitan genera dominate.

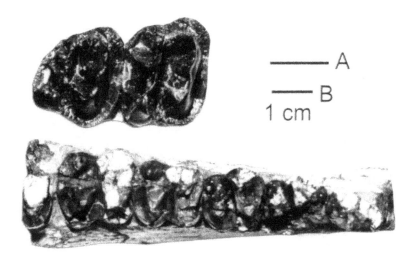

Figure 10-10 *Dentitions of selected Nongshanian phenacolophids: occlusal view of upper molars of Tienshanilophus lianmuqinensis from the Turpan basin Xinjiang (above) and lower molars of Minchenella grandis from the Nanxiong basin, Guangdong (below).*

However, the index taxa of the Bumbanian, the coryphodontid pantodonts *Asiocoryphodon* (see figure 10-12) and *Heterocoryphodon*, are endemic to Asia. These coryphodontids are much larger than *Coryphodon* and have more advanced bilophodont cheek teeth. They are endemic to China and represent an episode of early Eocene endemism of the Asian mammal fauna.

Hyracotherium ("*Eohippus*") is the oldest equid, *Homogalax* the oldest tapiroid, and *Hyopsodus* is a small, omnivorous "condylarth." Bumbanian primates (Beard et al. 1993) are carpolestids, which are also known from the Paleocene of North America.

The Bumbanian mammal fauna of China is not diverse and needs to be collected further. Still, what taxa are present indicate an episode of Holarctic cosmopolitanism that encompassed China during the latest Paleocene–earliest Eocene (Ting 1998).

Arshantan and Irdinmanhan Mammals

Arshanto and Irdin Manha (see figure 10-13), in Nei Monggol, are classic localities for Eocene mammals first collected by the Central Asiatic Expeditions of the American Museum of Natural History in 1922–1923. They lend their names to the middle Eocene LMAS used in China. Other Arshantan and Irdinmanhan

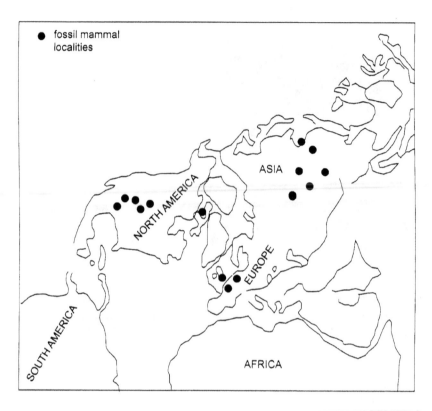

Figure 10-11 *The north polar projection of the continents shows the Holarctic distribution of cosmopolitan latest Paleocene–earliest Eocene mammals.*

mammal-bearing strata in China are: the Dongjun Formation in the Bose basin of Guangxi; the Chuankou Formation in the Lingbao basin, Henan; the Lushi Formation in the Lushi basin of Henan; the Lizhuang Formation in the Pingchangguan basin of Henan; the Lishigou and Maojiapo formations in the Wucheng basin of Henan; the Dacangfang and Hetaoyuan formations of the Xichuan basin, Henan; the Pailoukou Formation in Hubei; the Honghe Formation in Shaanxi; the Guanzhuang Formation in Shandong; the Xiangshan Formation in the Lijiang basin of Yunnan; the Lumeiyi Formation in the Lunan basin of Yunnan; the Honglishan Formation in the Junggur basin of Xinjiang; the Liankan Formation in the Turpan basin of Xinjiang; the Changxindian Formation near Beijing; and the Ulan Shireh and "Tukum" formations in Nei Monggol. Arshantan and Irdinmanhan mammals thus have a broader known geographic distribution in China than do mammal assemblages of the other Paleogene LMAs.

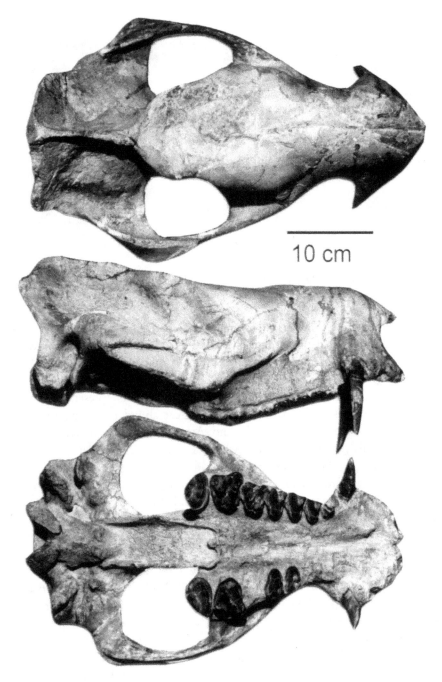

Figure 10-12 *Skull of Asiocoryphodon, a hippopotamus-like Bumbanian pantodont. Dorsal (top), lateral (middle), and ventral (bottom) views.*

Figure 10-13 *Outcrops of the Eocene Irdin Manha Formation at Irdin Manha, Nei Monggol (courtesy of R.J. Emry).*

The Arshantan and Irdinmanhan mark a dramatic change in the composition of the Chinese Paleogene mammalian fauna. Those of the Paleocene-early Eocene were dominated by archaic and wholly extinct eutherian orders—Anagalida, Pantodonta, Dinocerata, Notoungulata and Mesonychia. During the Arshantan, representatives of some of these orders are still present, but new and modern orders of mammals appear in great abundance—Lagomorpha, Rodentia, Carnivora, Perissodactyla and Artiodactyla. The replacement of "paleoplacentals" by "neoplacentals" (discussed below) is a fundamental turnover in the history of eutherian mammals.

The Arshantan-Irdinmanhan mammal fauna represents the first appearance and immediate dominance of the neoplacentals. Leftover paleoplacentals, some of which persisted into the subsequent Sharamurunian, are the pantolambdodontid and coryphodontid pantodonts (*Eudinoceras* is the most widespread coryphodontid genus); the trogosine tillodonts *Trogosus* (= *Kuanchuanius*) and *Chungchienia*; the bizarre uintathere *Gobiatherium* (see figure 10-14) and the more "typical" *Uintatherium;* a diversity of oxyaenid and hyaenodontid creodonts, not seen in China before, but well known from older horizons in North America; and diverse mesonychids, which include *Andrewsarchus*, the largest terrestrial meat-eating mammal to have ever lived. Neoplacentals, especially perissodactyls, dominate Irdinmanhan mammal assemblages in China. More than 100 species of these Arshantan and Irdinmanhan perissodactyls have been named, and they represent one of the most significant records of early perissodactyl (especially ceratomorph) evolution. Most of the Arshantan-

Irdinmanhan perissodactyls are tapiroids assigned to the families Depere-tellidae, Helaletidae and Lophialetidae (Radinsky 1965; Reshetov 1979; Schoch 1989). The deperetellids are medium- to large-sized tapiroids distinguished by their high crowned and very bilophodont molars. The helaletids are smaller and have less bilophodont cheek teeth, whereas the similar-sized lophialetids have even less bilophodont molars (see figure 10-15).

Rhinocerotoids are the second most common Arshantan-Irdinmanhan perissodactyls. They are either hyracodontids (hornless, cursorial rhinos) or amynodontids (hornless, amphibious rhinos). The hyracodontids range in size

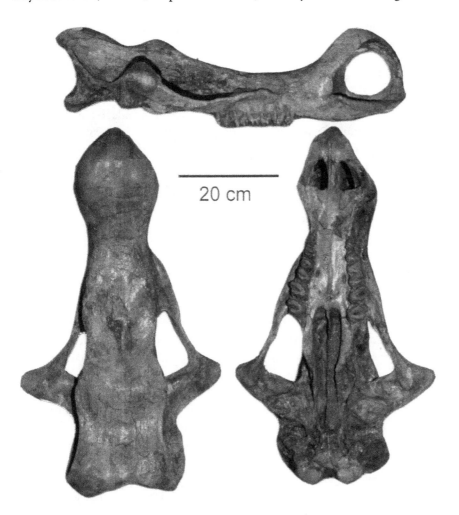

Figure 10-14 *Skull of the bizarre Arshantan uintathere Gobiatherium; lateral (top), dorsal (bottom left) and ventral (bottom right) views.*

from tiny *Rhodopagus*, about the size of a beagle dog, to horse-sized *Forstercoo-peria*, the oldest indricothere (see figure 10-15). The amynodonts were diverse but relatively large sized; they are assigned to the genera *Sharamynodon*, *Lushia-mynodon*, *Sianodon* and *Caenolophus*.

Brontotheres were rhinoceros-like in overall body build, but during Arshan-tan-Irdinmanhan time mostly lacked horns. A few paleotheres (*Propalaeothe-rium*) are known from Arshantan-Irdinmanhan strata, as are the oldest Asian chalicotheres—*Eomoropus*, *Grangeria*, and *Litolophus*.

China's oldest true primate, *Lushius*, is of Irdinmanhan age. Also present are the first true leporid lagomorphs in China (*Shamolagus*, *Lushilagus*) and a diversity of rodents, mostly ctenodactylids. Arshantan-Irdinmanhan carnivores are miacids (*Miacis*), canids (*Cynodictis*), and felids (*Eusmilus?*). Artiodactyls are diverse, and include dichobunids (*Dichobune*), entelodontids (*Eoentelodon*), anthracotheres (*Anthracokeryx*), leptomerycids (*Archaeomeryx*), and helohyids (*Gobiohyus*). The genus *Gobiohyus* is particularly well-known (Coombs and Coombs 1977). *Archaeomeryx* is the oldest ruminant artiodactyl in Asia, appar-ently an immigrant from North America (Webb and Taylor 1980).

Most of the neoplacentals in the Chinese Arshantan-Irdinmanhan faunas appear to be immigrants (or descendants of immigrants) from Europe and North America because they lack Asian antecedents. This massive immigration over-hauled the Chinese mammalian fauna in one stroke, converting it from paleo-placental-dominated to neoplacental rich.

Sharamurunian Mammals

At Ula Usu near Shara Murun in Nei Monggol, the Central Asiatic Expeditions of the American Museum of Natural History discovered Eocene mammals in 1922. The mammal assemblage from the Shara Murun Formation has given its name to the youngest middle Eocene LMA used in China. Sharamurunian mammals are widespread in China, occurring in Henan (Jiyuan, Hunshuihe, Chugouyu, and Wulidui formations), Nei Monggol (the type assemblage), Shanxi (Heti Formation), and Xinjiang ("lower green" formation), but they are not as abundant as those of Arshantan and Irdinmanhan age.

The general character of the Sharamurunian land mammals of China closely resembles that of the Arshantan-Irdinmanhan. However, there are less Shara-murunian paleoplacentals (pantodonts and dinoceratans were extinct by Shara-murunian time, and the other paleoplacental orders are much less diverse). New immigrants have appeared (such as cricetid and dipodid rodents), perisso-dactyl diversity is still very high (especially of amynodontids and brontotheri-

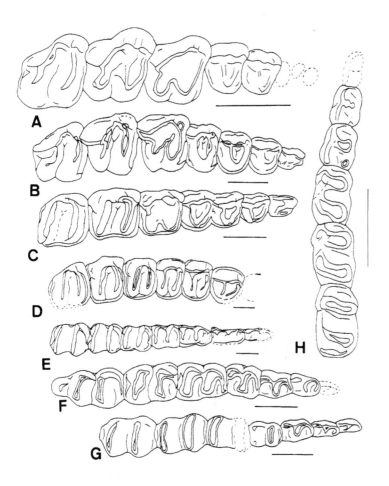

Figure 10-15 *Occlusal views of the cheek teeth of representative Arshantan-Irdinmanhan tapiroids (after Radinsky, 1965). A, H - Rhodopagus; B, F – Lophialetes; C, G – Teleolophus; D, E – Deperetella. A- D are upper teeth, E-H are lower teeth; scale bars = 1 cm.*

ids), and southern China has a rich anthracothere fauna during the Sharamurunian.

The anthracotheres from southern China may represent the first indication of the development of two zoogeographic zones in China during the Cenozoic. The difference is best seen by comparing the Shara Murun fauna of Nei Mong-gol, which has two species of anthracothere, and the approximately contemporaneous fauna of the Pondaung Sandstone in Burma (just south of Yunnan, China), with at least seven anthracothere species.

Ergilian Mammals

Fossil mammals from the Ergilian-Dzo Formation of Mongolia are the type fauna of the Ergilian LMA (Russell and Zhai 1987). Ergilian mammals were long considered to be of early Oligocene age, but are now technically of late Eocene age (see above). The mammalian faunas of the Ulan Gochu and Urtyn Obo (= Ardyn Obo) formations of Nei Monggol are the best-known Chinese mammalian assemblages of Ergilian age. Other assemblages of this age in China come from Guangxi (Naduo, Yongning, and Gongkang formations), Yunnan (unnamed strata in Xuanwei County, Xiaotun and Cajiachong Formations), Guizhou (Shinao Formation), Nei Monggol (Houldijin, Baron Sog, and Chaganbulage Formations), Shaanxi (Bailuyuan Formation), Shanxi (Baishuicun Formation), and Xinjiang (unnamed unit in the Hami basin) (see table 10-1).

The Ergilian mammal fauna resembles the earlier Arshantan, Irdinmanhan, and Sharamurunian mammal faunas in overall composition. Perissodactyls dominate, especially brontotheriids and amynodontids (the other perissodactyls are not very diverse, especially the tapiroids). Rodents are very diverse during the Ergilian, and include cylindrodontids, cricetids, ctenodactylids and dipodids. Lagomorphs are both rabbits (*Gobiolagus*) and pikas (*Procaprolagus*). Bats (vespertilionoids), shrews (soricids), and hedgehogs (erinaceids) first appear in China, as do marsupials. No Ergilian carnivores or hyaenodontid creodonts are known from the Chinese Ergilian, but they are known from nearby Ergilian assemblages in Mongolia. Mesonychids are rare in the Ergilian. Diverse Ergilian artiodactyls include enteledonts, anthracotheres, leptomerycids (*Miomeryx*), cervids, and gelocids (*Lophiomeryx*). The first true rhinoceroses (Rhinocerotidae) appear during the Ergilian, and the last anagalid (*Anagale*, ironically the first anagalid described, see figure 10-6) is Ergilian.

The Ergilian fauna represents the virtual completion of neoplacental hegemony. It also is in many ways a transitional fauna. Ergilian faunas are very similar to earlier Arshantan, Irdinmanhan, and Sharamurunian faunas, but contain the first representatives of the later, rather different mammalian faunas of the Oligocene.

Shandgolian Mammals

The early Oligocene LMA in China is the Shandgolian. Named from the Shandgol Formation in Mongolia, this "age" was considered middle Oligocene until realignment of the Eocene-Oligocene boundary (see above) placed it in the early Oligocene. The St. Jacques mammal locality in Nei Monggol has produced the most extensive Shandgolian mammal assemblage from China. The mammal fauna of the nearby Wulanbulage Formation (see figure 10-16) is equally diverse, as is the Ulantatal mammal fauna, also from Nei Monggol. Other Chinese Shandgolian mammals are also from northern China (Ningxia). This geographic restriction of Shandgolian mammals to northern China stands in stark contrast to the more widespread mammalian assemblages of the earlier Paleogene.

The Shandgolian mammal fauna is much more diverse than that of the earlier Ergilian. Groups that first appeared during the Ergilian are now much more abundant and more speciose: erinaceid insectivores; ochotonid lagomorphs; cylindrodontid, castorid, cricetid, and ctenodactylid rodents; canid carnivores; rhinocerotid perissodactyls; and ruminant artiodactyls. Perissodactyls, artiodactyls and rodents dominate the Shandgolian mammal faunas. Hyaenodontid creodonts are the only paleoplacentals left; they survived until the Miocene in Pakistan.

New rodents appeared and proliferated during the Shandgolian. These were the tsaganomyids, endemic to Asia, and the aplodontids and tachyoryctoids, immigrants from North America. The first Chinese talpids also appeared during the Shandgolian, as did the amphicyonids (bear dogs), mustelids, and

Figure 10-16 Outcrops of the Wulanbulage Formation in Nei Monggol (courtesy of R.J. Emry).

viverrids. Among the perissodactyls, indricothere hyracodontids reached truly gigantic size (see below). Ruminants (leptomerycids, cervids, and geolocids) dominate the artiodactyls, and didymoconids are more common than during the Ergilian.

Close contact of the Chinese Shandgolian mammals with contemporaries in Europe is indicated by a large number of shared Chinese-European genera at this time, including *Schizotherium, Eucricetodon, Ronzotherium, Hyaenodon, Entelodon, Bothriodon,* and *Lophiomeryx.*

Tabenbulukian Mammals

Like the Shandgolian mammals, Tabenbulukian mammal localities are restricted to northern China. These mammals are of late Oligocene age, and the LMA takes its name from the mammal assemblage found at Taben Buluk and Yindirte in western Gansu by the Swedish paleontologist Birger Bohlin in the 1930s, and first monographed by him (Bohlin, 1942, 1946). Bohlin also collected Tabenbulukian mammals at Sharagaltein Gol in Gansu and at Shih-ehr-ma-cheng, nearby. The Yikebulage Formation in Nei Monggol contains Tabenbulukian mammals, as do the "brown" and Taoshuyuanzi formations of Xinjiang.

Tabenbulukian mammals much resemble those of the Shandgolian in being mostly ochotonid lagomorphs, rodents, rhinocerotoid perissodactyls, and ruminant artiodactyls. Most Tabenbulukian insectivores are erinaceids; records of soricids and talpids are fragmentary. All the lagomorphs are ochotonids (*Desmatolagus, Sinolagomys*), and most of the rodents are tachyoryctoids (*Tachyoryctiodes*) and ctenodactylids (*Tataromys*). Sciurid?, tsaganomyid, castorid, cricetid, and dipodid (*Parasminthus*) rodents are less diverse. Carnivores are little known from the Chinese Tabenbulukian, but felids, amphicyonids, and mustelids are present in correlative units in Mongolia and Soviet Middle Asia. Hyaenodontid creodonts (*Hyaenodon*) persist, as do chalicothere and tapiroid perissodactyls.

Most perissodactyl diversity is in hyracodontids, especially the giant indricotheres. True rhinoceroses and the last amynodontid are present. Most artiodactyls are gelocids and cervids, but entelodonts and anthracotheres still are present, as are didymoconids. Most of the Tabenbulukian mammal fauna is endemic, but it shares a number of genera with the late Oligocene of Europe, including *Eucricetodon, Hyaenodon, Entelodon, Lophiomeryx,* and *Amphitragulus.*

Paleogene Rodent Evolution

Today, rodents are the most diverse and abundant terrestrial mammals. Their complex evolutionary history began during the late Paleocene when their first fossils are known in North America. The relationships of rodents to lagomorphs—the rabbits, hares, and pikas—have long been disputed. Linnaeus (1758) included rodents and lagomorphs in a single order, Glires. This grouping was maintained, with Rodentia and Lagomorpha as separate orders in a higher category (Cohort or Superorder Glires), at least until the 1960s.

Chinese Paleogene mammals have recently been used to bolster the concept Glires. These mammals are the eurymylids, especially Nongshanian *Heomys* (see figure 10-17), which are very similar to primitive rodents and lagomorphs. The eurymylids are usually placed in a separate order, Mixodontia (Sych 1971; Hartenberger 1996). Some workers, however, (C. Li and Yan 1979) have included the mixodonts in the Rodentia. Recent classifications divide Glires into Duplicidentata (lagomorphs) and Simplicidentata (rodents and mixodonts) (C. Li et al. 1987; Korth 1994; Meng et al. 1994; McKenna and Bell 1997). The mixodonts include the eurymylids, rhombomylids and the mimotonids and have gliriform lower teeth that place them close to the ancestry of Glires. Mimotonids (see figure 10-17) may even be ancestral lagomorphs (C. Li and Ting 1985).

The oldest Chinese rodents are of Eocene age. Unlike the Eocene rodent record of North America and Europe, ischyromyids are rare, and ctenodactyloids dominate the Eocene record of rodents in China (C. Li 1963; Dawson 1968; Dawson et al. 1984; Qi 1987). The first cricetid appeared in China during the middle Eocene (Tong 1992). The early Asian cricetids gave rise to the subsequent cricetid radiations in Europe and North America (B. Wang and Dawson 1994). A zapodid and the probable eomyid (or sciuravid) *Zelomys* also appeared in the Chinese Eocene (C. Li and Ting 1983; B. Wang and Li 1990). All Chinese Eocene rodents, except the ctenodactyloids, are probable immigrants from North America (Korth 1994).

Ctenodactyloids continue to dominate the Chinese rodent record during the Oligocene (B. Wang 1997). All other rodents were immigrants: cylindrodontids, aplodontids, zapodids and eomyids (Bohlin 1946; Rensberger and Li 1986; B. Wang 1987; B. Wang and Emry 1991). Castorids also first appeared in China during the Oligocene, but their origin is unclear, as is the case withthe tsaganomyids, an endemic Asian group of Oligocene rodents (Wood 1974).

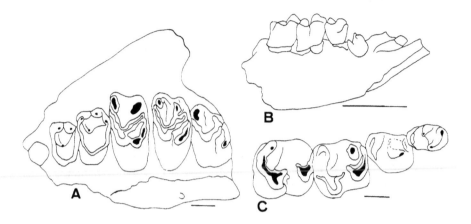

Figure 10-17 *Dentition of the mimotonid Mimotona wana: A – occlusal view of upper cheek teeth; B, C – lower jaw, lateral (B) and occlusal (C) views. After C. Li and Ting (1985). Scale bars = 1 cm.*

Indricothere Evolution

The evolution of the giant rhinoceroses (indricotheres) took place in Eurasia between the middle Eocene (Irdinmanhan) and the late Oligocene (Tabenbulukian). (Previous reports of the oldest indricothere *Forstercooperia* from North America [Radinsky 1967; Lucas et al. 1981; Lucas and Sobus 1989] are erroneous [Holbrook and Lucas 1997].) This evolution began with the Irdinmanhan appearance of *Forstercooperia*, a small-horse-sized hyracodontid that represents the first evolutionary step toward the giant indricotheres of the Oligocene. These first steps are seen in the facial skeleton where the nasal bones form a robust shelf that begins above the upper canines and a high, flattened eminence that terminates above the orbits, and a preorbital fossa on the maxillary bones runs parallel to most of the posterior tooth row. Thus began the evolutionary modification of the nasal and maxillary bones for the support of an elaborated muscular snout, which probably supported a short proboscis in the largest and most advanced indricotheres.

The succeeding evolution of indricotheres is an almost orthogenetic trend to produce the largest land-mammals of all time—*Paraceratherium*—5 m tall at the shoulder and weighing at least three tons (see figure 10-18). These giant rhinoceroses of the Oligocene are known from Chinese localities throughout the country; their two evolutionary antecedents, *Juxia* of the Sharamurunian

and Ergilian *Urtinotherium*, are known only from China. The giant indricotheres had only two large cropping, tusk-like incisors in each jaw and deeply
retracted nasal incisions to support huge snout muscles. They were treetop
browsers who roamed Eurasia (fossils are known from Serbia to Nei Monggol)
during the late Oligocene. Most of what we know about indricothere evolution
is based on the group's Chinese fossil record.

Paleogene Lower Vertebrates

After the terminal Cretaceous extinctions, which removed the dinosaurs, nonmarine lower vertebrates of the Cenozoic are mostly teleost fishes, lissamphibians, turtles, lizards, snakes, and crocodilians. The Chinese Paleogene has
yielded a modest and much understudied record of lower vertebrates. A comparison with the age-equivalent, much more extensive Paleogene lower vertebrate record in western North America reveals that this is a potentially rich and
unexplored area for future research in China.

No lissamphibians have been reported from the Chinese Paleogene; the oldest Chinese lissamphibians are of middle Miocene age (see chapter 11). Only
five teleost fishes of Paleogene age have been described from China, all of

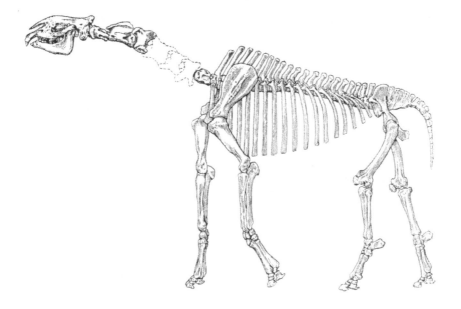

Figure 10-18   *The skeleton of the giant rhinoceros Paraceratherium is approximately 8
meters long (after Gromova 1954).*

Eocene age: *Knightia yuhanga* Liu; *Osteochilus hunanensis* Cheng; *Aoria lacus* Cheng; *Tungtingichthys gracilis* Liu, Liu & Tang; and *T. hsiawanpuensis* Cheng.

Most of the record of Chinese Paleogene lower vertebrates is of turtles (see figure 10-19). This record, however, is much less than is known of the Mesozoic, especially Cretaceous, turtles of China. Chinese Paleogene turtles (Ye 1994) are:

1. *Mongolemys australis* Ye, a dermatemydid from the Early Paleocene of the Nanxiong basin, Guangdong

2. Another species of *Mongolemys*, *M. turfanensis* Ye, from the Paleocene of Xinjiang

3. The dermatemydid *Hokouchelys chenshuensis* Ye, from the early Paleocene of the Nanxiong basin, Guangdong

4. *Adocus orientalis* Gilmore, a middle Eocene dermatemydid from Irdin Manha, Nei Monggol

5. *Anhuichelys siaochihensis,* an early Paleocene emydid from Anhui. Other Paleocene species of *Anhuichelys* are *A. tsienshanensis* Ye, also from Anhui, and *A. xinzhouensis* Chen from Hubei

6. *Isometremys lacuna* Chow & Ye, an emydid from the Eocene of Guangdong

7. Two other emydids from the Eocene Ulan Gochu Formation of Nei Monggol are ?*Palaeochelys elongata* and *Sharemys hemispherica,* both described by Gilmore (1931)

8. China's oldest tortoise (Testudininae) is *Sinohadrianus sichuanensis* Ping, from the middle Eocene of Henan

9. Younger Paleogene tortoises: *Kansuchelys chiayukuanensis* Ye from Gansu; *K. ovalis* Ye from Shanxi; *K. tsiyuanensis* Ye from Henan; *Testudo ulanensis* Gilmore and *T. sharanensis* Ye, both from Nei Monggol; and *T. yunnanensis* Ye, and *T. lunanensis* Ye, both from Yunnan

10. Five Paleogene species of the carettochelyid genus *Anosteira*: *A. mongoliensis* Gilmore from the middle Eocene of Nei Monggol, *A. manchuriana* Zangerl from the late Eocene of Liaoning, *A. maomingensis* Chow & Liu from the late Eocene of Guangdong, *A. shantungensis* Cheng from the late Eocene of Shandong, and *A. lingnanica* Young & Chow from the early Paleocene of Guangdong

11. Two species of the trionychid genus *Aspideretes: A. muyuensis* Li & Yefrom the early Eocene of Hubei and *A. impressus* Ye from the late Eocene of Guangdong

12. The trionychid genus *Amyda*, which still has living representatives in China, is known from four Paleogene species: *?A. linchuensis* Ye, from the early Eocene of Shandong, *A. neimenguensis* Ye and *A. johnsoni* Gilmore from the middle Eocene of Nei Monggol, and *A. gregaria* Gilmore from the late Eocene? of Nei Monggol

13. The trionychid *Platypeltis subcircularis* Chow & Ye from the middle Eocene of Henan

Virtually all Cenozoic fossil lizards from China are of Paleogene age. *Arretosaurus ornatus* Gilmore is the sole basis for the family Arretosauridae, a poorly known group of Iguania from the upper Eocene of Nei Monggol. Chinese

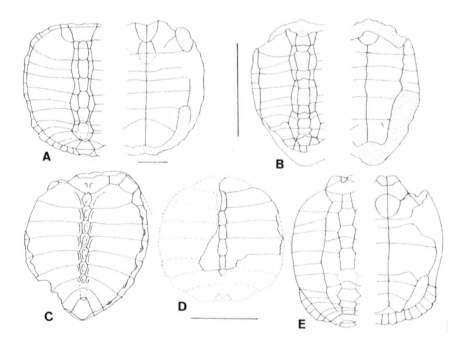

Figure 10-19 *Selected Chinese Paleogene turtles: A – Anhuichelys siaoshinensis, carapace (left) and plastron (right); B – Sinohadrianus sichuanensis, carapace (left) and plastron (right); C – Anosteira maomingensis, carapace; D – Platypeltis subcircularis, carapace; E – Kansuchelys tsiyuanensis, carapace (left) and plastron (right). Bar scales = 8 cm. After Ye (1994).*

Paleogene agamids are three species of *Tinosaurus*, a poorly known but widely distributed Northern Hemisphere form of the Eocene: *T. asiaticus* Gilmore from the middle Eocene of Nei Monggol, *T. lushihensis* Dong from the middle Eocene of Henan, and *T. doumuensis* Hou from the Paleocene of Anhui.

Only one Paleogene chamaeleonid has been reported from China: *Anguingosaurus brevicephalus* Hou from the Paleocene of Anhui. Known from a nearly complete skull and lower jaws whether or not *Angingosaurus* actually is a chamaeleonid is uncertain (Estes 1983).

A frontal bone from the middle Eocene of Nei Monggol belongs to an anguid and has been referred to *Glyptosaurus* by Gilmore (1943), as the type of the species *Helodermoides mongoliensis* Sullivan and most recently as *Placosaurus mongoliensis* by Estes (1983). Other anguid material has been reported from the Eocene of Henan (Chow 1957).

Other Chinese Paleogene lizards are very poorly known. The "agamid" *Agama sinensis* Hou from the Paleocene of Anhui is known from only part of a maxillary and may be a *nomen dubium* or a specimen of *Mimeosaurus* (Estes 1983). *Anhuisaurus huainanensis* Hou is known from very poorly preserved skulls, lower jaws, and vertebrae from the Paleocene of Anhui. Originally considered to be an agamid (L. Hou 1974), Estes (1983), argued that *Anhuisaurus* is too poorly known to be assigned with certainty to any lacertilian family. Estes (1983) offered a similar view of the affinities of *Changjiangosaurus huananensis* Hou from the Paleocene of Anhui. This taxon is known only from incomplete lower jaws and part of a quadrate. The lower jaw is unique among lizards in having a flange that projects posteriorly from the angular bone. In this feature and others, *Changjiangosaurus* resembles *Qianshanosaurus huangpuensis*, also described by L. Hou (1974) from the Paleocene of Anhui. L. Hou (1974) considered *Qianshanosaurus* an iguanid, and L. Hou (1976) created the new family Changjiangidae for *Changjiangosaurus*. Estes (1983), however, considered both taxa Lacertilia, *incertae sedis*. No Paleogene snakes have been described from China.

Crocodilia have been studied more than Chinese Paleogene Squamata. Four subfamilies are represented: crocodylines, alligatorines, pristichampsines and thoracosaurines. *Asiatosuchus* from the Paleocene-Eocene of Guangdong, and Nei Monggol (Mook 1940; Young 1964b) and *Dzungarisuchus* from the late Eocene of Xinjiang (Dong 1974) are the crocodylines. Neither taxon is well known; fossils are confined to jaw fragments and isolated postcrania. *Lianghusuchus* from the Eocene of Hunan is equally poorly known from a few skull fragments, isolated vertebral, and scutes.

Eoalligator is the Chinese Paleogene alligatorine (Young 1964b). It is known from skull and jaw fragments found in Paleocene strata of Anhui and Guangdong. The best known Chinese Paleogene crocodilian is *Planocrania* (see figure

10-20), a pristichampsine known from nearly complete skulls collected in the Paleocene-Eocene of Guangdong and Hunan (J. Li 1976, 1984). Pristichampsines are unique among crocodilians in their serrated teeth, which are also known from the Eocene of Henan (Chow et al. 1973a).

Thoracosaurines were marine crocodilians with extremely long rostra. Two have been reported from eastern China. *Tienosuchus hsiangi* Young is based on a single tooth and some postcrania from the Eocene of Hunan, whereas a partial skull and lower jaw are the holotype of *Tomistoma petrolica* Ye from the Eocene of Guangdong. *Wanosuchus atresus* Zhang is based on a crocodilian lower jaw from the Paleocene of Anhui and is of uncertain affinities

Paleogene Birds

China's fossil record of Paleogene birds is very limited and much less extensive than the record in nearby Mongolia and Kazakstan (Kurochkin 1976). No

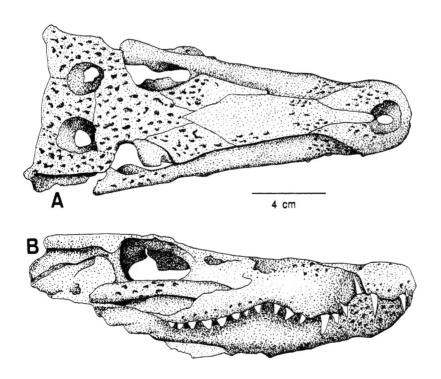

Figure 10-20 *Skull of the crocodilian Planocrania, dorsal (A) and lateral (B) views (after J. Li 1976).*

Paleocene or Oligocene bird fossils have been described from China; the Paleogene avian record is wholly Eocene and almost entirely from Nei Monggol.

The Middle Eocene Irdin Manha Formation at Ulan Shireh in Nei Monggol yielded a falcon coracoid and a femur of a numidid that Wetmore (1934) named *Telecrex grangeri*. This is the oldest Asian numidid. Many specimens of the crane *Eogrus aeola* were also described from the Irdin Manha Formation at various localities in Nei Monggol (Wetmore 1934; Kurochkin 1976).

Approximately age-equivalent birds are known from Henan, Hubei, and Xinjiang. The middle Eocene Yuhuangding Formation at Xichuan, Henan, yielded a single tibiotarsus of a large, flightless diatrymiform bird, *Zhongyuanus xichuanensis* Hou. The upper Eocene Lizhuang Formation in Henan bore a very small threskiornithid, *Minggangia changgouensis* Hou. The lower Eocene Yangxi Formation in Hubei yielded the skull and partial skeleton of *Songzia heidangkouensis* Hou, sole representative of a new subfamily, Songzidae. In Xinjiang, the ciconid *Eociconia sangequanensis* Hou is known from the Yixibaila Formation of middle Eocene age, and an indeterminate bird has been reported from the late Eocene of the Junggur Basin (Zhou et al. 1982).

Paleoplacentals and Neoplacentals

Osborn (1894) distinguished two groups of placental mammals, the Mesoplacentalia and the Cenoplacentalia. Osborn and Earle (1895: 3–4) further noted that "the difference between these two groups consists mainly in the lower state of evolution and apparent incapacity for higher development exhibited by the mesoplacentals in contrast with the capacity for rapid development shown by the cenoplacentals." They identified "amblypods" (pantodonts + uintatheres), "condylarths," creodonts, tillodonts, insectivores, and "lemuroid" primates as mesoplacentals and proboscideans, artiodactyls, perissodactyls, carnivores, rodents, and "anthropoid" primates as cenoplacentals. Because the taxa Mesoplacentalia and Cenoplacentalia do not refer to monophyletic groups, they have never been used by paleomammalogists since Osborn. However, these concepts may have some utility when stripped of their formal taxonomic meaning. I propose to recast them as the terms paleoplacentals and neoplacentals to refer to two distinct, adaptive radiations of eutherians.

Thus, Paleocene-Eocene eutherians present us with a broad dichotomy into which most eutherian orders are readily placed (see figure 10-21). Paleoplacentals had their ancestry during the Late Cretaceous or early Paleocene and were mostly evolutionary dead ends that did not provide ancestry to neoplacentals. Paleoplacentals had lower encephalization quotients and more primitive brains than neoplacentals, and more generalized limb structures. Although they con-

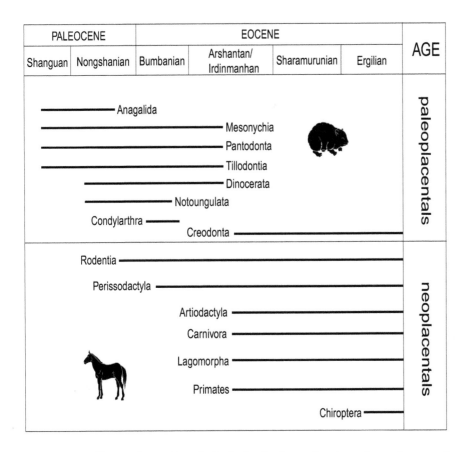

PALEOCENE		EOCENE				AGE
Shanguan	Nongshanian	Bumbanian	Arshantan/ Irdinmanhan	Sharamurunian	Ergilian	

Figure 10-21 *Temporal succession of principal paleoplacental and neoplacental mammal orders in the Chinese Paleogene.*

verged on many dental structures of neoplacentals, they evolved these dentitions from different, mostly zalambdodont, starting points.

Neoplacentals originated during the late Paleocene-Eocene and encompass most of the extant mammalian orders (see figure 10-21). Neoplacentals relatively higher encephalization quotients and more sophisticated brains, specialized limb structures and more "advanced" dentitions have usually been thought to have given them a competitive edge over contemporaneous paleoplacentals. This is supposedly why neoplacentals survived the Eocene and flourished, whereas paleoplacentals did not. Yet, paleoplacentals and neoplacentals coexisted throughout the Eocene for some 20 million years or more, suggesting a complex and prolonged replacement. Indeed, the Chine record of Paleogene mammals well documents the Eocene replacement of paleoplacentals by neoplacentals.

Chapter 11

Miocene-Pliocene

The early-middle Neogene encompasses two epochs, Miocene and Pliocene, and lasted about 23 million years, from 24 Ma to 1.8 Ma (Berggren et al. 1995; Berggren 1998). China has a rich and extensive fossil record of Miocene-Pliocene vertebrates, dominated by mammals. Like the Paleogene record, this record is most easily reviewed by organizing it into a succession of land-mammal "ages." It documents a complex pattern of endemism overlain by immigration events from Africa, southern Asia (the Indo-Pakistani subcontinent), Europe and North America. Questions of paleozoogeography thus dominate analysis of China's Miocene-Pliocene mammals.

Miocene-Pliocene strata have a very broad distribution in China (see figure 11-1) and are particularly widely exposed in the northern and western portions of the country. These rocks are mostly relatively coarse-grained and oxidized fluvial deposits. Miocene-Pliocene mafic volcanic rocks have a limited distribution in eastern China.

Miocene-Pliocene Vertebrate-Bearing Strata

During the Miocene-Pliocene, fluvial and lacustrine deposition took place across China in several large basins with which we are already familiar and in numerous smaller basins. Limited epicontinental seas of the Paleogene disappeared completely from the Chinese mainland during the Miocene-Pliocene. Further uplift of the Himalayas and Tibetan Plateau altered the Chinese climate so that western China was relatively dry, whereas eastern China had a much wetter, monsoonal climate. The effects of this climate are reflected by Miocene-Pliocene nonmarine sediments in China (see figure 11-2). Those in western China and as far east as the Ordos basin are fluvial and lacustrine red beds with extensive paleosols. Those of the eastern China basins (Jianghan, north China, and Songlia basins, etc.) contain variegated sequences of mudstone and sandstone and locally, are coal bearing. The Miocene-Pliocene was a time of active compressional tectonism in China (Himalayan orogeny), and this is reflected by the extremely thick (some more than 4000 m) accumulations of nonmarine strata in many of the Miocene-Pliocene sedimentary basins.

It is beyond the scope of this book to review all the Miocene-Pliocene vertebrate-bearing deposits of China (see figure 11-1). Instead, the sequence of vertebrate-bearing deposits in the Ordos basin—certainly the classic section for Chinese Miocene-Pliocene vertebrates—can be considered representative.

Figure 11-1 *Distribution of Neogene rocks and Neogene vertebrate fossil localities of China (after Qiu 1990): 1 – Lanzhou; 2 – Suosuoquan; 3 – Xiejia; 4 – Anjihai; 5 – Wafongyongzi; 6 – Zhangjiaping; 7 – Jiaozigou; 8 – Taben Buluk; 9 – Sihong; 10 – Fangshan; 11 – Shanwang; 12 – Xiaodian; 13 – Tongxin; 14 – Koujiacun; 15 – Quantougou; 16 – Jiulongkou; 17 – Lengshuigou; 18 – Tunggur; 19 – Halamagai; 20 – Erlanggang; 21 – Lingyanshan; 22 – Xiaolongtan; 23 – Qaidam; 24 – Amuwusu; 25 Bahe; 26 – Hezheng; 27 – Biru; 28 – Wuzhong; 29 – Zhongning; 30 – Wangdaifuliang; 31 – Baode; 32 – Lufeng; 33 – Ertemte; 34 – Gaozhuang; 35 – Jingle; 36 – Youhe; 37 – Dongyaozitou; 38 – Shagou.*

During the Miocene-Pliocene, the Ordos basin was a north-south graben system divided by the Ordos Plateau (see figure 11-3). The grabens were actually formed during the Late Cretaceous-Paleogene, so most of their thick basin fill is older than Miocene-Pliocene. The Miocene-Pliocene succession is best exposed along the Huang He and Fen He drainages, where a total thickness of 3000–4000 m of red-bed sandstones, conglomerates, and mudstones are exposed that yield a succession of vertebrate faunas of Miocene and Pliocene age.

Also worthy of mention is the Shanwang vertebrate-bearing deposit of central Shandong. This early Miocene locality is in a lacustrine deposit of paper shales and marls reminiscent of the Green River Formation fossil beds of Wyoming (X. Wang 1996). A wealth of fossil vertebrates, insects and plants come from these

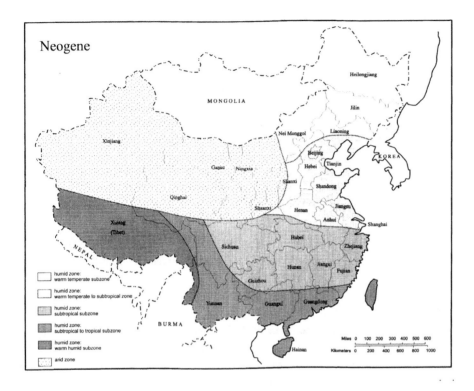

Figure 11-2 *This map of the Neogene climatic belts of China shows the arid zone caused largely by the uplift of the Himalayan Mountains (after H. Wang 1985).*

strata, which are part of a 1900 m thick succession of MiocenePliocene lacustrine shales and fluvial sandstones deposited in the north China basin.

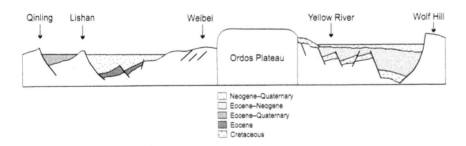

Figure 11-3 *The Neogene graben system of the Ordos basin is representative of the extensional basins of the Chinese Neogene.*

Miocene-Pliocene Land-Mammal "Ages"

As is the case with the Paleogene, Chinese Miocene-Pliocene land-mammal assemblages provide the basis for a sequence of Land-Mammal "Ages" (LMAs) by which the Chinese Miocene-Pliocene vertebrate record can be organized. This organization (Chiu et al. 1979; C. Li et al. 1984; Qiu and Qiu 1990 1995; Tong et al. 1995) identifies ten LMAs of Miocene-Pleistocene age, five corresponding to the Miocene and five to the Pliocene-Pleistocene (see figure 11-4).

Xiejian Mammals

The oldest Miocene fossil vertebrate assemblages from China are of Xiejian age. These are the Xiejia fauna of Qinghai, its correlatives, and some assemblages that may be slightly older.

The best known of these older assemblages is the Lanzhou fauna of Gansu (Qiu and Gu 1988). The assemblage is only of small mammals—the brachyericine shrew *Metaxallerix* and the rodents *Tataromys, Leptotataromys,* and *Tsaganomys*. These taxa are either Tabenbulukian (late Oligocene) in age elsewhere or endemic to Lanzhou (*Metaxallerix*). Qiu (1990) suggests the stage of evolution of the Lanzhou representatives of these taxa indicates they are younger than Tabenbulukian, but this is uncertain. A similar stage-of-evolution argument is used by Qiu (1990) to assign to the early Xiejian a small mammal fauna that includes the rabbit *Sinolagomys* and the mole *Tachyoryctoides* from the Suosuoguan Formation of the Junggur basin in Xinjiang (Tong 1987). Problems thus exist in China, as elsewhere in Eurasia, differentiating earliest Miocene mammals from latest Oligocene mammals. There simply is no easily identified event in mammalian evolution with which to benchmark the Paleogene-Neogene boundary.

The characteristic Xiejian fauna from the Xiejia Formation of the Xining basin in Qinghai (see figure 11-1) is mostly of small mammals. These mammals are a rabbit (*Sinolagomys*), rodents (sciurid, *Eucricetodon*, "*Plesiosminthus*," *Tataromys*), a mole (*Tachyoryctoides*), a mustelid, a rhinoceros (*Brachypotherium*) and a bovid (*Oioceros?*) (C. Li and Qiu 1980). The taxa in this fauna include *Eucricetodon* and *Brachypotherium*. The lack of characteristic Oligocene taxa and the stage of evolution of the others support correlation with early Miocene mammals of Europe. Correlative mammal faunas are known from Anjihai in the Junggur basin of Xinjiang (Chiu 1965 1973) and Wafangyingzi in Hebei (C. Li 1962).

Epoch		land-mammal "age"
Pleistocene		Salawusuan
		Zhoukoudianian
Pliocene	Late	Nihewanian
		Youhean
	Early	Jinglean
Miocene	Late	Baodean
		Bahean
	Middle	Tunggurian
	Early	Shanwangian
		Xiejian

Figure 11-4 *Neogene LMAs of China divides the Miocene, Pliocene, and Pleistocene into 10 intervals.*

The Paleogene-Neogene (Oligocene-Miocene) transition in the northern continents was a profound and complex evolutionary turnover in which the warmer, wetter, densely forested Paleogene world gave way to the cooler, drier grassland of the Miocene-Pliocene world. Primitive mammalian herbivores of the Paleogene disappeared, and new, hypsodont herbivores, especially among the Rodentia and Artiodactyla, evolved. No one extinction or evolutionary event marks this transition. We see this clearly in the earliest Miocene and Xie-jian mammals of China. They are mostly Oligocene holdovers, only slightly advanced evolutionarily over their Paleogene ancestors. They predate a massive immigration and mixing of the European and Asian mammal faunas during the Shanwangian

Shanwangian Mammals

Shanwang is a diatomaceous lake deposit about 20 km east of Lingu in Shandong. This deposit is one of the great Cenozoic Lagerstätte, comparable to the Solnhofen limestones of Bavaria in the diversity and quality of its fossil record, including plants, insects, and vertebrates. The nearly two dozen mammalian species (Yan et al. 1983) recovered from the Shanwang diatomites are the basis of the Shanwangian LMA.

A mammalian fauna older than the Shanwang, the Zhangjiaping fauna of Gansu, has also been included in the Shanwangian (Qiu 1990). This fauna is from coarse-grained sandstones of the Xianshuihe Formation (Young and Bien 1936), 30 km north of Lanzhou. Qiu (1990) provided preliminary identifications of the mole *Tachyoryctoides*, the creodont *Hyaenodon*, tusk fragments of a proboscidean, a chalicothere (*Phyllotillon?*), and two rhinoceroses, *Aprotodon* and a large indricothere. Particularly significant are the proboscidean fossils, which mark the first appearance of proboscideans in China as immigrants from Europe. This *Gomphotherium* datum event (Madden and Van Couvering 1976; Tassy 1990, 1996; Lucas and Bendukidze 1997) took place about 19 Ma in China and can be used to define the beginning of the Shanwangian.

Other faunas of early Shanwangian age—correlatives of the Zhanjiapang fauna—are from Jiaozigou in Gansu and from near Taben-buluk, also in Gansu (Bohlin 1946; Qiu 1990). These faunas also include proboscidean remains. The fauna near Taben-buluk should not be confused with the classic Taben-buluk faunas of late Oligocene age (Conroy and Bown 1974). The Shanwangian fauna from the Taben-buluk area comes from a thick and structurally complex section that encompasses localities near Hsishui, Tienchiangtziku, and Yindirte. The fossils are of a proboscidean (aff. *Gomphotherium*), the rodent *Sayimys*, cervids, bovids, rhinocerotids, and the anthropoid primate "Kansupithecus" (a *nomen nudum*).

Bohlin (1946) proposed the generic name "Kansupithecus" without a species name (so the genus name is a *nomen nudum*) for an edentulous lower jaw fragment and isolated lower molar fragment. The specimen appears to belong to an anthropoid primate, one of the earliest known from Asia, but it is too poorly preserved to allow more definite conclusions.

The next faunal assemblage of Shanwangian age is the Sihong fauna, which is thought to be intermediate in age between the Zhanjiaping and Shanwang faunas. The Sihong fauna comes from conglomeratic sandstones of the Xiaocaowan Formation in Jiangsu, 150 km northwest of Nanjing. C. Li et al. (1983) listed 65 vertebrate taxa from the fauna, 47 of which are mammals. But the list has been considerably reduced. Important taxa include a wide diversity of rodents, such as aplodontids (*Ansomys*), sciurids (*Parapetaurista, Shuanggouia, Eutamias, Plesiosciuris*), castorids (*Youngofiber*), cricetids (*Sayimys, Diatomys, Megacricetodon, Democricetodon, Spanocricetodon*), glirids (*Microdyromys*), anthropoid primates (*Dinysopithecus, Platodontopithecus*), carnivores (*Semigenetta, Pseudaelurus*), the horse *Anchitherium*, the tragulid *Dorcatherium*, the palaeomerycid *Lagomeryx*, and the cervid *Stephanocemas*.

The appearance of *Anchitherium* (see figure 11-5) represents an important immigration from North America. Some other taxa (*Democricetodon, Megacricetodon, Dorcatherium* and *Lagomeryx*) are also known from Europe, and thus indicate a cosmopolitanism of Eurasian mammal faunas during the middle Shanwangian. Another *Anchitherium*-bearing locality correlative to the Sihong fauna is at Fangshan south of Nanjing (Chow and Hu 1956; C. Li 1977b). This fauna comes from a horizon above a basalt with a radiometric age of about 14 Ma.

The Shanwang mammal fauna, characteristic of the LMA, lacks *Anchitherium* and is slightly younger than the Sihong fauna. It contains a bat (*Shanwangia*), aplodontid (*Ansomys*), sciurid (*Plesiosciurus*), petauristid (*Meinia*) and sciurid (*Diatomys*) rodents, carnivores (*Hemicyon, Amphicyon, Ursavus*), indeterminate proboscideans, a tapir ("*Palaeotapirus*," based on a metapodial previously misidentified as *Anchitherium*), rhinoceroses (*Plesiaceratherium, Brachypotherium*), a chalicothere (*Chalicotherium*), a pig, and palaeomerycids (*Palaeomeryx, Lagomeryx*). Most interesting here are the hemicyonids, or "beardogs." These large, probable scavengers, are also known from Europe (*Amphicyon*) and North America (*Hemicyon*) and clearly underscore the evidence of

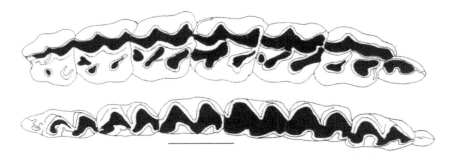

Figure 11-5 *Occlusal view of the cheek teeth of Anchitherium, upper premolars-molars (above) and lower premolars-molars (below). Scale bar = 2 cm (after Handbook of Chinese Vertebrate Fossils Editorial Group 1979).*

cosmopolitanism seen in the earlier Sihong fauna. *Hemicyon* has also been found near Xiaodian in Hubei. At this site, a supposed molar of a macaque monkey (Gu 1980) probably was misidentified and actually belongs to a fossil pig (Qiu 1990).

The Shanwangian mammal faunas of China thus indicate great cosmopolitanism of the Eurasian mammal fauna with some connections to North America. The arrival of proboscideans in China from Europe marks the beginning of the Shanwangian. The emigration of the horse *Anchitherium* from North America to Asia is an important datum during the middle Shanwangian.

Tunggurian Mammals

Tunggur is one of the most famous Miocene-Pliocene mammal localities in China. Located in Nei Monggol, fossil mammals were discovered at Tunggur in 1928 by the Central Asiatic Expeditions of the American Museum of Natural History. The shovel-tusker proboscidean *Platybelodon* (see figure 11-6) found at Tunggur caused quite a sensation. The Tunggur mammal fauna now is the basis of the LMA that represents most of the middle Miocene. Nevertheless, like the older Shanwang fauna, the Tunggur mammal fauna is the youngest mammal assemblage in the LMA bearing its name.

The oldest Tunggurian mammal assemblage is the Tongxin fauna from Ningxia (Guan et al. 1981; Guan 1988). This assemblage includes the monkey *Pliopithecus*, ochotonids, the proboscidean *Platybelodon*, the pig *Kubanochoerus*, carnivores (*Percrocuta*, *Sansanosmilus*, and *Gobicyon*), a chalicothere, two types of rhinoceros, and artiodactyls (*Eotragus*). Key here is the endemism (to China) of these taxa, in contrast to the cosmopolitanism of the preceding Shanwangian.

Platybelodon is the Tunggurian index taxon, and the listriodont pig *Kubanochoerus* (see figure 11-7) also is characteristic. Possible correlatives of the Tongxin fauna include the Koujiacun fauna in Shaanxi (Liu and Li 1963; M. Zhou 1978), the Quantougou locality in Gansu and possibly the Jiulongkou fauna of Hebei. Liulongkou, however, lacks *Platybelodon* and *Kubanochoerus* as well as many other taxa of the Tongxin fauna (G. Chen and Wu 1976), so its correlation is somewhat problematic. Fossil mammals from the Chetougou Formation in the Xining basin of Qinghai (Qiu et al. 1981) may also be early Tunggurian.

The Lengshuigou fauna comes from the formation of the same name at a locality 20 km northeast of Xian in Shaanxi. The presence of *Listriodon*, not *Kubanochoerus*, and other advanced aspects of the fauna suggest it is slightly younger than the Tongxin fauna, but older than the Tunggur type fauna.

Figure 11-6 *Note the long, scoop-like tusks on this skeleton of the shovel-tusker* Platybelodon *from Tongxin, Ningxia.*

Indeed, the proboscidean genus *Selenolophodon* is based on nothing more than a species of *Platybelodon* morphologically intermediate between the Tongxin species *P. tongxinensis* and the Tunggur species *P. grangeri* (Ye and Jia 1986; Tobien et al. 1986). Other Lengshuigou mammals are the rhinoceros *Hispanotherium* and the artiodactyls *Palaeomeryx, Palaeotragus,* and "*Oioceros.*"

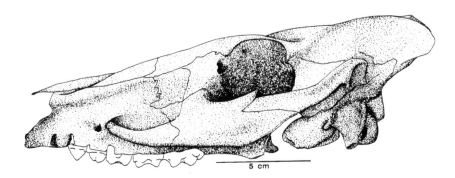

5 cm

Figure 11-7 *Skull of the listriodont pig* Kubanochoerus *in lateral view (after* Handbook of Chinese Vertebrate Fossils Editorial Group *1979).*

The most extensive Tunggurian mammal assemblage (59 taxa) is, not surprisingly, from Tunggur. It includes erinaceid, talpid, and soricid insectivores, a chiropteran, a great diversity of aplodontid, sciurid, castorid, eomyid, glirid, zapodid, dipodid, and cricetid rodents, lagomorphs (*Alloptox*, *Bellatona*), a great diversity of carnivores (*Gobicyon*, hemicyonids, many mustelids), proboscideans (*Platybelodon*, *Serridentinus*, *Zygolophodon*), the horse *Anchitherium*, the chalicothere *Chalicotherium*, rhinocerotids (*Acerorhinus* and *Hispanotherium*), the listriodonts *Kubanochoerus* and *Listriodon*, and a variety of ruminant artiodactyls (*Dicrocerus*, *Micromeryx*, *Lagomeryx*, *Euprox*, *Palaeotragus*, "*Oioceros*") (Osborn and Granger 1932; Colbert 1936, 1939; Qiu et al. 1988; Qiu 1996; Cerdeño 1996). This fauna is mostly Asian endemics, well demonstrating the endemism characteristic of the Tunggurian. However, *Zygolophodon* and *Serridentinus* are immigrants from Europe (Tobien et al. 1984).

The Halamagai Formation of Xinjiang (Tong 1987), Erlanggang locality in Hubei (Yan 1979), the Lingyanshan locality in Jiangsu (Bi et al. 1977) and the Xiaolongtan locality of Yunnan (G. Dong 1987) yield mammals correlative with the Tunggur fauna. The Xiaolongtan fauna may actually be slightly younger (G. Dong 1987). It contains fossils of the early orangutan *Sivapithecus*.

Bahean Mammals

The Bahean LMA begins with the first appearance in China of the horse *Hipparion* (see figure 11-8), an immigrant from North America. The importance of this immigration event to understanding the Miocene-Pliocene mammalian record of Eurasia cannot be overstated and is discussed at greater length below. The Bahe Formation along the Bahe River in Shaanxi produces the fossil mammal assemblage typical of the LMA.

The Qaidam (Tsaidam) mammalian fauna from the upper Youshashan Formation in western Gansu is the oldest, Chinese *Hipparion*-bearing fauna (Bohlin 1937; C. Li et al. 1984). It shows a mixture of survivors of the Tunggurian (e.g., *Stephanocemas*, *Lagomeryx*, *Dicrocerus*) and mammals characteristic of the Bahean mammal fauna (*Hipparion*, *Tetralophodon*, *Ictitherium*). Unique taxa are the bovids *Qurlignoria*, *Olonbulukia*, and *Tsaidamotherium*. *Hipparion* teeth occur at Amuwusu, 200 km southwest of Tunggur together with some typical Tunggurian mammals, and this is at least as old an occurrence as the Qaidam fauna (Qiu 1990).

The Bahe Formation fauna is the most diverse Bahean mammal assemblage. It includes a hedgehog (*Erinaceus*), carnivores (*Miomachairodus*, *Dinocrocuta*), the proboscidean *Tetralophodon*, primitive *Hipparion*, rhinoceroses (*Acerorhinus* and *Dicerorhinus*), and a diversity of artiodactyls (*Chleuastochoerus*, *Palaeotragus*,

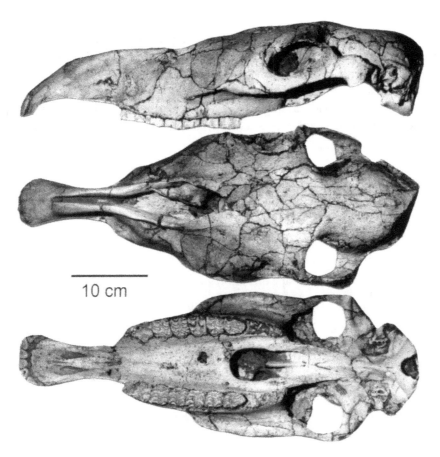

Figure 11-8 *The hipparionine horse Proboscidipparion sinense from the Plio-Pleistocene of Henan, lateral (top), dorsal (middle), and ventral (bottom) views. Some paleontologists consider Proboscidipparion to be a subgenus of Hipparion.*

Samotherium, Gazella, and *Shaanxispira*) (Liu et al. 1978; Qiu 1990). This fauna thus is very similar to the classic *Hipparion* fauna of the younger Baodean LMA. Bahe fauna correlatives are also known from Gansu, Tibet, Ningxia, and Shaanxi (G. Chen 1977; Qiu et al. 1987, 1988; Qiu 1990; Qiu and Qiu 1995). Fossils of *Hipparion, Dinocrocuta,* giraffids, gazelles, and cervids characterize these early *Hipparion* faunas.

Baodean Mammals

The classic *Hipparion* fauna of northern China is well represented by the Baode fauna of Shaanxi. First discovered by the Swedish geologist Johan Gunnar

Andersson in 1916 (see chapter 2), the *Hipparion* fauna of northern China is widespread in red beds of late Miocene age, especially on the so-called "great loess plateau" (see figure 11-1). The fossils provided many of the "dragon bones" of the Chinese pharmacopeia and thus include most of the first Chinese vertebrate fossils studied in the West (see chapter 2).

Rather than provide a single generic list of this rich and diverse fauna (which, incidentally, badly needs revision), the key elements can be reviewed as follows:

1. Baodean insectivores are of limited diversity, either erinaceids, talpids, or soricids.

2. Ochotonids (*Ochotonoides, Ochotona*) are the common lagomorphs.

3. Rodent diversity is very high, especially of castorids (*Sinocastor*), alactagids (*Paralactaga*), and cricetids (*Heterosminthus*).

4. Primates include the ape *Sivapithecus*.

5. Carnivores are very diverse, especially musteloids and ictitheres.

6. Mastodonts are *Tetralophodon, Gomphotherium, Anancus* and *Stegotetrabelodon*, all descendants of earlier immigrant mastodonts.

7. *Hipparion*, of course, is very abundant and diverse (Bernor et al. 1990).

8. Chalicotheres are rare, but the acerathere rhinoceroses (*Chilotherium, Aceratherium, "Diceratherium," Sinotherium*) are diverse and abundant (see figure 11-9).

9. Pigs are present (*Chleuastochoerus, Propotamochoerus*).

10. Ruminant artiodactyl diversity is very high, especially of giraffids, cervids, and gazelles.

Highly significant is the fact that the *Hipparion* fauna is also known from southern China. The most famous locality is the "*Ramapithecus*" site at Lufeng in Yunnan. Here, the fauna includes rhizomyid rodents (*Brachyrhizomys*); the rabbit *Alilepus*; the primates *Sinoadapis, Lufengopithecus,* and *Sivapithecus*; a variety of carnivores (*Ursavus, Indarctos, Ictitherium, Machairodus,* etc.); and the ruminant artiodactyls *Dorcabune* and *Yunnanotherium. Lufengopithecus* is one of the oldest fossil apes, and its occurrence in Yunnan is the only Chinese record. Undescribed elements of the fauna include rodents, tapirs, chilotheres, *Hipparion*, suids and bovids (Zhang et al. 1981).

Andersson also discovered the youngest Baodean mammal assemblage, at Ertemte in Nei Monggol, in 1919. Schlosser (1924) first described the assemblage. About 51 mammal taxa are now known (Fahlbusch et al. 1983; Storch

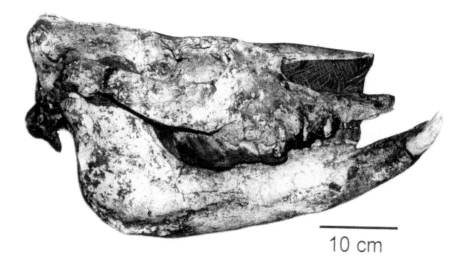

10 cm

Figure 11-9 *Lateral view of the skull and lower jaw of the hornless rhinoceros (acerathere) Chilotherium, a characteristic Baodean mammal.*

and Qiu 1983; Qiu 1985, 1987; Fahlbusch 1987; Storch 1987), and the overall character of the fauna is very similar to that just outlined. Stage of evolution, however, suggests the Ertemte fauna is slightly younger than the Baode fauna.

The Baodean thus begins with the immigration of *Hipparion* from North America and encompasses the time of the classic *Hipparion* fauna of northern China.

Jinglean Mammals

Fossil mammals from near Hefeng, Jingle County, Shanxi are the type fauna of the Jinglean LMA. This fauna is of low diversity, but significantly includes the canid *Nyctereutes* (see figure 11-10), an immigrant from North America. Other mammals are the carnivore *Metailurus*; the cricetid rodents *Chardinomys*, *Prosipheneus* and *Ungaromys*; a mammoth ("*Elephas*"); rhinoceros; *Hipparion*; a cervid; and gazelles (*Gazella*, *Antilospira*).

A much better known mammalian fauna of Jinglean age is that from the Yushe basin in eastern Shanxi. The mammal assemblages here are from the Gaozhuang Formation. Immigrants from North America in the Gaozhuang fauna are the camelid *Paracamelus* and the canid *Nyctereutes*. The fauna also includes large mammals evolutionarily advanced over those of Baodean age, including the mastodonts *Mammut* and *Stegodon*, and the carnivores *Chasmoporthetes*, *Pliohyaena*,

Agriotherium, and *Ursus.* Hipparionine horses still are common (*Hipparion, Proboscidipparion*), as are cervids (*Paracervulus*). The small mammal fauna is extensive; it includes talpid, soricid, and erinaceid insectivores; ochotonid lagomorphs; and a large number of rodent taxa (Flynn et al. 1991).

Recent detailed study of the Jinglean mammal succession in the Yushe basin by Flynn et al. (1991, 1997) identifies a complex succession of immigration events to China from North America, Africa, Europe, and northern and southern Asia (see figure 11-11). Mammalian turnover events in this succession appear to correspond to global climatic shifts. An important early Pliocene interchange between North America and Asia (see figure 11-12) seems to be the most significant biotic event affecting mammal evolution in the Yushe basin.

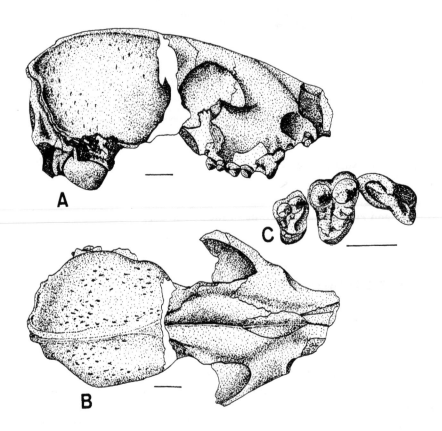

Figure 11-10 *Skull of the canid Nyctereutes; lateral (A) and dorsal (B) views of skull, and occlusal view of upper cheek teeth (C). Scale bars = 1 cm (after* Handbook of Chinese Vertebrate Fossils Editorial Group *1979).*

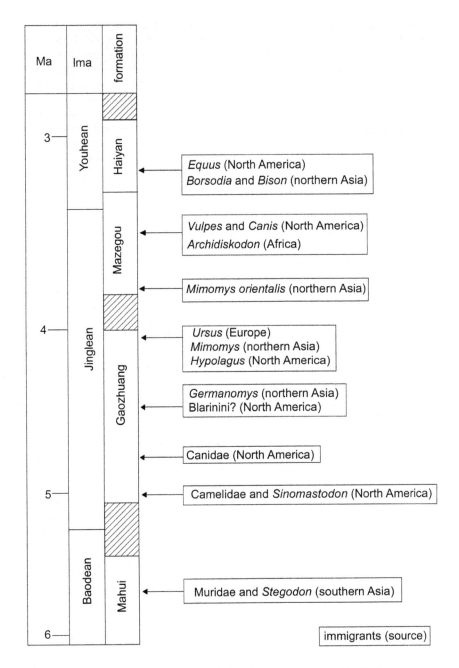

Figure 11-11 *Immigration events recorded in the Mio-Pliocene of the Yushe area, eastern Shanxi (after Flynn et al. 1991); lma = land-mammal "age" and Ma = millions of years.*

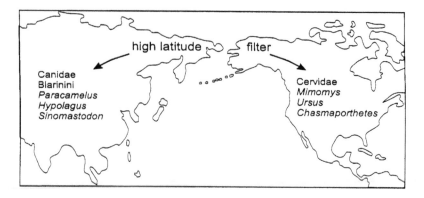

Figure 11-12 *Map of early Pliocene Asian-North American mammal interchange through a high-latitude filter (after Flynn et al. 1991).*

Youhean Mammals

The youngest Miocene-Pliocene LMA recognized in China is the Youhean, named for the Youhe fauna from Gansu. (The younger Nihewanian land-mammal "age" begins during the latest Pliocene but is mostly Pleistocene, so it is discussed in the next chapter.) The immigration to China of true elephantids ("*Elephas*") and equids (*Equus*), the first from Europe, the second from North America, is distinctive of Youhean time. Other key Youhean mammals are *Hipparion, Sus, Cervavitus, Nyctereutes, Ochotonoides,* and *Mimomys* (Xue 1981). A correlative of the Youhe fauna is from Dongyaozitou west of Beijing (Tang 1980; Tang and Ji 1983; Cai 1987).

Forest and Steppe Faunas

Schlosser (1903), followed more explicitly by Kurtén (1952) classified Chinese Miocene-Pliocene mammal localities into "forest" and "steppe" faunas. They applied this distinction particularly to Baodean sites; the forest faunas were to the south (southeastern Shanxi and northern Henan), whereas the steppe faunas were to the north (northern Shanxi, Gansu), with intermediate ("mixed") faunas between them. Cervids (*Cervocerus*), giraffids (*Honanotherium*) and a brachydont bovid (*Gazella gaudryi:* see figure 11-13) characterize the forest fauna (also sometimes called the *gaudryi* fauna). In contrast, the steppe fauna includes a hypsodont rhinoceros (*Sinotherium*), a diversity of bovids, the giraffids *Palaeotragus* and *Samotherium,* and the hypsodont *Gazella dorcadoides* (see figure 11-13) (thus, the fauna is also called the "*dorcadoides* fauna"). These distinctions still are accepted (Watabe 1992) and identify two ecologically separate mammalian communities in the Chinese Miocene-Pliocene.

Proboscidean Evolution

The fossil record points to the origin of the Proboscidea in North Africa during the Oligocene, from whence they spread out to live, at one time or another, on all the continents except Australia and Antarctica. During the early Miocene, mastodont proboscideans emigrated to Asia and first appear in the Chinese fossil record (Tassy 1996). This reflects the so-called "proboscidean datum event" (Madden and Van Couvering 1976) in the lower Miocene deposits of Eurasia. This event now appears to have been diachronous, with gomphotheres and mammutids arriving in China about 18 Ma (Tassy 1990, 1996), so Lucas and Bendukidze (1997) have renamed it the "*Gomphotherium* datum event."

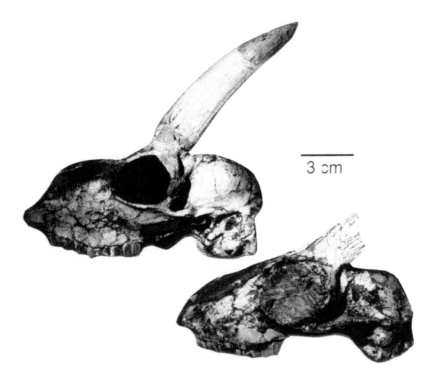

3 cm

Figure 11-13 *Lateral views of skulls of the gazelles Gazella gaudryi (left) and Gazella dorcadoides (right).*

The ensuing Miocene-Pliocene (and Quaternary) fossil record of proboscideans in China is rich and diverse. Tobien et al. (1984, 1986, 1988) have recently restudied what they term the mastodonts (gomphotheres + mammutids) and have divided their Chinese record into three phases: an *"Anchitherium* fauna" of middle Miocene age, a *"Hipparion* fauna" of late Miocene age, and a "last hipparions-*Equus* fauna" of Plio-Pleistocene age (this actually represents an alternative, broader subdivision of the Chinese Miocene-Pliocene land-mammal faunas than the LMAs used here) (see figure 11-14).

The proboscideans of the *"Anchitherium* fauna" were immigrants from the west. These are the bunodont taxa *Gomphotherium* (Lucas and Bendukidze 1997), *Choerolophodon* and *Synconolophus*, the shovel tuskers *Amebelodon* and *Platybelodon* (Guan 1991), and the stegodontid *Zygolophodon*. Characteristic mastodonts of the *Hipparion* fauna are the bunodont *Tetralophodon*, an advanced species of *Zygolophodon*, *Choerolophodon*, and an immigrant from North America, the brevirostrine *Sinomastodon*. The "last hipparion-*Equus* fauna" still includes *Sinomastodon* as well as the widespread *Anancus* and *Mammut*.

These mastodonts well reflect the combination of immigration and *in situ* evolution characteristic of Chinese Miocene-Pliocene mammals. The proboscideans of the *Anchitherium* fauna immigrated to China from the west, either from Europe or the Indo-Pakistani subcontinent (*Synconolophus*). Some of these proboscideans (e.g., *Gomphotherium*, *Platybelodon*) continued eastward to

land-mammal "ages"	fauna	mastodonts
Nihewanian	last hipparions - *Equus* fauna	*Anancus, Mammut, Sinomastodon*
Youhean		
Jinglean		
Baodean	*Hipparion* fauna	*Sinomastodon, Tetralophodon, Zygolophodon, Choerolophodon*
Bahean		
Tunggurian	*Anchitherium* fauna	*Amebelodon, Synconolophus, Zygolophodon, Choerolophodon, Gomphotherium, Platybelodon*
Shanwangian		

Figure 11-14 *The temporal distribution of mastodonts in the Chinese Neogene defines three phases (after Tobien et al. 1984, 1986, and 1988).*

North America via the Bering land bridge. *Sinomastodon* of the *Hipparion* fauna came the other way, immigrating from North America to China during the late Miocene.

Other mastodonts in the Chinese Miocene-Pliocene show clear *in situ* evolution in two lineages: *Zyglophodon* to *Mammut* and *Gomphotherium* to *Tetralophodon* to *Anancus*. Chinese Miocene-Pliocene mastodonts have relatively high crowned, cement-covered molars (see figure 11-15). They probably mostly inhabited savannas and grasslands.

Hipparion First-Appearance Datum

The hipparionine horses were grazing, three-toed horses that originated in North America (MacFadden 1984). The isolated protocone of the upper molars is one of the key shared, derived features of hipparionines (see figure 11-16). The oldest North American hipparionines are from the USA and are of middle Miocene age, about 15 Ma. They emigrated to the Old World soon thereafter, first appearing in China during the late Miocene (Bahean), about 11-12 Ma.

Thereafter, the hipparionines underwent an extensive evolutionary diversification in Eurasia (e.g., Woodburne and Bernor 1980; Bernor et al. 1990; MacFadden 1992; Watabe 1992) and also invaded Africa during the late Miocene. In China, the hipparionines replaced the anchitheres, large browsing horses that had emigrated from North America to Asia at the beginning of the Miocene. The emigration of large, monodactyl grazing horses, the equines (*Equus*), from North America to Asia during the late Pliocene (Youhean) witnessed the demise of the hipparionines.

Chinese Miocene-Pliocene Apes

Other than the large adapid *Sinoadapis* from Lufeng (R. Wu and Pan 1985), all Chinese Miocene-Pliocene fossil primates are apes. They represent pliopithecids and dryopithecines.

The pliopithecid *Laccopithecus* is also from the uppermost Miocene deposits at Lufeng. *Laccopithecus* is dentally identical to *Pliopithecus*, the archetypal pliopithecid from the Miocene of Europe. It was a large ape (estimated weight of 12 kg) and is one of the last pliopithecids. It has been identified as a fossil gibbon (R. Wu and Pan 1984, 1985). Similar, gibbonlike forms are *Dionysopithecus* and *Platydontopithecus*, also from Lufeng.

Chinese fossils long placed in the genera *Sivapithecus* and *Ramapithecus* have recently been renamed *Lufengopithecus* (R. Wu 1987). These fossils are from

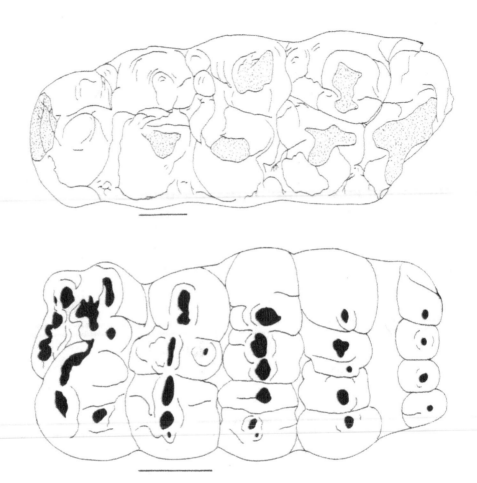

Figure 11-15 *Representative molars of Chinese Neogene mastodonts (after Chow and Zhang 1974): Gomphotherium (above, scale bar = 1 cm), Stegodon (below, scale bar = 2 cm).*

the late Miocene Lufeng locality (e.g., R. Wu 1983, 1985; R. Wu et al. 1983). Specimens previously called *Ramapithecus* may be females; those called *Sivapithecus* may be males.

Miocene-Pliocene Lower Vertebrates

Lower vertebrate fossils are abundant at many Chinese Miocene-Pliocene localities but have received much less study than the mammals. The most remarkable Miocene-Pliocene lower vertebrate locality in China is the lacustrine deposits at Shanwang in Shandong. These lower Miocene strata have produced a prolific record of fossil fishes and lissamphibians (see figure 11-17). The fishes (Young and Tchang 1936) were originally identified as extant genera of cyprinids (*Leuciscus, Barbus, Pseudorasbora*) and an ariid (*Rhineastes*). They occur as complete, articulated specimens. However, a recent reappraisal of the Shanwang fishes suggests they belong to extinct genera and are very similar to coeval fish assemblages from Japan (M. Chang et al. 1996; P. Chen et al. 1999). Similarly, the Miocene-Pliocene fishes from Yushe, Shanxi closely resemble Japanese fish assemblages, though the early Pliocene fish fossils from China belong to extant genera (M. Chang et al. 1996). The obvious conclusions are that the same freshwater fishes inhabited eastern China and Japan during the Miocene-Pliocene, and that by the Pliocene modern genera and species had appeared.

All important specimens of Chinese Miocene-Pliocene lissamphibians are also from Shanwang. They are: (1) two nearly complete skeletons of the salamander *Procynops* (Young 1965a); (2) a nearly complete skeleton of the pelobatid anuran *Macropelobates* (Gao 1986); (3) the somewhat less completeskeleton of the bufonid *Bufo* (Young 1977; Gao 1986); and (4) two ranid skeletons of the genus *Rana* (Young 1936a; H. Liu 1961). The Shanwang lissamphibian record is one of the best Miocene-Pliocene records in the world.

Outside of Shanwang, Miocene-Pliocene lower vertebrate collecting in China has largely been of isolated elements, mostly obtained by screenwashing.   No

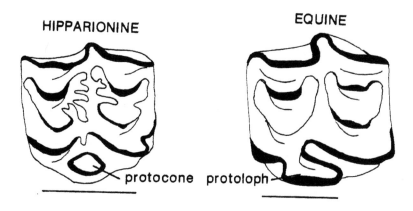

Figure 11-16 *Protocone morphology in hypsodont horse upper molars (after MacFadden 1992). Scale bars = 1 cm.*

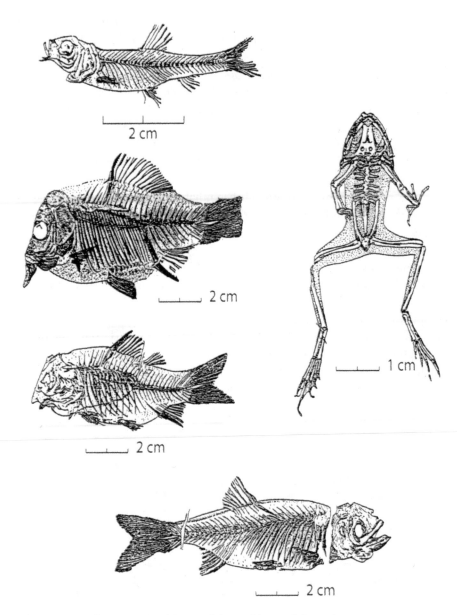

Figure 11-17 *Representative Miocene fishes and lissamphibians from Shanwang, Shandong (after Young and Tchang 1936).*

systematic review of this record can be made because so little of it has been studied and described. A characteristic record is the middle Miocene Sihongfaunas, which contain isolated elements of cyprinid and bagrid fishes, a ranidanuran, emydid, and trionychid turtles, varanid lizards, colubroid snakes and crocodilians. This record needs to be developed and studied in detail.

These kinds of occurrences and isolated finds give us our only insight into Chinese Miocene-Pliocene reptiles. Turtles are emydids (*Epiemys, Shansiemys, Clemmys, Chinemys, Cuora*); testudinids (*Testudo*); and trionychids (*Aspideretes, Amyda*) (Wiman 1930; Ye 1963, 1994) (see figure 11-18). The tortoises are especially well-known from Baodean fossil assemblages.

No good fossil lizard material has been described from the Chinese Miocene-Pliocene, and only a single snake from Shanwang, *Mionatrix* Sun, has been adequately published. Shanwang has also produced the only well-known Pliocene crocodilian, a skull of *Alligator* (J. Li and Wang 1987).

Miocene-Pliocene Birds

Relatively few Miocene-Pliocene fossil birds have been described from China, and as is the case with the lower vertebrates, the Shanwang locality dominates this record. A list of Chinese Miocene-Pliocene birds is as follows:

1. The Baodean ostrich *Struthio wimani* Lowe from Shanxi (see figure 11-19)

2. The large threskiornithid *Platalea tiangangensis* (Hou) from the late Shanwangian of Jiangsu

3. The anatid *Aythya shihuibas* Hou from the Baodean at Lufeng, Yunnan

4. Another duck, *Sinanus diatomas* Ye from Shanwang

5. A falcon, *Mioaegypius gui* Hou from the Baodean of Jiangsu

6. Two phasianids from Shanwang, *Shandonggornis shanwanensis* Ye and *Linquornis gigantis* Ye

7. Another phasianid from the Baodean of Jiangsu, *Palaeoalectoris songlinensis* Hou

8. The phasianids *Phasianus lufengia* Hou and *Diangallus mious* Hou from the Baodean at Lufeng

9. A rallid, *Youngornis gracilis* Ye from Shanwang

10. A passeriform bird of uncertain affinities, *Yunnanus gaoyuansis* Hou from Lufeng

This limited record of Chinese Miocene-Pliocene birds thus comes almost entirely from three localities: Sihong in Jiangsu, Lufeng in Yunnan and Shan-

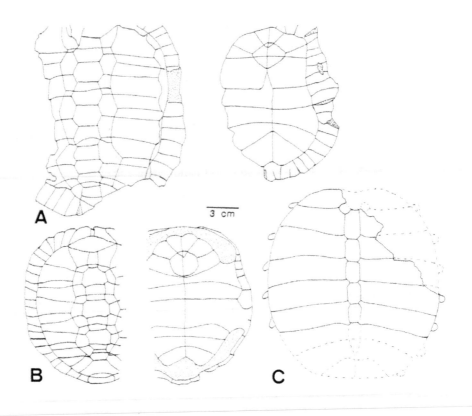

Figure 11-18 *Representative Neogene fossil turtles from China. A, Cuora pitheca Ye, carapace (left) and plastron (right). B, Testudo hipparionum Wiman, carapace (left) and plastron (right). C, Aspideretes sinuosus Chow & Ye, carapace.*

wang in Shandong. Clearly, much work needs to be undertaken to develop a more extensive record.

Paleozoogeography

China's Miocene-Pliocene record reveals a complex history of alternating episodes of endemism and cosmopolitanism of the mammal faunas of the northern continents. Clearly, Eurasia was a single zoogeographic province during most of the Miocene-Pliocene, with regular and frequent immigration events across the vast continent. North America's connection to Eurasia, via a Bering land bridge, was more intermittent and less extensive. Four immigration events from North America to Eurasia are the high points of this connection: (1) early Miocene *Anchitherium* immigration, (2) late Miocene *Hipparion* immigration; (3) early Pliocene camelid and canid immigration, and (4) latest Pliocene elephantid and equid immigration. African immigrants to China during the Miocene-Pliocene came by way of Europe or the Indo-Pakistani subcontinent and were proboscideans.

The origin of what modern zoogeographers term the Palaearctic and Oriental zoogeographic regions of Asia (see figure 11-20) can be seen in the Chinese Miocene-Pliocene. The boundary between these two regions runs from the Himalayas in the west and the Huai River in the east, so that much of southern China is in the Oriental region (see figure 11-20). A distinctive mammal fauna in this region can be traced back to about the middle Miocene. No doubt, the uplift of the Himalayas and the Tibetan Plateau, which altered climates (and hence, vegetation) throughout Asia during the Miocene-Pliocene, played a strong role in creating these two zoogeographic regions. Not only plants, but also invertebrates and lower vertebrates responded quickly to these changes, and mammals developed into two distinct faunas—north and south—during the Miocene.

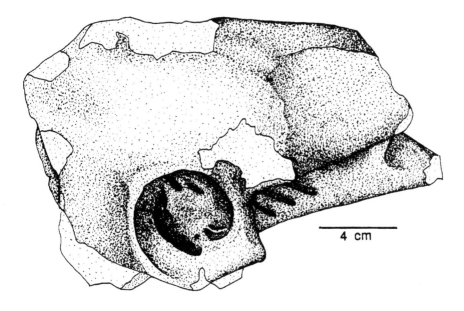

Figure 11-19 *Lateral view of the pelvis of Struthio wimani Lowe.*

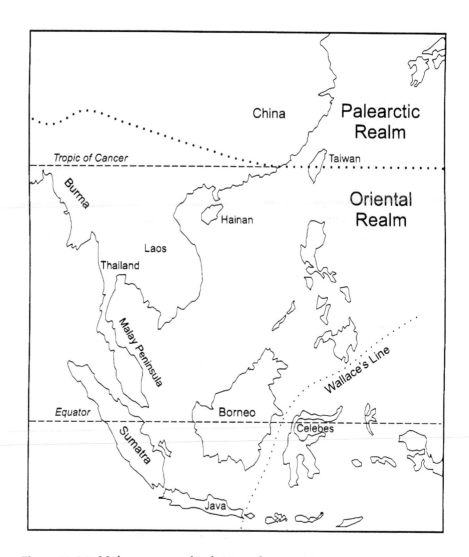

Figure 11-20 *Modern zoogeographic divisions of eastern Asia.*

Chapter 12

Pleistocene

The Pleistocene is the next to last Neogene epoch. The last is the Holocene. By international agreement, the Pleistocene began 1.8 Ma and ended 10,000 years ago (Berggren et al. 1998). However, in China the beginning of the Pleistocene is still a subject of debate, many workers wanting to place it earlier than 1.8 Ma to coincide with the onset of the first late Cenozoic glacial age evident in China. In this book, the international definition of the Pleistocene Epoch is employed.

Pleistocene deposits are widely distributed in China (see figure 12-1) and are mostly loess, fluvio-lacustrine deposits, alluvial fan sediments, glacial deposits, and cave and other karst fillings. Deposition took place during the Pleistocene in a tectonic and climatic setting that began to develop during the Neogene. Indeed, Pleistocene deposition in China represents little more than the culmination of processes that began during the Neogene.

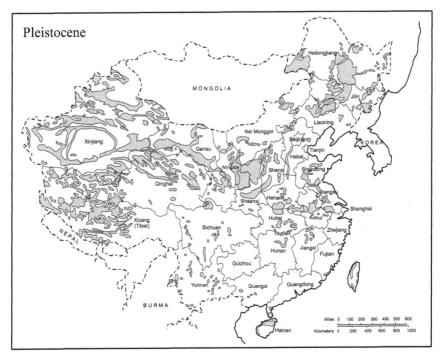

Figure 12-1 *Distribution of Pleistocene rocks in China (after H. Wang 1985).*

Thus, the Himalayas and Tibetan Plateau continued to rise, roughly 3,000 to 4,000 m during the Pleistocene. This rise cut off the Chinese interior from Indian monsoonal moisture, producing very dry climates in the Tarim and Qaidam basins to the north, both of which subsided about 1,000 m during the Pleistocene. In these, and nearby basins of northern China, large alluvial fans at least 500 m thick developed, and basin floors were (and continue to be) blanketed with active sand dunes dotted with evaporitic playa lakes (e.g., Zhou and Chen 1992).

Pleistocene glaciation was confined to mountainous regions of China, leaving glacial till and fluvial outwash deposits in most of China's higher elevations, as well as on some of its central and eastern plains. The Baitou Shan, an active Pleistocene volcano, roughly 2,744 m high, was present in eastern Jilin. Other volcanic rocks of Pleistocene age are confined to eastern China. In southern and eastern China, extensive karstification of Paleozoic limestones took place during the Pleistocene, resulting in numerous cave and karst deposits.

The thick and extensive loess deposits of northern China are classic (see figure 12-2). These are the largest loess deposits on earth. They formed during the Pleistocene, as wind-blown silts and clays from the Gobi Desert of Mongolia and Nei Monggol were dropped in northern China in front of the rising Tibetan Plateau.

China's Pleistocene vertebrate record is one of the most extensive records known. It broadly resembles the Chinese record of Neogene vertebrates in its abundance of mammals and less well-studied material of lower vertebrates and birds. China's Pleistocene mammals show both the influence of immigration from North America and other parts of Eurasia as well as endemic evolution. They also include an important record of the evolution of the genus *Homo*, and even encompass a widespread record of Pleistocene cetacean fossils (Cao 1993). China's Pleistocene mammals reveal much about the origin of the extant Chinese mammalian fauna, but they so far have told us little about the terminal Pleistocene extinction of large mammals.

Pleistocene Vertebrate-Bearing Deposits

Chinese geologists have developed a local nomenclature for the late Cenozoic glacial ages evident in their local rock and fossil record. These ages have been correlated to the classic Alpine glacial ages of western Europe (see figure 12-3). Pleistocene glaciation in China occurred mainly at high altitudes, especially in western and northeastern China. Glacial strata record four principal glacial ages that correlate well to North American and European glacial ages. It is the

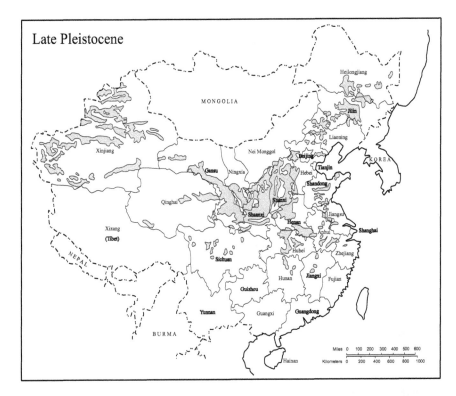

Late Pleistocene

Figure 12-2 *Distribution of late Pleistocene loess deposits of China (after Liu and Ding 1983).*

interglacial deposits that contain fossil vertebrates, especially at Zhoukoudian, the "Peking man" cave (see figure 12-3).

Glacial tillites and outwash only provide very coarse resolution of the Pleistocene climatic history of China. The loess deposits of northern China contain a much more sensitive climatic record (Kukla 1987) as well as most of China's Pleistocene vertebrate record (see figure 12-4). The loess covers an area of more than 440,000 km^2 and in places is more than 150 m thick (T. Liu et al. 1985; Kukla 1987). The loess deposits are usually divided into four units (ascending order): (1) the Wucheng Formation (loess) overlying red-bed mudstones of Baodean age is as much as 50 m thick and composed of yellowish and reddish-brown compact loess interbedded with soils and carbonate concretions; (2) the lower Lishi Formation is similar to the Wucheng but has a lighter color and less common horizons of carbonate concretions; (3) the upper Lishi Formation is 20 to 30 m thick and consists of four paleosol and four loess members; and (4) the Malan Formation is about 10 m thick and composed of grayish yellow, porous,

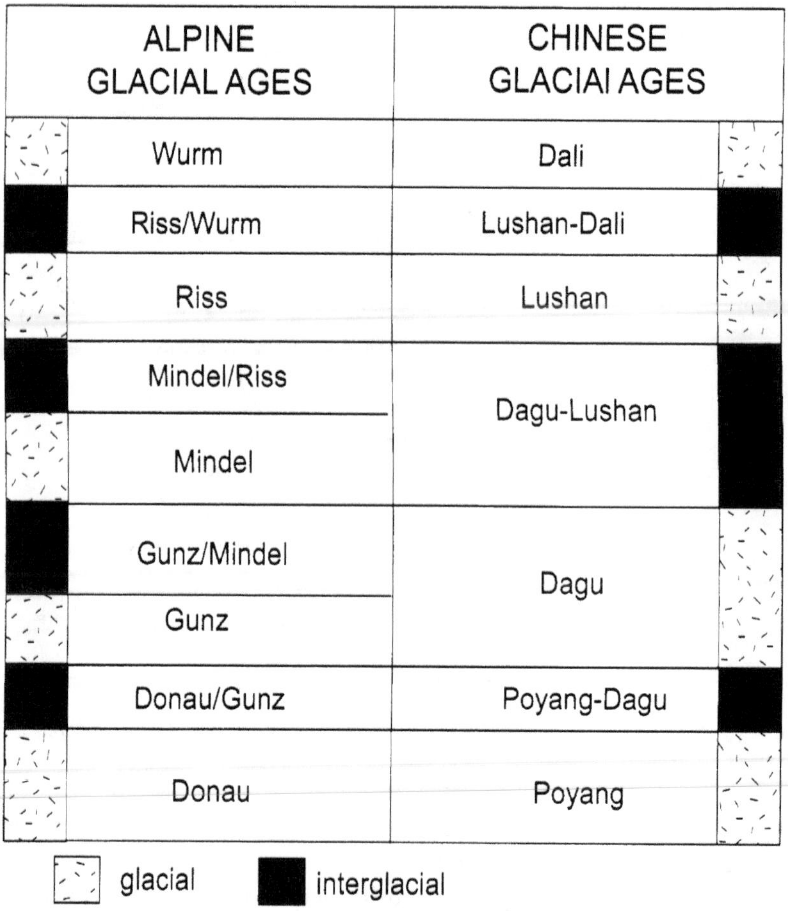

ALPINE GLACIAL AGES		CHINESE GLACIAI AGES	
	Wurm	Dali	
	Riss/Wurm	Lushan-Dali	
	Riss	Lushan	
	Mindel/Riss	Dagu-Lushan	
	Mindel		
	Gunz/Mindel	Dagu	
	Gunz		
	Donau/Gunz	Poyang-Dagu	
	Donau	Poyang	

glacial   ■ interglacial

Figure 12-3 *Correlation of Alpine and Chinese nomenclature for the late Cenozoic glacial ages.*

loosely cemented and bioturbated loess. It is overlain by sediments and soils of latest Pleistocene and Holocene referred to as the "black loam formation."

Fossil mammals are abundant in the Chinese loess. The fauna of the Wucheng Formation includes voles (*Myospalax*) and hamsters (*Kowalskia*) and is of Nihewanian age (Xue 1984). Micromammal successions in the Lishi and Malan Formations indicate middle and late Pleistocene ages (Xue 1982, 1984), and recent work has provided a magnetostratigraphic calibration of the important mammalian faunas from the loess (Yue and Xue 1996) (see figure 12-4).

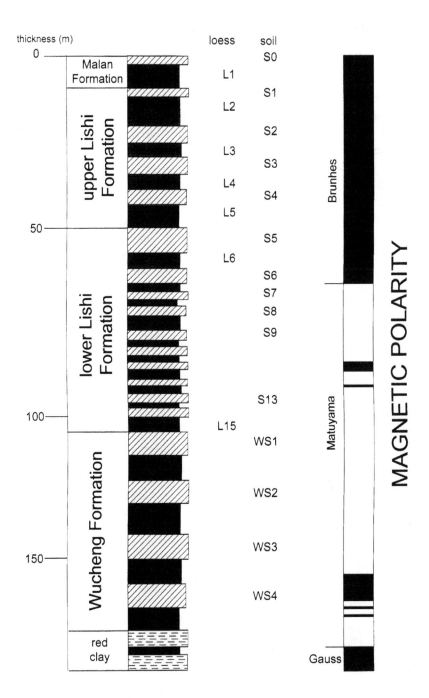

Figure 12-4 *Stratigraphy of the Pleistocene loess in the Xifeng area (after Kukla 1987).*

Nihewanian Land-Mammal "Age"

The Nihewanian LMA is recognized for China's early Pleistocene fossil mammals (see figure 11-4). Two LMA's, Zhoukoudianian and Salawusuan, have recently been proposed for China's middle and late Pleistocene mammals (S. Zheng and Han 1991) (see figure 11-4; table 12-1). Teilhard de Chardin and Piveteau (1930) originally described the type Nihewan (Shagou) fauna from Hebei (see figure 12-5). The Nihewanian as now construed actually encompasses a succession of mammal faunas that range in age from late Pliocene (about 2.4 Ma) to the latest early Pleistocene, about 0.7 Ma (Qiu 1990; Xue and Zhang 1991; Tong et al. 1995) (see figure 12-6). The type Nihewan fauna is characteristic of early Pleistocene faunas that occur throughout northern China (especially Gansu, Shaanxi, Shanxi, and Qinghai provinces, where the "Sanmen Formation" produces correlative mammals). About 20% of its genera are Neogene survivors such as the carnivores *Megantereon* and *Nestoritherium* and the horses *Hipparion* and *Proboscidipparion* (see figure 11-8). Typical early Pleistocene mammals in the Nihewan fauna include the horse *Equus sanme-*

Figure 12-5 *Photograph of type Nihewan outcrops (courtesy of J. Li).*

niensis, the dog *Canis chiliensis,* the bison *Bison palaeosinensis,* and the wooly rhinoceros *Elasmotherium.* Cervids are diverse and are mostly forms with simple branching antlers such as *Euctenoceras, Elaphurus, Axis,* and *Rusa.*

An important zoogeographical distinction can be made between the Pleistocene mammal faunas of northern and of southern China. Northern China here refers to the area north of a line from Xinjiang to Shandong (33.5–42° N latitude, 100–122° E longitude). This is within the Palearctic zoogeographic region. Southern China is south of a line from Sichuan to Taiwan (20–33.5° N latitude, 100–122° E longitude). This is largely in the Oriental zoogeographic region (see figure 11-20). A transitional region exists between these two regions, especially in eastern China (see figure 12-6). Because the characteristics of the Pleistocene mammal faunas of these three regions are distinctive, they can be discussed separately.

Pleistocene Mammals of Northern China

The Nihewan mammal fauna just discussed is characteristic of the early Pleistocene mammal assemblages of northern China. Other important early Pleistocene faunas from northern China are from the Wucheng Formation (Yanggou fauna) from Weinam County, Shaanxi (Ji 1975), the Gonghe fauna of Qinghai, and locality 18 at Zhoukoudian in the Beijing suburbs.

The Yanggou fauna is thought to be somewhat younger than the type Nihewan fauna because it lacks as many Neogene survivors and some characteristic early Pleistocene taxa, such as *Canis chiliensis* (see figure 12-7) and *Elasmotherium.* Nevertheless, the two faunas are difficult to correlate with each other because of facies differences. The Nihewan fauna comes from fluviolacustrine gravels and sands, whereas the Yanggou fauna is from lithic loess and associated concretions. The Nihewan fauna has many forest browsers, whereas open-country grazers dominate the Yanggou fauna.

Locality 18 at Zhoukoudian represents yet another facies: cave deposits. Neogene survivors such as *Proboscidipparion* apparently are extinct at this level; they account for only 7% of the genera. Only about 26% of the total genera are extinct, and many early-middle Pleistocene genera are gone, including cervids such as *Axis rugosus* and the canid *Canis chiliensis.* For this reason, the locality 18 fauna may be slightly younger than the Yanggou fauna.

Middle Pleistocene (Zhoukoudianian) faunas of northern China begin with the Chenjiawo fauna from Lantian County, Shaanxi, which includes the mandible of "Lantian Man" (Hu and Qi 1978; Zhou 1964, 1965). The cranium from the same site has now been made the type of a new subspecies, *H. erectus gongwanglingensis,* whereas the mandible is still termed *H. erectus lantianensis*

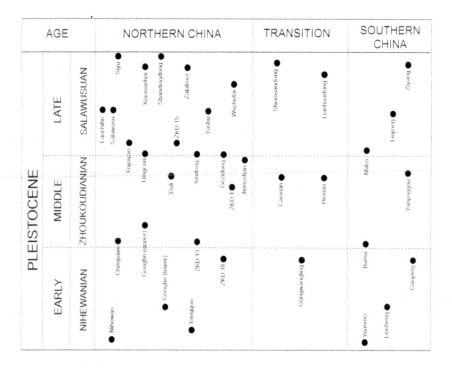

Figure 12-6 *Correlation of Pleistocene fossil mammal localities of China recognizes three land-mammal "ages" and three geographic regions (after Xue and Zhang 1991).*

(Xue and Zhang 1991). (Simply referring to both specimens as *H. erectus* should be sufficient.)

A lack of Pliocene survivors and a virtual lack of characteristic early Pleistocene taxa (see table 12-1) characterize the Chenjiawo fauna and its correlatives (especially Zhoukoudian locality 13). The presence of hominid fossils also is characteristic, as is the appearance of characteristic middle Pleistocene mammals, including the giant elk *Megaloceros* (see figure 12-8), the deer *Pseudaxis,* the rabbit *Lepus wongi,* and the carnivore *Cuon alpinus.*

Younger middle Pleistocene mammal assemblages of northern China include the classic locality 1 at Zhoukoudian—the "Peking man cave," discussed below. This locality has abundant fossils of *Homo erectus,* and only 11% of its genera are now extinct. Characteristic mammals include the giant elk *Megaloceros,* the saber-toothed tigers *Homotherium ultima* and *Megantereon inexpectatus,* and early Pleistocene survivors such as *Equus sanmeniensis, Sus lydekkeri,* and *Paracamelus gigas.* Newly appearing mammals include progres-

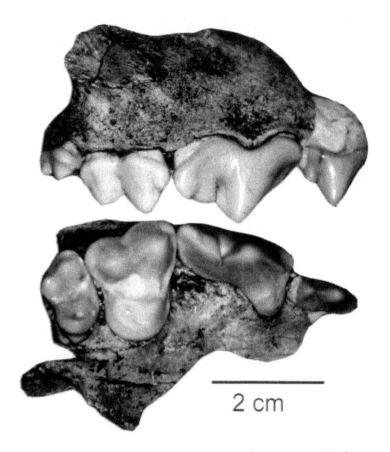

Figure 12-7 *Maxillary fragment of Canis chiliensis, a characteristic early Pleistocene dog from China, lateral (above) and occlusal (below) views.*

Table 12-1 Pleistocene Mammal Genera of China Divided into Early (Nihewanian), Middle (Zhoukoudianian), and Late (Salawusuan) Pleistocene Occurrences (Based on Data in Xue and Zhang 1991)

	Pleistocene		
	Early	Middle	Late
Insectivora			
Erinceidae:			
Erinaceus	X	X	X
Neotetracus	—	—	X
Soricidae:			
Crocidura	—	X	X

Table 12-1 Pleistocene Mammal Genera of China Divided into Early (Nihewanian), Middle (Zhoukoudianian), and Late (Salawusuan) Pleistocene Occurrences (Based on Data in Xue and Zhang 1991) (Continued)

	Pleistocene		
	Early	Middle	Late
Nectogale	—	X	—
Anourosorex	—	X	X
Blarinella	—	X	X
Chodsigoa	—	X	—
Soriculus	—	—	X
Chimmarogale	—	—	X
Neomys	—	X	—
Sorex	—	—	X
Peisorex	X	—	—
Talpidae:			
Scaptochirus	X	X	X
Scaptonys	—	—	X
Parascaptor	—	—	X
Mogera	—	—	X
Chiroptera			
Rhinolophidae:			
Rhinolophus	X	X	X
Hipposideridae:			
Hipposideros	—	X	X
Vespertillionidae:			
Miniopterus	—	X	—
Murina	—	X	—
Eptesicus	—	—	X
Laio	—	X	X
Plecotus	—	—	X
Myotis	X	X	X
Hesperopternus	—	X	—
Pipistrellus	—	X	—
Cercopithecidae:			
Macaca	X	X	X

Table 12-1 Pleistocene Mammal Genera of China Divided into Early (Nihewanian), Middle (Zhoukoudianian), and Late (Salawusuan) Pleistocene Occurrences (Based on Data in Xue and Zhang 1991) (Continued)

	Pleistocene		
	Early	Middle	Late
Szechuanopithecus	—	X	—
Rhinopithecus	—	X	X
Procynocephalus	X	—	—
Pongidae:			
Gigantopithecus	X	X	—
Hylobates	X	X	X
Pongo	X	X	X
Hominidae			
Australopithecus	X	—	—
Homo	X	X	X
Lagomorpha			
Leoporidae:			
Altelepus	X	—	—
Hypolagus	X	—	—
Lepus	X	X	X
Ochotonidae			
Ochotona	X	X	X
Ochotonoides	X	X	—
Rodentia	—	—	—
Sciuridae:			
Marmota	X	X	X
Petaurista	X	X	X
Sciurotamias	X	—	X
Cittelus	—	X	X
Tamias	—	X	—
Rupestes	—	—	X
Sciurus	X	—	X
Eutamias	—	—	X
Spermophilus	—	—	X
Primates			

Table 12-1 Pleistocene Mammal Genera of China Divided into Early (Nihewanian), Middle (Zhoukoudianian), and Late (Salawusuan) Pleistocene Occurrences (Based on Data in Xue and Zhang 1991) (Continued)

	Pleistocene		
	Early	Middle	Late
Castoridae:			
Castor	—	X	X
Trogontherium	X	X	X
Sinocaster	X	X	—
Rhizomyidae:			
Brachyrhizomys	—	X	—
Rhizomys	X	X	X
Cricetidae:			
Cricetulus	X	X	X
Allocricetus	X	—	—
Clethrionomys	—	X	X
Microtus	X	X	X
Mimomys	X	X	—
Prosiphneus	X	X	—
Myospalax	X	X	X
Arvicola	X	X	X
Eothenomys	—	X	X
Ellobius	—	—	X
Meriones	—	X	X
Sinocricetus	X	—	—
Kowalskia	X	—	—
Bahomys	—	X	—
Gerbillus	X	X	—
Pitymys	—	X	—
Alticola	—	X	X
Allophaiomys	X	—	—
Phaiomys	—	X	—
Muridae:			
Apodemus	X	X	X
Micromys	—	X	X

Table 12-1 Pleistocene Mammal Genera of China Divided into Early (Nihewanian), Middle (Zhoukoudianian), and Late (Salawusuan) Pleistocene Occurrences (Based on Data in Xue and Zhang 1991) (Continued)

	Pleistocene		
	Early	Middle	Late
Vernaya	—	—	X
Bandicota	—	X	X
Rattus	X	X	X
Mus	—	X	X
Orientalomys	X	—	—
Hyperacrius	X	—	—
Stephanomys	X	—	—
Dipodidae:			
Alactaga	—	—	X
Dipus	—	—	X
Sminthoides	—	X	—
Hystricidae:			
Hystrix	X	X	X
Atherurus	X	X	X
Hyracoidea			
Postschizotherium	X	—	—
Proboscidea			
Gomphotheriidae	—	—	—
Gomphotherium	X	X	—
Sinomastodon	X	—	—
Tetralophodon	X	—	—
Pentalophodon	X	—	—
Stegodontidae			
Stegodon	X	X	X
Elephantidae			
Archidiskodon	X	X	—
Palaeoloxodon	X	X	X
Mammuthus	—	—	X
Elephas	X	X	X
Carnivora			

Table 12-1 Pleistocene Mammal Genera of China Divided into Early (Nihewanian), Middle (Zhoukoudianian), and Late (Salawusuan) Pleistocene Occurrences (Based on Data in Xue and Zhang 1991) (Continued)

	Pleistocene		
	Early	Middle	Late
Canidae:			
Canis	X	X	X
Nyctereutes	X	X	X
Cuon	X	X	X
Cynalopex	—	X	—
Vulpes	X	X	X
Procyonidae:			
Ailurus	X	X	X
Ailuropoda	X	X	X
Ursidae:			
Ursus	X	X	X
Mustelidae:			
Martes	X	X	X
Charronia	—	X	—
Mustela	X	X	X
Gulo	—	X	X
Putorius	—	X	—
Arctonys	X	X	X
Meles	X	X	X
Parameles	—	X	—
Lutra	X	X	X
Vormela	—	X	—
Viverridae:			
Viverra	X	X	X
Viverricula	X	X	X
Paguma	X	X	X
Hyaenidae:			
Hyaena	X	X	X
Crocuta	—	X	X
Felidae:			

Table 12-1 Pleistocene Mammal Genera of China Divided into Early (Nihewanian), Middle (Zhoukoudianian), and Late (Salawusuan) Pleistocene Occurrences (Based on Data in Xue and Zhang 1991) (Continued)

	Pleistocene		
	Early	Middle	Late
Machairodus	X	X	—
Megantereon	X	X	—
Homotherium	X	X	—
Epimachairodus	X	—	—
Metailurus	X	—	—
Felis	X	X	X
Neofelis	—	—	X
Lynx	X	X	X
Panthera	X	X	X
Acinonyx	X	X	X
Perissodactyla			
Equidae			
Hipparion	X	—	—
Proboscidipparion	X	—	—
Equus	X	X	X
Chalicotheriidae:			
Nestoritherium	X	X	—
Circotherium	X	—	—
Tapiridae:			
Megatapirus	X	X	X
Tapirus	X	X	X
Rhinocerotidae:			
Rhinoceros	X	X	X
Dicerorhinus	X	X	X
Coelodonta	X	X	X
Elasmotherium	X	X	—
Artiodactyla			
Suidae:			
Dicoryphochoerus	X	X	—
Potamochoerus	X	X	—

Table 12-1 Pleistocene Mammal Genera of China Divided into Early (Nihewanian), Middle (Zhoukoudianian), and Late (Salawusuan) Pleistocene Occurrences (Based on Data in Xue and Zhang 1991) (Continued)

	Pleistocene		
	Early	Middle	Late
Sus	X	X	X
Camelidae:			
Camelus	—	—	X
Paracamelus	X	X	—
Tragulidae:			
Dorcabune	X	—	—
Cervidae:			
Moschus	X	—	—
Eostyloceros	X	—	—
Metacervulus	X	—	—
Paracervulus	X	—	—
Muntiacus	X	X	X
Cervulus	—	X	—
Elaphodus	X	X	—
Cervocerus	X	—	—
Eucladoceros	X	—	—
Megaloceros	X	X	X
Axis	X	—	—
Rusa	X	X	X
Pseudaxis	X	X	X
Elaphus	—	X	X
Cervus	—	X	X
Elaphurus	X	—	X
Procapreolus	—	X	X
Alces	X	—	—
Capreolus	—	X	X
Bovidae:			
Antilospira	X	—	—
Spirocerus	X	X	X
Gazella	X	X	X

Table 12-1 Pleistocene Mammal Genera of China Divided into Early (Nihewanian), Middle (Zhoukoudianian), and Late (Salawusuan) Pleistocene Occurrences (Based on Data in Xue and Zhang 1991) (Continued)

	Pleistocene		
	Early	Middle	Late
Budorcas	—	—	X
Boopsis	X	X	—
Naemorhedus	—	X	X
Ovis	X	X	X
Megavolis	X	X	—
Capricornis	—	X	X
Capra	—	—	X
Pseudovis	—	X	—
Leptobos	X	—	—
Bubalus	X	X	X
Bison	X	X	X
Bos	X	—	X
Bibos	X	X	X

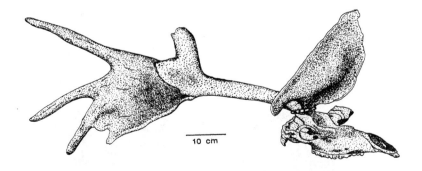

10 cm

Figure 12-8 *Skull of Megaloceros, the giant elk (after Handbook of Chinese Vertebrate Fossils Editorial Group 1979).*

sive species of *Myospalax*, the cave bear *Ursus spelaeus*, the fox *Vulpes vulpes*, *Crocuta ultima*, and the deer *Cervus canadensis*.

Mammal fossils from fluvial deposits in Dali County, Shaanxi, probably are somewhat younger than those from locality 1 at Zhoukoudian. They include the "Dali man," a complete skull interpreted to be the earliest stage of evolution of *Homo sapiens* in China (Y. Wang et al. 1979). Associated nonhuman mammals include castorids, the stegodontid *Palaeoloxodon naumannii* (see figure 12-9), horse (*Equus*), wooly rhinoceros (*Coelodonta antiquitatis*), giant elk (*Megaloceros*), Indian rhinoceros (*Dicerorhinus mercki*), deer (*Pseudaxis*), and *Bubalus*. Cypriniform and siluroid fishes and the ostrich *Struthio anderssoni* also are known from Dali. The presence of early *Homo sapiens* and the advanced dentition of the Dali *Equus* (it is not *E. sanmeniensis*) suggest the fauna is younger than Zhoukoudian locality 1.

The largest concentrations of Pleistocene vertebrate localities in China are those of late Pleistocene (Salawusuan) age in northern China (see figure 12-6). The Dingcun fauna of Shanxi is characteristic of the older of these localities, which include Xujiayao in Shanxi, Xindong near Zhoukoudian, and Gezidong in Liaoning. New mammals characteristic of the late Pleistocene first appear at these localities—the Przewalsky horse *Equus przewalskyi*, the hemione *E. hemionus*, the pig *Sus scrofa,* and the Old World bison *Bos primigenius*. About 45 to 60% of all mammal species in these faunas are extinct, and radioisotopic ages indicate they are between 100,000 and 200,000 years old.

The Salawasu River fauna from Nei Monggol, from which the Late Pleistocene LMA takes its name, typifies somewhat younger late Pleistocene faunas from northern China. This fauna has a radioisotopic age of 36,000–45,000

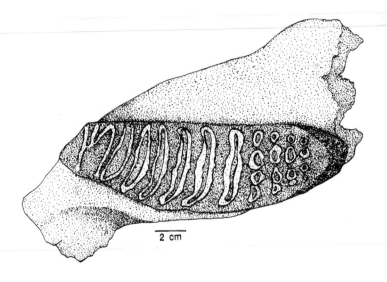

Figure 12-9 *Occlusal view of lower jaw of Palaeoloxodon naumanni (after* Handbook of Chinese Vertebrate Fossils Editorial Group *1979).*

years old and has 32% extinct species. It contains characteristic late Pleistocene mammals of the slightly older assemblages, but also marks the first appearance of other, extant species, including *Pseudaxis horturolum, Elaphurus menziestanus,* and *Nyctereutes procyonoides.*

One interesting aspect of assemblages of this age is the distribution of elephantids. The wooly mammoth *Mammuthus primigenius* is almost exclusively found east of longitude 116°E, whereas the contemporaneous stegodont *Palaeoloxodon namadicus* is found mostly west of that longitude (which is a north-south line, almost intersecting Beijing). This suggests much colder climates in northeastern China than to the west during the late Pleistocene.

Latest Pleistocene faunas from northern China are from cave deposits. The Shandingdong fauna from the Zhoukoudian area is characteristic. Its radiocarbon age is 10,470 year B.P., and it only has 12% extinct species. Fully modern *Homo sapiens* associated with abundant artifacts are found at Shandingdong.

Pleistocene Mammals of Southern China

Southern China has a much less extensive Pleistocene record of fossil vertebrates than does northern China, and all the southern Chinese localities are in cave deposits or fissure fills. These assemblages typically contain fossils of the giant panda *Ailuropoda* and the stegodont proboscidean *Stegodon* (see figure 12-10). For this reason, they were long referred to as the "*Ailuropoda-Stegodon* fauna," which also can be recognized in Southeast Asia and Indonesia.

The Yuanmo locality in Yunnan yields the oldest Pleistocene fauna from southern China. It is the only southern China occurrence of *Homo erectus.* The Yuanmo fauna has many Pliocene survivors and numerous taxa characteristic of the early Pleistocene of northern China (Zhou et al. 1978). However, it also has several endemics, including the dog *Canis yuanmouensis,* the mastodonts *Stegodon yuanmouensis* and *S. elephantoensis,* and the horse *Equus yunnanensis.* Ironically, it lacks *Ailuropoda,* the tapir *Tapirus,* and the giant pongid primate *Gigantopithecus,* characteristic Pleistocene mammals of southern China. Paleomagnetic data initially indicated an age of about 1.7 Ma for the Yuanmo mammals, but more recent data indicate an age of less than 1.0 Ma (T. Liu and Ding 1984).

The Liucheng locality from Guangxi marks the first appearance in southern China of *Ailuropoda, Stegodon, Tapirus,* and *Gigantopithecus.* This is the famous "*Gigantopithecus* cave," which contained most of the known fossils of this giant ape (see figure 12-11). About 60% of the species at Liucheng are extinct. The taxa present include carnivores (*Hyaena, Felis, Cuon*), proboscideans (*Gomphotherium, Stegodon*), perissodactyls (*Nestoritherium, Tapirus, Equus*), artiodactyls (*Sus,*

Figure 12-10 *Skeleton of Stegodon, a characteristic Pleistocene proboscidean of southern China.*

Dorcabuna, Cervoceros, Rusa), and *Gigantopithecus*. This large mammal fauna appears to be somewhat younger than the Yuanmo fauna yet older than fauna from the "Dragon bone cave" at Gaoping in Hubei.

The Gaoping fauna is one of the classic localities of the "*Ailuropoda-Stegodon* fauna." Its mammalian fauna includes *Ailuropoda, Stegodon, Tapirus,* and *Gigantopithecus* as well as gomphotheres (*Trilophodon*), machairodontine saber tooths, hyenas (*Hyaena licenti*), a rhinoceros (*Rhinoceros sinensis*), and a horse (*Equus yunnanensis*).

A cave in Bama County, Guangxi, yielded a very similar fauna and may be slightly younger than the Gaoping fauna. Definitely younger is the cave fauna from Yanjinggou in Sichuan. No *Gigantopithecus* is present, no hominids are known either, and the *Ailuropoda* and *Tapirus* fossils are much larger and more evolutionarily advanced than those are at Gaoping and Bama. Correlative cave faunas are well known from Yunnan and Guangxi. The Bama and Yanjinggou faunas thus represent the characteristic and widespread (in southern Asia) "*Ailuropoda-Stegodon* fauna."

This fauna continued into the late Pleistocene in southern China. These late Pleistocene faunas thus also contain giant pandas, stegodonts, tapirs, and rhinoceroses. Their chronology is based not only on radioisotopic (especially C^{14}) dating, but also on a perceived chronocline in the evolution of *Homo sapiens* and associated artifacts. Maba cave in northern Guangdong, at about 100,000 years old, yielded early *Homo sapiens*. Correlative occurrences are at Jiande in Fujian and Tongzi in Guizhou. Liujiang in Guangxi represents somewhat younger localities with an age range of 30,000 to 70,000 years B.P. These include remains of more progressive *Homo sapiens*, including the Liujiang (67 Ka), Suicheng (52 Ka), and Chengong (31 Ka) occurrences. Ziyang in Sichuan, at about 20 Ka, is one of the youngest Pleistocene vertebrate faunas of southern China.

Pleistocene Mammals of the Transition Zone

Between the northern and southern Chinese Pleistocene localities we can recognize a transition zone of localities with a clear mixture of northern and southern species (see figure 12-6). The oldest such fauna is from Gongwangling in loess along the northern foothills of the Qinling Mountains in Shaanxi. *Homo erectus* is present here, and about one-third of the mammal species are members of the *Ailuropoda-Stegodon* fauna of southern China, including *Ailuropoda*, *Stegodon*, and *Tapirus*.

The Hexian fauna of Anhui is a somewhat younger transitional fauna. Its northern faunal component looks very similar to the mammals from locality 1 at Zhoukoudian. As at Gongwangling, about one-third of the mammals at Hexian are characteristic of the *Ailuropoda-Stegodon* fauna. Advanced *Homo erectus* is present at Hexian, and the fauna's age is about 150 Ka. At Caoxian, only 50 km away, a similar fauna occurs.

The youngest transitional faunas of the Pleistocene are well represented by the Shenxiandong and Lianhuadong faunas of Jiangsu. These faunas include mostly southern species of the *Ailuropoda-Stegodon* fauna with only a few northern species.

Clearly, China has a complex Pleistocene record of mammals that documents two great zoogeographic provinces—north and south—and a clear trend toward modernization through time.

Gigantopithecus

The largest primate to have ever lived, with an estimated body weight of 300 kg, was *Gigantopithecus*. A latest Miocene species, *G. giganteus*, is known from India-Pakistan, and the Pleistocene species *G. blacki* is known from cave deposits in southern China and Vietnam (Simons and Chopra 1969; Von Koenigswald 1983). The Dutch paleontologist G. H. R. von Koenigswald originally bought teeth of *G. blacki* from a Chinese druggist.

Known only from the lower jaw and isolated teeth (see figure 12-11), *Gigantopithecus* has small, vertical incisors, thick and short canines, and broad cheek teeth with thick enamel and low flat cusps. These features, plus the extremely thick and deep lower jaws, suggest *Gigantopithecus* ate hard fibrous plant matter. In part because of its very large size, *Gigantopithecus* is considered to have been a specialized dead-end in pongid evolution (Simons and Ettel 1970).

Fossil *Homo*

China has yielded one of the most extensive fossil records of Pleistocene *Homo erectus*, particularly from the "Peking man" cave at Zhoukoudian. The most famous *H. erectus* sites (see figure 12-12) in China are:

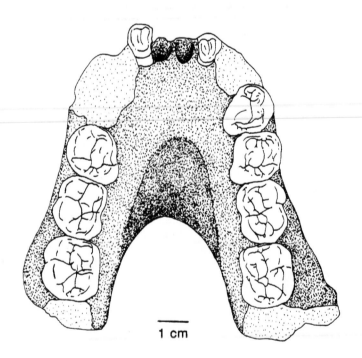

1 cm

Figure 12-11 *Occlusal view of the lower jaw of Gigantopithecus.*

1. Zhoukoudian, Bejing, where fossils representing more than 40 individuals of middle Pleistocene age have been collected (see figure 12-13).

2. Yuanmo, Yunnan is the site of so-called "Yuanmo man," based on two upper central incisors. This site was first believed to be about 1.7 Ma, but now is considered to be less than 1.0 Ma (T. Liu and Ding 1983; Aguirre 1997). Palynology and associated mammals suggest the site represented a subtropical broad-leaf forest.

3. Lantian, Shaanxi is the site of "Lantian man," an early Pleistocene skull cap and mandible of middle Pleistocene age. Associated mammals also indicate warm climates, and the skullcap was found in association with some crude stone tools.

4. In middle Pleistocene strata at He Xian, Anhui, most of a skull, bone fragments and a tooth of *H. erectus* were recovered. Associated fossil mammals indicate a warm climate, as they do with other, less well-documented *H. erectus* occurrences in Hubei (Yun Xian, Yunxi), Henan (Nanzhao, Xichuan), and Anhui (Xiao Xian).

Clearly, the record of *Homo erectus* in China is confined to interglacial deposits. All *Homo erectus* incisors from China are shovel shaped, a feature characteristic of modern Mongoloids among *Homo sapiens*.

Weidenreich (1943) drew attention to these and other features that "regionalize" the characteristics of *H. erectus* fossils. He argued, as have some later authors (e.g., Thorne and Wolpoff 1981), for a direct ancestry of Mongoloid *H. sapiens* from a Chinese population of *H. erectus*. The evidence to support such an ancestry is impressive. Not only is there temporal overlap of *H. erectus* and *H. sapiens* in China (e.g., T. Chen et al. 1994), but Chinese *Homo erectus* share several unique skeletal features with modern Mongoloids (X. Wu 1990; X. Wu and Poirier 1995). These include midsagittal elevation, flatness of the nasal saddle, a forward facing antero-lateral flange of the frontal process of the zygomatic, a lower upper face, shovel-shaped incisors, rounded orbital margins, and presence of a lambdoidal ossicle. These features strongly support a direct ancestry of Mongoloid *Homo sapiens* from Chinese *Homo erectus*.

However, this view remains a point of controversy. The alternative argument is that *H. sapiens* arose from *H. erectus* in Africa and that the new species spread from there and replaced *H. erectus* throughout the Old World. Besides the bearing Chinese *H. erectus* have on this controversy, they also provide us with a much more holistic view of *H. erectus* than any of its non-Chinese fossils because of the unique record preserved at Zhoukoudian.

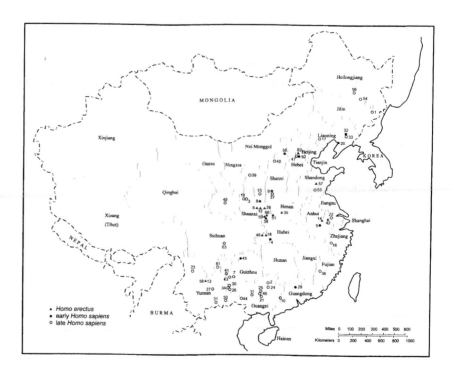

Figure 12-12 *Chinese fossil hominid sites: 1 – Antu, 2 – Baojiyan, 3 – Changwu, 4 – Changyang, 5 – Chaoxian, 6 – Chenjiawo, 7 – Chuandong, 8 – Dali, 9 – Dingcun, 10 – Dongzhongyan, 11 – Duan, 12 – Gongwangling, 13 – Guojiabao, 14 – Hexian, 15 – Huanglong, 16 – Jiande, 17 – Jianpin, 18 – Jianshi, 19 – Jinchuan, 20 – Jinniushan, 21 – Laibin, 22 – Lianhua, 23 – Lijiang, 24 – Lipu, 25 – Liujiang, 26 – Longlin, 27 – Longtanshan, 28 – Luonan, 29 – Maba, 30 – Maomaodong, 31 – Mengzi, 32 – Miaohoushan, 33 – Miaohoushan Dongdong, 34 – Nalai, 35 – Nanzhao, 36 – Qingliu, 37 – Quwo, 38 – Quyan River Mouth, 39 – Salawusu, 40 – Shiyu, 41 – Shuicheng, 42 – Taohua, 43 – Tiandong, 44 – Tongzi, 45 – Tubo, 46 – Upper Cave, 47 – Wushan, 48 – Yuanyang, 49 – Xichou, 50 – Xichuan, 51 – Xingdong, 52 – Xintai, 53 – Xuetian, 54 – Xujiayao, 55 – Yanjiagang, 56 – Yiyuan, 57 – Yuanmou, 58 – Yunxi, 59 – Yunxian, 60 – Zhaotong, 61 – Zhoukoudian, 62 – Ziyang, 63 – Zuozhen.*

Zhoukoudian

Zhoukoudian (= Choukoutien) is a village just southwest of Beijing. In this area, cave deposits and fissure fillings characterize extensive karst developed in Ordovician limestones. For vertebrate paleontology, the most famous such deposit is the "Peking man cave," more precisely referred to as locality 1 at Zhoukoudian (see figure 12-13). This deposit has yielded most of what we know about Chinese Pleistocene lower vertebrates, birds (see table 12-2) and *Homo erectus*. It also has yielded a diverse and rich mammalian fauna of middle Pleistocene age. Zdansky (1928) first described the mammals, which include diverse insectivores, rodents, lagomorphs, carnivores, proboscideans, perissodactyls and artiodactyls. The deposits that contain these and the fossils of *Homo erectus* encompass 10 distinct layers with a total thickness of 40 m.

Two interpretations of the cave deposits have been offered. The traditional view is that *H. erectus* occupied the cave for nearly 200,000 years during the Dagu-Lushan interglacial (e.g., Chia 1975; Jia 1980; R. Wu and Lin 1983; Jia and Huang 1990; Xu et al. 1996). Most of the bones found in the cave were

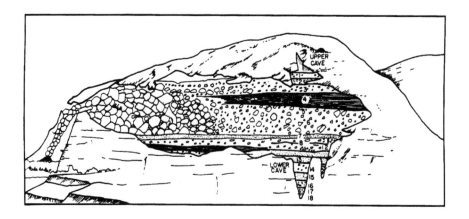

Figure 12-13 *This cross section of locality 1 at Zhoukoudian shows the reconstruction of the cave filling and the numbered layers of the excavation (after M. Ren et al. 1981). Compare to figure 2-7.*

Table 12-2 Pleistocene Birds from Locality 1 at Zhoukoudian

Order Struthioniformes	
Family Struthionidae	Ostriches
Struthio anderssoni	Extinct
Order Gruiformes	
Family Rallidae	Rails
Rallus aquaticus	Water Rail (across N China, widespread Palearctic, Oriental region)
Gallinula chloropus	Common Gallinule (cosmopolitan, except for Australasian region)
Order Falconiformes	Birds of Prey
Family Accipitiridae	Hawks

Table 12-2 Pleistocene Birds from Locality 1 at Zhoukoudian (Continued)

Buteo buteo	Common Buzzard (N China, winter S China & Tibet)
Accipiter nisus	Northern Sparrowhawk (NE and Central China; winter SW China
Family Falconidae	Falcons
Falco chowi	Extinct
Order Galliformes	Fowl-like Birds
Family Phiasianidae	Pheasants
Alectoris graeca	Rock Partridge (now found in mountains of France, Italy, Austria; not in China)
Perdix dauricae	Daurian Partridge (Across N China)
Coturnix coturnix	Common Quail (NE and W China, winters S China, Tibet)
Pucrasia macrolopha	Kolelass Pheasant (E and central China)
Phasianus colchius	Common Pheasant (throughout China ; 19 subspecies)
Order Columbiformes	Pigeons and Doves
Family Pteroclididae	Sandgrouse
Syrrhaptes paradoxus	Pallass Sandgrouse (China)
Family Columbidae	
Columba livia	Rock Pigeon (Far W China & Tibet)
Streptopelia chinensis	Spotted Dove (east-central, S China)
Order Strigiformes	Owls
Family Strigidae	
Asio flammeus	Short-eared Owl (NE, central China)
Ninox scutulata	Brown Hawk Owl (NE, central China)
Order Apodiformes	Swifts
Family Apodidae	Swifts
Apus apus	Eurasian Swift (N China)
Order Piciformes	
Family Picidae	Woodpeckers
Dendrocopos major	Great Spotted Woodpecker (throughout China)
Picus canus	Gray-headed Woodpecker (NE, central China)
Dryocopus	Great Black Woodpecker (north-central China)

Table 12-2 Pleistocene Birds from Locality 1 at Zhoukoudian (Continued)

Order Passeriformes		
Family Alaudidae	Larks	
Alauda arvenis	Common Skylark (east-central, W China)	
Melamocorypha mongolica	Mongolian Lark (N, north-central China)	
Eremophila alpestris	Horned Lark (widespread Palearctic, Nearctic)	
Calandrella cinerea	Greater Short-toed Lark (NE, north-central China)	
Galerida cristata	Crested Lark (N China, Tibet)	
Family Hirudinidae	Swallows	
Hirundo daurica	Red-rumped Swallow (E, central China)	
Hirundo rustica	Barn Swallow (throughout China)	
Riparia riporia	Sand Martin (Bank Swallow) (N China, winters in S China)	
Family Motacillidae	Pipits, Wagtail	
Motacilla flava	Yellow Wagtail (N China, winters in S China)	
Anthus spinoletta	Water Pipit (NW China, winters in S China)	
Family Laniidae	Starlings	
Sturnus cineraceus	White-cheeked Starling (NE and north-central China)	
Family Corvidae	Crows and Jays	
Cyanopica cyana	Azure-winged Magpie (E and central China)	
Pyrrhocorax pyrrhocorax	Red-billed Chough (N, central, W China)	
Corvus monedula	Eurasian Jackdaw (W China)	
Corvus torquattus	Collared Crow (E, central, & S China)	
Urocissa erythrorhyncha	Red-billed Magpie (east-central, SW China)	
Family Musciapidae		
Subfamily Turdinae		
Erithacus calliope	Siberian Rubythroat (NE, central China)	
Erihtacus svecieus	Bluethroat (N, W China)	
Hodgsonius phoenicuroides	White-bellied Redstart (NE, central China)	
Zoothera dixoni	Long-tailed Thrush (south-central China)	
Rhyacornis fuliginosus	Plumbeous Redstart (throughout China)	

Table 12-2 Pleistocene Birds from Locality 1 at Zhoukoudian (Continued)

Subfamily Panuridae or Paradox-ornithidae	
Paradoxornis webbianus	Vinous-throated Parrotbill (E and central China)
Subfamily Sylviinae	Old World Warblers
Acrocepahlus	Thick-billed Warbler (NE, north-central China)
Prinia polychroa	Brown Prinia (S China)
Family Paridae	Tits
Parus major	Tit
Family Ploceidae	Weavers
Passer domesticus	House Sparrow (throughout China)
Family Fringillidae	
Fringilla montifringilla	Brambling (north-central China in water)
Carpodacus roseus	Pallas's Rosefinch (NE, east-central China in winter)
Loxia curvirostra	Red Crossbill (north-central China)
Coccothraustes occo-thraustes	Hawfinch (NE, NW, central China)
Family Emberizidae	
Emberiza leucocephala	Pine Bunting (NE, NW China)
Emberiza spodocephala	Black-faced Bunting (NE China)
Emberiza cia	Eurasian Rock Bunting (throughout China)
Emberiza pusilla	Little Bunting (winters in SE China)
Emberiza shyrcophrys	Yellow-browed Bunting (winters in SE China)

accumulated by human hunting (stone and bone tools have been found), and fire was used by *H. erectus* (see figure 12-14).

An alternative point of view (Binford 1981; Binford and Ho 1985) argues that: (1) many of the bones in the cave, including at least some of those of hominids, were brought there by carnivores; (2) many of the so-called bone tools actually are bone modified by gnawing rodents, including porcupines; and (3) the geological evidence indicates burning in place of deposits in the Zhoukoudian cave, not necessarily hominid use of fire. Strong evidence lends credence to at least part of this alternative view of the Zhoukoudian deposit.

Figure 12-14 *Statue of Homo erectus using fire in a Beijing museum exhibit.*

Pleistocene Mammoths

More than 150 Pleistocene mammoth localities are known in China, most of them restricted to the northeastern portion of the country. The fossils are part of what Pei (1957) termed the *Mammuthus-Coelodonta* fauna of the late Pleistocene in northeastern China (T. Liu and Li 1984). This was a time of cold glacial climate, but no Chinese mammoth is known that is younger than about 20,000 years old; the last glacial maximum in China occurred about 18,000 years ago.

There is clear and unambiguous association between mammoth fossils and human artifacts in northeastern China. However, to what extent the extinction of *Mammuthus* (and other large late Pleistocene mammals) was influenced by Paleolithic hunters in China has never really been analyzed (Martin 1984).

Pleistocene Lower Vertebrates

Remains from the Zhoukoudian cave site dominate China's Pleistocene record of lower vertebrates. Pleistocene fossil fishes are very few in number (Bien 1934). Locality 3 at Zhoukoudian contained *Ctenopharyngodon* (see Bien 1934). Only two localities have yielded lissamphibians: (1) locality 3 of Zhoukoudian yielded limb bones of *Rana nicromaculata* and *R. asiatica* (Bien 1934); and (2) Pleistocene deposits at Ertemte, Nei Monggol, yielded remains of at least 60 individuals of *R. hipparionum* (Schlosser 1924).

Chinese Pleistocene turtles are emydids (*Ocadia, Chinemys, Cuora*), testudinids (*Testudo*) and trionychids (*Amyda*) (Chow 1961; Ye 1963, 1994; Sun et al. 1992). The Chinese Pleistocene lower vertebrate record is little studied and this is well demonstrated by the fact that no lizard or snake records of Pleistocene age have been published. Clearly, much work remains to be done to develop a Pleistocene ichthyofauna and herpetofauna from China.

Pleistocene Birds

Locality 1 at Zhoukoudian, the famous "chicken-bone hill" bird-bone locality that initially drew attention to the area, has yielded the only extensive avifauna of Pleistocene age from China (Hou 1985; table 12-2). Most of these birds are still extant and live today in China, but a few, such as the ostrich *Struthio* and the extinct falcon *Falco chowi* Hou, are peculiar to the Chinese Pleistocene.

Indeed, eggs and eggshell fragments of *Struthio anderssoni* are common in the Pleistocene loess of northern China. Today, ostriches live only in southern

Africa, but their fossil record begins in the early Pliocene of Eurasia. During the Pleistocene, ostriches were diverse and spread over the Old World from China, Russia and Mongolia, to India across Europe and throughout Africa (Feduccia 1996).

Other Pleistocene records of Chinese fossil birds are few in number: (1) *Buteo* from Wanxian in Sichuan; (2) *Tetrastes* and *Scolopax* from Fuxian in Liaoning (Zhou et al. 1990); and (3) *Phasianus* from Yunshui cave near Beijing (Huang and Hou 1984). Clearly, much potential to develop a more extensive Chinese Pleistocene avifauna exists and should be pursued.

Origin of China's Extant Vertebrates

China's record of Neogene and Pleistocene lower vertebrates is too poorly known to allow inferences about when the modern Chinese ichthyofauna and herpetofauna first evolved. The avifauna of middle Pleistocene locality 1 at Zhoukoudian is almost totally modern. However, so little is known of Chinese Neogene birds, it is impossible to say if the modernization of the avifauna took place before the Pleistocene. Mammals, however, are well enough known from the Chinese Pleistocene (and Miocene-Pliocene) to allow meaningful discussion of the origin of China's extant mammals.

China's Pleistocene mammals segregate into northern and southern zoogeographic regions that parallel today's Palearctic and Oriental zoogeographic regions (see figure 11-20). As was clear in the last chapter, the demarcation of these two zoogeographic realms began in the early Neogene. This demarcation was intensified during the Pleistocene, when a northern mammalian fauna can be readily distinguished from an *Ailuropoda-Stegodon* fauna to the south.

Pliocene survivors and taxa unique to the Pleistocene dominate early Pleistocene mammal faunas of China. By the late Pleistocene, however, most extant mammals had appeared, Pliocene survivors were extinct, and the last occurrences of Pleistocene taxa are recorded.

Chapter 13

Summary

This chapter briefly summarizes the preceding chapters, presenting a concise overview of Chinese fossil vertebrates.

History of Vertebrate Paleontology in China

Fossil vertebrates have been known from China for more than 2000 years, but scientific study of them did not begin until the 1800s. This study can be organized to fit three phases of a model proposed by Basalla (1967) to explain the introduction of a science into any non-European nation.

During the first phase, from about the 1870s until the 1920s, only foreign vertebrate paleontologists collected and studied Chinese vertebrate fossils, sending their collections to the West. The 1920s and 1930s saw foreign vertebrate paleontologists living in China and beginning to train Chinese vertebrate paleontologists. The most significant student they trained was C.C. Young (Yang Zhungjian), who can truly be called the "father" of Chinese vertebrate paleontology. Young implemented the third phase between the 1930s and 1950s by establishing an independent Chinese vertebrate paleontology that thrives today.

Cambrian-Ordovician

The oldest fossil vertebrates are jawless fishes of Early Cambrian age from southern China. No other vertebrates of Cambrian or Ordovician age are yet known from China, but extensive marine deposits of these systems hold great potential for a record of fossil vertebrates.

Silurian

China contains an extensive fossil record of vertebrates of Silurian age. Southern China contains fossils of galeaspid agnathans and "acanthodians" from Lower-Upper Silurian strata, thelodont agnathans and antiarch placoderms from middle-Upper Silurian strata, and its oldest fossil records of Osteichthyes and Chondrichthyes are from the Upper Silurian. China's Silurian vertebrate record can be organized into four time intervals: Early Silurian (*Dayongaspis* biochron), early middle Silurian (*Hanyangaspis* biochron), late middle Silurian (*Sinogalaeaspis* biochron) and Late Silurian. The Silurian vertebrates of China emphasize the clear isolation of the south China microplate. Endemic galeaspid agnathans and "acanthodians" dominate the early-middle Silurian vertebrate-fossil assemblages of China. The first hint of cosmopolitanism comes in the Late Silurian when some wide-ranging thelodonts and chondrichthyans first appear in the Chinese vertebrate-fossil record.

Devonian

China's extensive fossil record of Devonian fishes comes mostly from marine deposits in the southern part of the country. Fossils represent agnathans, placoderms, acanthodians, chondrichthyans and osteichthyans; no tetrapods are known. China's Devonian vertebrate-fossil record can be organized into three time intervals: Early Devonian (*Yunnanolepis* biochron), Middle Devonian (*Bothriolepis* biochron) and Late Devonian (*Remigolepis* biochron). Eugaleaspid agnathans and antiarch placoderms dominated highly endemic Early Devonian fish faunas of southern China. By Middle Devonian time this endemism began to breakdown with the appearance of cosmopolitan taxa such as the antiarch placoderm *Bothriolepis*. Almost all Chinese Devonian agnathans represent a distinct, endemic clade of jawless fishes, the Galeaspida.

Carboniferous

China's Carboniferous vertebrate fossil record is extremely limited. Ichthyoliths of *Acanthodes* and chondrichthyans from the Lower Carboniferous of Guizhou are the bulk of this record. Ironically, Carboniferous marine and terrestrial strata are widely distributed in China and have yielded extensive fossil assemblages of marine invertebrates and nonmarine plants. The dearth of Chinese Carboniferous vertebrate fossils almost certainly reflects a lack of discovery.

Permian

Permian fossil vertebrates from China are only known from Middle-Upper Permian strata in northern China. These vertebrates are fishes and tetrapods (principally indeterminate pareiasaurs and *Dicynodon*) very similar to correlative vertebrate assemblages in the Karoo basin of South Africa and the Russian

Urals. Chinese Permian vertebrates serve as a compelling argument for a complete assembly of Pangea by the Middle Permian. A lack of older (Early) Permian vertebrates from China represents a significant gap in the Chinese vertebrate fossil record.

Triassic

China's Triassic record of terrestrial vertebrates is essentially restricted to Lower-Middle Triassic strata in northern China. A diverse array of marine reptiles—especially nothosaurs, huhpesuchians and ichthyosaurs—come from Triassic marine strata in southern China.

Northern China's terrestrial Triassic tetrapod succession is very similar to that seen in the Lower-Middle Triassic strata of the Karoo basin in South Africa and in the Russian Urals. The Chinese succession is dicynodont-dominated, with the classic Pangean succession *Lystrosaurus-Kannemeyeria-Shansiodon* well established. However, China also has its own large, endemic Triassic dicynodonts—*Parakannemeyeria* and *Sinokannemeyeria*. Other important groups of Chinese Triassic terrestrial tetrapods include procolophonids, proterosuchians and erythrosuchids. Labyrinthodont amphibians are rare. The lack of a Late Triassic tetrapod fauna from China is a significant gap in its vertebrate fossil record.

Jurassic

China's Jurassic vertebrate fossils are much more widely distributed geographically and more densely cover their geologic time period than do China's Permian or Triassic vertebrate fossils. The most significant occurrences are in the Sichuan basin, where a 3000 m thick section of fluvial and lacustrine strata contains superposed fossil vertebrates of Early, Middle and Late Jurassic age. This is the most complete, single succession of Jurassic vertebrate assemblages on earth.

The Early Jurassic Lufeng Formation vertebrate fauna from Yunnan yields prosauropod dinosaurs, tritylodont therapsids and primitive mammals (especially *Sinoconodon*) that have had a major influence on paleontological understanding of the evolution of these groups. Chinese Middle Jurassic dinosaurs—especially sauropods and the earliest stegosaur, *Huayangosaurus*—also occupy key roles in deciphering Jurassic dinosaur evolution. Chinese Late Jurassic dinosaurs are part of a rather cosmopolitan dinosaur fauna found across the diverging Pangean continents.

Cretaceous

China's Cretaceous fossil vertebrates have a broader geographic distribution than do its Jurassic fossil vertebrates. However, the Cretaceous vertebrate fossil record has a significant temporal gap, most of the "middle" Cretaceous (Albian-Santonian), from which almost no vertebrate fossils are known.

Because of this gap, two basic vertebrate fossil assemblages can be recognized in the Chinese Cretaceous. The older assemblage is of late Neocomian-early Albian age and is characterized by the primitive ceratopsian dinosaur *Psittacosaurus*. The younger assemblage is of Campanian-Maastrichtian age and is characterized by hadrosaurs and *Tarbosaurus*. Despite this disjunction, the Chinese Cretaceous vertebrate fossil record documents important aspects of the evolution of ceratopsian and hadrosaurid dinosaurs. It also contains significant Early Cretaceous birds and the most extensive assemblages of dinosaur eggs on earth.

Paleogene

Throughout China, Paleogene strata (mostly lacustrine and fluvial red-beds) contain mammal-dominated assemblages of vertebrate fossils. The mammals represent one of the most significant records of Paleogene mammals known, and can be organized into nine land-mammal "ages." A variety of eutherian orders of mammals probably arose in eastern Asia during the Paleogene (or Late Cretaceous) and include the Anagalida, Notoungulata, Dinocerata, Pantodonta, Rodentia, Lagomorpha, Mesonychia and Carnivora. Chinese Paleocene mammals are mostly endemics; some early Paleocene and Eocene immigration took place between Asia and North America, and during the Oligocene more European forms appear in the Chinese mammal record. The turnover from archaic placental orders of the Paleocene to the modern placental orders of the later Cenozoic is well documented by Chinese Paleogene mammals. This turnover took place mostly during the middle Eocene (Arshantan-Irdinmanhan). China has a diverse but very incomplete record of Paleogene lower vertebrates dominated by turtles. The fossil record of Paleogene birds is very limited.

Miocene-Pliocene

Mammals dominate China's rich and extensive Miocene-Pliocene vertebrate-fossil record and have been organized into eight land-mammal "ages." These fossil mammals document a complex pattern of endemism overlain by immigration events from Africa, southern Asia (India-Pakistan), Europe and North America. Particularly significant Miocene immigrations involved: (1) proboscideans, who first appeared in China in the early Miocene as immigrants from the west; and (2) horses, including key taxa that came to China from North America, *Anchitherium* (an early Miocene immigrant) and *Hipparion* (a late Miocene immigrant). During the Pliocene, immigrations of camelids and canids (early Pliocene) and of elephantids and *Equus* (late Pliocene) took place. Chinese Neogene ape fossils represent an important record of early gibbons and related forms. The Neogene uplift of the Himalayas and Tibetan Plateau altered the Chinese climate so that western and northern China were relatively dry, whereas southeastern China had a much wetter, monsoonal climate. This may in part explain the appearance of two separate zoogeographic regions—Palearctic and Oriental—in eastern Asia that can be detected in China's fossil mammals as far back as the middle Miocene.

Pleistocene

China has very limited fossil records of Pleistocene lower vertebrates and birds, but an extensive record of Pleistocene mammals. These mammals have classically been organized into northern and southern faunas (zoogeographic provinces) that show clear modernization of the mammal fauna through time. The northern mammals are characteristic of the Nihewanian, Zhoukoudianian and Salawusuan land-mammal "ages"; the southern mammals belong to the "*Ailuropoda-Stegodon* fauna." This southern fauna includes the giant extinct ape *Gigantopithecus*. The extensive fossil record of *Homo* (especially *Homo erectus* from Zhoukoudian) plays an important role in current debate about the origin of modern Mongoloids among *Homo sapiens*. Little is known about late Pleistocene extinctions in China. Pleistocene mammals indicate that the modern Palearctic and Oriental zoogeographic regions are readily recognized in eastern Asia by the Pleistocene.

References

Adams, A. L., 1868, Has the Asiatic elephant been found in a fossil state? Quarterly Journal of the Geological Society of London, v. 24, pp. 496–498.

Aguirre, E., 1997, Human evolution in the Plio-Pleistocene interval; *in* Van Couvering, J. A., ed., The Pleistocene and the beginning of the Quaternary: Cambridge, Cambridge University Press, pp. 129–137.

Amalitzky, V., 1922, Diagnoses of the new forms of vertebrates and plants from the Upper Permian of North Dvina: Bulletin Academy of Science St. Petersburg, v. 16, pp. 329–340.

Anderson, J. M., 1980, World Permo-Triassic correlations: their biostratigraphic basis; *in* Cresswell, M. M. and Vella, P., eds., Gondwana five: Rotterdam, A. A. Balkema, pp. 3–10.

Anderson, J. M. and Cruickshank, A. R. I., 1978, The biostratigraphy of the Permian and the Triassic. Part 5. A review of the classification and distribution of Permo-Triassic tetrapods: Palaeontologica Africana, v. 21, pp. 15–44.

Andersson, J. G., 1922, Current palaeontological research in China: Bulletin of the American Museum of Natural History, v. 46, pp. 727–737.

———, 1923, Essays on the Cenozoic of northern China: Memoirs of the Geological Survey of China (A), v. 3, pp. 1–152.

Andrews, R.C., 1926, On the trail of ancient man. New York, G.P. Putnam's Sons, p. 375.

———, 1929a, Ends of the earth: New York, G.P. Putnam's Sons.

———, 1929b, On the trail of ancient man: New York, G.P. Putnam's Sons.

———, 1932, Natural history of Central Asia, v. 1, the conquest of Central Asia. New York, American Museum of Natural History, p. 677.

Bandyopadhyay, S., 1988, A kannemeyeriid dicynodont from the Middle Triassic Yerrapalli Formation: Philosophical Transactions of the Royal Society of London B, v. 320, pp. 185–233.

Barrett, P. M., 1999, A sauropod dinosaur from the lower Lufeng Formation (Lower Jurassic) of Yunnan Province, People's Republic of China: Journal of Vertebrate Paleontology, v. 19, pp. 785–787.

———, 2000, Evolutionary consequences of dating the Yixian Formation: Tree, v. 15, pp. 99–103.

Barrett, P. M., You, H., Upchurch, P., and Burton, A. C., 1998, A new ankylosaurian dinosaur (Ornithischia: Ankylosauria) from the Upper Cretaceous of Shanxi Province, People's Republic of China: Journal of Vertebrate Paleontology, v. 18, pp. 376–384.

Barsbold, R., 1974, Saurornithoididae, a new family of small theropod dinosaurs from Central Asia and North America: Palaeontologica Polonica, v. 30, pp. 5–22.

Basalla, G., 1967, The spread of western science: Science, v. 156, pp. 611–622.

Battail, B., Dejax, J., Richir, P., Taquet, P., and Ve'ran, M., 1995, New data on the continental Upper Permian in the area of Luang-Prabang, Laos: Geological Survey of Vietnam Journal of Geology B, v. 5–6, pp. 11–15.

Beard, C., Wang, J., and Tong, Y., 1993, The first undoubted Asian plesiadapoids: Journal of Vertebrate Paleontology, v. 13, supplement to no. 3, pp. 25A–26A.

Bennet, S. C., 1989, A pteranodontid pterosaur from the Early Cretaceous of Peru; with comments on the relationships of Cretaceous pterosaurs: Journal of Paleontology, v. 63, pp. 669–677.

Berggren, W. A., 1998, The Cenozoic Era: Lyellian (chrono) stratigraphy and nomenclatural reform at the millennium; *in* Blundell, D. J. and Scott, A. C., eds., Lyell: The past is the key to the present: Geological Society, London, Special Publication 143, pp. 111–132.

Berggren, W. A., Kent, D. V., Swisher, C. C. III, and Aubry, M.-P., 1995, A revised Cenozoic geochronology and chronostratigraphy: SEPM Special Publication 54, pp. 129–212.

Berman, D. J., Sumida, S. S. and Lombard, R. E., 1997, Biogeography of primitive amniotes; *in* Sumida, S. S. and Martin, K. L. M., eds., Amniote origins: San Diego, Academic Press, pp. 85–139.

Bernor, R. L., Qiu, Z. and Hayek, L. C., 1990, Systematic revision of Chinese *Hipparion* species described by Sefve, 1927: American Museum Novitates, no. 2984, p. 60.

Bi, Z., Yu, Z. and Qiu, Z., 1977, First discovery of mammal remains from Upper Tertiary deposits near Nanking: Vertebrata PalAsiatica, v. 15, pp. 126–138.

Bien, M., 1934, On the fossil Pisces, Amphibia and Reptilia from Choukoutien localities 1 and 3: Palaeontologica Sinica C, v. 10(1), pp. 1–32.

Bien, M. N., 1940, Discovery of Triassic saurischian and primitive mammalian remains at Lufeng, Yunnan: Bulletin of the Geological Society of China, v. 20, pp. 225–234.

Binford, L. R., 1981, Bones: Ancient men and modern myths. New York, Academic Press.

Binford, L. R. and Ho C. K., 1985, Taphonomy at a distance: Zhoukoudian, "the cave home of Beijing Man"?: Current Anthropology, v. 26, pp. 413–442.

Black, D., 1926, Tertiary man in Asia: the Chou K'ou Tien discovery: Nature, v. 118, pp. 733–734.

Blieck, A., Elliott, D. K., and Gagnier, P.-Y., 1991, Some questions concerning the phylogenetic relationships of heterostracans, Ordovician to Devonian jawless vertebrates; *in* Chang, M., Liu, Y. and Zhang, G., eds., Early vertebrates and related problems of evolutionary biology: Beijing, Science Press, pp. 1–17.

Bohlin, B., 1926, Die Familie Giraffidae: Palaeontologica Sinica C, v. 4(1).

——, 1937, Oberoligozäne Säugetiere aus dem Shargaltein-Tal (Western Kansu): Palaeontologia Sinica, New Series, C, v. 3, pp. 1–66.

——, 1942, The fossil mammals from the Tertiary deposit of Taben-buluk, west Kansu. Pt. I, Insectivora and Lagomorpha: Palaeontologia Sinica, New Series, C, v. 8a, pp. 1–113.

——, 1946, The fossil mammals from the Tertiary deposit of Taben-buluk, west Kansu. Pt. II, Simplicidentata … : Palaeontologia Sinica, New Series, C, v. 8b, pp. 1–259.

——, 1953, Fossil reptiles from Mongolia and Kansu: Reports from the Scientific Expedition to N. W. Provinces of China, under the leadership of Dr. Sven Hedin, The Sino-Swedish Expedition Publication 37, pp. 1–113.

Boonstra, L. D., 1938, A report on some Karoo reptiles from the Luangwa Valley, northern Rhodesia: Quarterly Journal of the Geological Society of London, v. 94, pp. 371–384.

——, 1953, A report on a collection of fossil reptilian bones from Tangyanika Territory: Annals of the South African Museum, v. 42, pp. 5–18.

Brett-Surman, M. K., 1979, Phylogeny and palaeobiogeography of hadrosaurian dinosaurs: Nature, v. 277, pp. 560–562.

Brinkman, D. B. and Dong, Z., 1993, New material of *Ikechosaurus sunailinae* (Reptilia: Choristodera) from the Early Cretaceous Laohongdong Formation, Ordos basin, Inner Mongolia, and the interrelationships of the genus: Canadian Journal of Earth Sciences, v. 30, pp. 2153–2162.

Brinkman, D. B. and Peng, J., 1993, *Ordosemys leios*, n. gen., n. sp., a new turtle from the Early Cretaceous of the Ordos basin, Inner Mongolia: Canadian Journal of Earth Sciences, v. 36, pp. 2128–2138.

Broom, R., 1905, Notes on the localities of some type specimens of the Karoo reptiles: Records of the Albany Museum, v. 1, pp. 275–278.

——, 1906, The classification of the Karoo beds of South Africa: Geological Magazine, New Series, Decade 5.

Buffetaut, E., 1982, Mesozoic vertebrates from Thailand and their paleogeological significance: Terra Cognita, v. 2, pp. 27–34.

——, 1993, Phytosaurs in time and space: Paleontologia Lombarda Nuova Serie, v. 2, pp. 39–44.

——, 1995, An ankylosaurid dinosaur from the Upper Cretaceous of Shandong (China): Geological Magazine, v. 132, pp. 683–692.

Buffetaut, E. and Suteethorn, V., 1992, A new species of the ornithischian dinosaur *Psittacosaurus* from the Early Cretaceous of Thailand: Palaeontology, v. 35, pp. 801–812.

Buffetaut, E. and Tong, H., 1995, The Late Cretaceous dinosaurs of Shandong, China: Old finds and new interpretations; *in* Sun, A. and Wang, Y., eds., Sixth symposium on Mesozoic terrestrial ecosystems and biota, short papers: Beijing, China Ocean Press, pp. 139–142.

Buffetaut, E. and Tong-Buffetaut, H., 1993, *Tsintaosaurus spinorhinus* Young and *Tanius sinensis* Wiman: A preliminary comparative study of two hadrosaurs

(Dinosauria) from the Upper Cretaceous of China: Compte Rendus Academie Science Paris II, v. 317, pp. 1255–1261.

Burke, J. J., 1941, New fossil Leporidae from Mongolia: American Museum Novitates, no. 1117.

Burrett, C. F., Long, J. A., and Strait, B., 1990, Early-middle Palaeozoic biogeography of Asian terranes derived from Gondwana; *in* McKerrow, W. S. and Scotese, C. R., eds., Paleozoic biogeography and palaeogeography: Geological Society of London Memoir, v. 12, pp. 163–174.

Cai, B., 1987, A preliminary report on the late Pliocene micromammalian fauna from Yangyuan and Yuxian, Hebei: Vertebrata PalAsiatica, v. 25, pp. 124–136.

Callaway, J. M. and Massare, J. A., 1989, Geographic and stratigraphic distribution of the Triassic Ichthyosauria (Reptilia, Diapsida): Neues Jahrbuch für Geologie und Paläontologie Abhandlungen, v. 178, pp. 37–58.

Camp, C. L., 1935, Dinosaur remains from the Province of Szechuan: Bulletin Department of Geology, University of California Publications, v. 23, pp. 467–471.

——, 1976, Vorläufige Mitteilung Über grosse Ichthyosaurier aus der oberen Trias von Nevada: Österreichische Akademie der Wissenschaften, Mathematisch-Naturwissenschafliche Klasse, Sitzungsberichte, Abteilung I, v. 185, pp. 125–134.

——, 1980, Large ichthyosaurs from the Upper Triassic of Nevada: Palaeontographica A, v. 170, pp. 139–200.

Campbell, K. S. W. and Barwick, R. E., 1990, Paleozoic dipnoan phylogeny: Functional complexes and evolution without parsimony: Paleobiology, v. , pp. 143-169.

Cao, K., 1993, A note on the findings of fossil Cetacea in China: Bulletin Museo Regina Scienze Naturale Torino, v. 11, pp. 365–376.

Cappetta, H., 1987, Chondrichthyes II Mesozoic and Cenozoic Elasmobranchii: Handbook of Paleoichthyology, v. 3B, p. 193.

Carpenter, K., 1982, Baby dinosaurs from the Late Cretaceous Lance and Hell Creek formations and a description of a new species of theropod: Contributions to Geology, University of Wyoming, v. 20, pp. 123–134.

——, 1992, Tyrannosaurids (Dinosauria) of Asia and North America; *in* Mateer, N. J. and Chen, P., eds., Aspects of nonmarine Cretaceous geology: Beijing, China Ocean Press, pp. 250–268.

Carpenter, K. and Alf, K., 1994, Global distribution of dinosaur eggs, nests, and babies; *in* Carpenter, K., Hirsch, K. F. and Horner, J. R., eds., Dinosaur eggs and babies: Cambridge, Cambridge University Press, pp. 15–30.

Carpenter, K., Hirsch, K. F., and Horner, J. R., 1994, Introduction; *in* Carpenter, K., Hirsch, K. F. and Horner, J. R., eds., Dinosaur eggs and babies: Cambridge, Cambridge University Press, pp. 1–11.

Carroll, R. L., 1988, Vertebrate paleontology and evolution. New York, Freeman.

Carroll, R. L. and Dong, Z., 1991, *Hupehsuchus*, an enigmatic aquatic reptile from the Triassic of China, and the problem of establishing relationships: Philosophical Transactions Royal Society of London B, v. 331, pp. 131–153.

Cerdeño, E., 1996, Rhinocerotidae from the middle Miocene of the Tung-gur Formation, Inner Mongolia (China): American Museum Novitates, no. 3184.

Chang, K., 1963, A new species of *Bothriolepis* from Kwangtung: Vertebrata PalAsiatica, v. 7, pp. 342–351.

——, 1965, New antiarchs from the Middle Devonian of Yunnan: Vertebrata PalAsiatica, v. 9, pp. 1–14.

Chang, M. and Chou, C., 1977, On late Mesozoic fossil fishes from Zhejiang Province, China: Memoirs Institute of Vertebrate Paleontology and Paleoanthropology, Academia Sinica, no. 12, pp. 1–59.

Chang, M. and Chow, C., 1986, Stratigraphic and geographic distributions of the late Mesozoic and Cenozoic fishes of China; *in* Uyeno, T., Arai, R., Taniuchi, T. and Matsuura, K., eds., Indo-Pacific fish biology: Tokyo, Ichthyological Society of Japan, pp. 529–539.

Chang M. and Jin, F., 1996, Mesozoic fish faunas of China; *in* Arratia, G. and Viohl, G., eds., Mesozoic fishes-systematics and paleoecology: Munich, Verlag Dr. Friedrich Pfeil, pp. 461–478.

Chang, M. and Wang, J., 1995, A new Emsian dinorhynchid (Dipnoi) from Guangan, southeastern Yunnan, China: Geobios Mémoire Spécial, no. 19, pp. 233–240.

Chang, M. and Yu, X., 1984, Structure and phylogenetic significance of *Diabolichthys speratus* gen. et sp. nov., a new dipnoan-like form from the Lower Devonian of eastern Yunnan, China: Proceedings of the Linnaean Society of New South Wales, v. 107, pp. 171–184.

Chang, M., Chen, Y. and Tong, H., 1996, A new Miocene xenocyprinine (Cyprinidae) from Heilongjiang Province, northeast China and succession of late Cenozoic fish faunas of east Asia: Vertebrata PalAsiatica, v. 34, pp. 165–183.

Chang, W., 1962, Ordovician of China. Beijing, Science Press.

Chao, S., 1962, Concerning a new species of *Psittacosaurus* from Laiyang, Shantung: Vertebrata PalAsiatica, v. 6, pp. 349–360.

Chao, T. and Chiang, Y., 1974, Microscopic studies on the dinosaurian egg-shells from Laiyang, Shantung Province: Scientia Sinica, v. 17, pp. 71–83.

Chen, G., 1977, A new genus of Iranotheriinae of Ningxia: Vertebrata PalAsiatica, v. 15, pp. 143–147.

Chen, G. and Wu, W., 1976, Miocene mammalian fossils of Jiulongkou, Cixian district, Hebei: Vertebrata PalAsiatica, v. 14, pp. 6–15.

Chen, P., 1982, Jurassic conchostracans from Mengying district, Shandong: Acta Palaeontologia Sinica, v. 21, pp. 133–139.

——, 1983, A survey of the non-marine Cretaceous in China: Cretaceous Research, v. 4, pp. 123–143.

——, 1986, On the Chinese non-marine Paleocene from the new discovery of the conchostracan fauna of Nanxiong, Guangdong: Acta Palaeontologia Sinica, v. 25, pp. 380–394.

——, 1987, Cretaceous paleogeography in China: Palaeogeography, Palaeoclimatology, Palaeoecology, v. 59, pp. 49–56.

——, 1988, Distribution and migration of Jehol fauna with reference to nonmarine Jurassic-Cretaceous boundary in China: Acta Palaeontologica Sinica, v. 27, pp. 659–683.

——, 1996, Nonmarine Jurassic strata of China: Museum of Northern Arizona Bulletin 60, pp. 395–412.

Chen, P. and Wang, Z., 1984, On the non-marine Cretaceous-Tertiary boundary of China: Developments in Geoscience, Academia Sinica Contributions to the 27th International Geological Congress, Moscow, pp. 129–134.

Chen, P., Chao, M., Pan, H., Ye, C., Li, W., Shen, Y., and Chen, J., 1980, Problems of Mesozoic continental stratigraphy in Shandong: Journal of Stratigraphy, v. 4, pp. 301–309.

Chen, P., Li, W., Chen, J., Ye, C., Wang, Z., Shen, Y.,and Sun, D., 1982a, Sequence of fossil biotic groups of Jurassic and Cretaceous in China: Scientia Sinica (Series B), v. 25, pp. 1011–1020.

——, 1982b, Stratigraphical classification of Jurassic and Cretaceous in China: Scientia Sinica (Series B), v. 25, pp. 1227–1248.

Chen, P[ingfu], Liu, H. and Yan, J., 1999, Discovery of fossil Coreoperca (Perciformes) in China: Vertebrata PalAsiatica, v. 37, pp. 212–227.

Chen, T., Yang, Q. and Wu, E., 1994, Antiquity of *Homo sapiens* in China: Nature, v. 368, pp. 55–56.

Cheng, Z., 1980a, Mesozoic stratigraphy and paleontology of the Shansi-Gansu-Ningxia basins (7). Vertebrate fossils. Beijing, Geological Publishing House, pp. 115–188.

——, 1980b, Vertebrates; *in* Permian and Triassic strata and fossil assemblages in the Dalongkou area of Jimsar, Xinjiang: Peoples Republic of China, Ministry of Geology and Mineral Resources, Geology Memoirs (2), v. 3, pp. 207–218.

——, 1981, Permo-Triassic continental deposits and vertebrate faunas of China; *in* Cresswell, M. M. and Vella, P., eds., Gondwana five: Rotterdam, A. A. Balkema, pp. 65–70.

Cheng, Z. and Li, J., 1996, First record of a primitive anteosaurid dinocephalian from the Upper Permian of Gansu, China: Vertebrata PalAsiatica, v. 34, pp. 123–134.

——, 1997, A new genus of primitive dinocephalian—the third report on Late Permian Dashankou lower tetrapod fauna: Vertebrata PalAsiatica, v. 35, pp. 35–43.

Cheng, Z. and Lucas, S. G., 1993, A possible nonmarine GSSP for the Permian-Triassic boundary: Albertiana, no. 12, pp. 39–44.

Cheng, Z., Hu, C., and Fang, X., 1995, Cretaceous stratigraphy and vertebrate faunal sequence in Laiyang-Zhucheng basin, Shandong, China—as a candidate stratotype of non-marine Cretaceous; *in* Sun, A. and Wang, Y., eds., Sixth symposium on Mesozoic terrestrial ecosystems and biota, short papers: Beijing, China Ocean Press, pp. 97–100.

Cheng, Z., Lucas, S. G., and Zidek, J., 1996, *Edestus* (Chondrichthyes, Elasmo-branchii) from the Upper Carboniferous of Xinjiang, China: Neues Jahrbuch für Geologie und Paläontologie Monatshefte, pp. 701–707.

Chi, Y., 1940, On the discovery of *Bothriolepis* in the Devonian of central Hunan: Bulletin of the Geological Society of China, v. 20, pp. 57–62.

Chia [Jia], L., 1975, The cave home of Peking man. Beijing, Foreign Languages Press.

Chiappe, L. M., 1995, The first 85 million years of avian evolution: Nature, v. 378, pp. 349–355.

Chiappe, L. M., Ji, S., Ji, Q., and Norrell, M. A., 1999, Anatomy and systematics of the Confuciusornithidae (Theropoda: Aves) from the late Mesozoic of north-eastern China: Bulletin of the American Museum of Natural History, v. 242, pp. 1–89.

Chinnery, B. J., Lipka, T. R., Kirkland, J. I., Parrish, J. M., and Brett-Surman, M. K., 1998, Neoceratopsian teeth from the Lower to Middle Cretaceous of North America: New Mexico Museum of Natural History and Science, Bulletin 14, pp. 297–302.

Chiu, C., 1965, First discovery of *Lophiomeryx* in China: Vertebrata PalAsiatica, v. 9, pp. 395–398.

——, 1973, A new genus of giant rhinoceros from Oligocene of Dzungaria, Sinki-ang: Vertebrata PalAsiatica, v. 11, pp. 182–191.

Chiu, C., Li, C., and Chiu, C., 1979, The Chinese Neogene—a preliminary review of the mammalian localities and faunas: Annales Géologique Pays Hellén Tome hors Serie 1979, pp. 263–272.

Chow, M., 1951, Notes on Late Cretaceous dinosaurian remains and the fossil eggs from Laiyang, Shantung: Bulletin Geological Society of China, v. 31, pp. 89–96.

——, 1954, Cretaceous turtles from Laiyang, Shantung: Acta Palaeontologia Sin-ica, v. 2, pp. 395–408.

——, 1957, Notes on some mammalian fossils from the late Cenozoic of Sinki-ang: Vertebrata PalAsiatica, v. 1, pp. 33–41.

——, 1958, Mammalian faunas and correlation of Tertiary and early Pleistocene of south China: Journal of the Paleontological Society of India, v. 3, pp. 123–130.

——, 1961, A new Pleistocene turtle from Hsiangfen, Shansi: Acta Paleontologia Sinica, v. 9, pp. 426–430.

——, 1962, A tritylodontid specimen from Lufeng, Yunnan: Vertebrata PalAsiat-ica, v. 6, pp. 365–367.

——, 1965, Mesonychids from the Eocene of Honan: Vertebrata PalAsiatica, v. 9, pp. 286–291.

Chow, M. and Chiu, C., 1963, New genus of giant rhinoceros from Oligocene of Inner Mongolia: Vertebrata PalAsiatica, v. 7, pp. 230–239.

Chow, M. and Hu, C., 1956, The occurrence of *Anchitherium aurelianense* at Fangshan, Nanking: Acta Palaeontologia Sinica, v. 4(4), pp. 525–533.

Chow, M. and Hu, C. C., 1959, A new tritylodontid from Lufeng, Yunnan: Verte-brata PalAsiatica, v. 3, pp. 9–12.

Chow, M. and Li, C., 1965, *Homogalax* and *Heptodon* of Shantung: Vertebrata PalAsiatica, v. 9, pp. 15–21.

Chow, M. and Qi, T., 1978, Paleocene mammalian fossils from Nomogen Forma-tion of Inner Mongolia: Vertebrata PalAsiatica, v. 16, pp. 77–85.

Chow, M. and Sun, A., 1960, A new procolophonid from northwestern Shanxi: Vertebrata PalAsiatica, v. 4, pp. 11–13.

Chow, M. and Tung, Y., 1962, Notes on some new uintathere materials of China: Vertebrata PalAsiatica, v. 6, pp. 368–374.

Chow, M. and Xu, Y., 1965, Amynodonts from the Upper Eocene of Honan and Shansi: Vertebrata PalAsiatica, v. 9, pp. 190–204.

Chow, M. and Zhang, Y., 1974, The Chinese fossil elephants: Beijing, Science Press.

Chow, M., Li, C., and Chang, Y., 1973[a], Late Eocene mammalian faunas of Honan and Shansi with notes on some vertebrate fossils collected therefrom: Vertebrata PalAsiatica, v. 11, pp. 165–181.

Chow, M., Chang, Y., Wang, B., and Ting, S., 1973[b], New mammalian genera and species from the Paleocene of Nanhsiung, N. Kwangtung: Vertebrata PalA-siatica, v. 11, pp. 31–35.

Chow, M., Zhang, Y., Wang, B., and Ding, S., 1977, Mammalian fauna from the Paleocene of Nanxiong basin, Guandong: Palaeontologia Sinica, New Series C, no. 20, p. 100.

Cifelli, R. L., Schaff, C. R., and McKenna, M. C., 1989, The relationships of the Arctostylopidae (Mammalia): New data and interpretation: Bulletin of the Museum of Comparative Zoology, v. 152, pp. 1–44.

Cluver, M. A., 1971, The cranial morphology of the dicynodont genus *Lystrosau-rus*: Annals of the South African Museum, v. 56, pp. 156–274.

Cluver, M. A. and Hotton, N., 1977, The dicynodont genus *Diictodon* (Reptilia, Therapsida) and its significance: Proceedings and Papers Fourth Gondwana Symposium, Calcutta, pp. 176–183.

——, 1981, The genera *Dicynodon* and *Diictodon* and their bearing on the classifi-cation of the Dicynodontia: Annals of the South African Museum, v. 83, pp. 99–146.

Colbert, E. H., 1936, *Palaeotragus* in the Tung Gur Formation of Mongolia: American Museum Novitates, no. 874, p. 17.

——, 1939, Carnivora of the Tung Gur Formation of Mongolia: Bulletin of the American Museum of Natural History, v. 76, pp. 47–81.

——, 1974, *Lystrosaurus* from Antarctia: American Museum Novitates, no. 2535.

——, 1986, Historical aspects of the Triassic-Jurassic boundary problem; *in* Padian, K., ed., The beginning of the age of dinosaurs: Cambridge, Cambridge University Press, pp. 9–18.

Conroy, G. C. and Bown, T. M., 1974, Anthropoid origins and differentiation: The Asian Question: Yearbook of Physical Anthropology, v. 18, pp. 1–6.

Coombs, M. C. and Coombs, W. P., Jr., 1977, Dentition of *Gobiohyus* and a reevaluation of the Helohyidae (Artiodactyla): Journal of Mammalogy, v. 58, pp. 291–308.

Cooper, M. R., 1980, The origins and classification of Triassic dicynodonts by A. W. Keyser and A. R. I. Cruickshank discussion: Transactions Geological Society South Africa, v. 83, pp. 107–110.

Cosgriff, J. W., Hammer, W. R. and Ryan, W. J., 1982, The Pangaean reptile, *Lystrosaurus maccaigi*, in the Lower Triassic of Antarctica: Journal of Paleontology, v. 56, pp. 371–385.

Cox, C. B., 1991, The Pangea dicynodont *Rechnisaurus* and the comparative biostratigraphy of Triassic dicynodont faunas: Paleontology, v. 34, pp. 767–784.

Cox, C. B. and Li, J., 1983, A new genus of Triassic dicynodont from East Africa and its classification: Palaeontology, v. 26, pp. 389–406.

Crompton, A. W. and Luo, Z., 1993, Relationships of the Liassic mammals *Sinoconodon*, *Morganucodon oehleri*, and *Dinnetherium*; in Szalay, F. S., Novacek, M. J. and McKenna, M. C., eds., Mammal phylogeny: Mesozoic differentiation, multituberculates, monotremes, early therians, and marsupials: New York, Springer Verlag, pp. 30–44.

Cruickshank, A. R. I., 1967, A new dicynodont genus from the Manda Formation of Tanzania (Tangyanika): Journal of zoology, v. 153, pp. 163–208.

Cui, G., 1976, *Yunnania*, a new tritylodont genus from Lufeng, Yunnan: Vertebrata PalAsiatica, v. 14, pp. 85–90.

——, 1981, A new genus of Tritylodontoidea: Vertebrata PalAsiatica, v. 19, pp. 5–10.

Cui, G. and Sun, A., 1987, Postcanine root system in tritylodonts: Vertebrata PalAsiatica, v. 25, pp. 245–259.

Danilov, M., 1971, A new dicynodont from the Triassic of Cisuralia: Paleontological Journal, v. 2, pp. 132–135.

Dashzeveg, D., 1993, Asynchronism of the main mammalian faunal events near the Eocene-Oligocene boundary: Tertiary Research, v. 14, pp. 141–149.

Dawson, M. R., 1968, Oligocene rodents (Mammalia) from East Mesa, Inner Mongolia: American Museum Novitates, no. 2324.

Dawson, M. R., Li, C. and Qi, T., 1984, Eocene ctenodactyloid rodents (Mammalia) of eastern and central Asia: Carnegie Museum of Natural History Special Publication 9, pp. 138–150.

Denison, R., 1978, Placodermi: Handbook of Paleoichthyology, v. 2.

——, 1979, Acanthodii: Handbook of Paleoichthyology, v. 5.

Ding, S., 1979, A new edentate from the Paleocene of Guang-dong: Vertebrata PalAsiatica, v. 17, pp. 57–64.

——, 1987, A Paleocene edentate from Nanxiong basin, Guangdong: Palaeontologia Sinica C, no. 24.

Ding, S. and Tong, Y., 1979, Some Paleocene anagalids from Nanxiong, Guangdong: Vertebrata PalAsiatica, v. 17, pp. 137–145.

Ding, S. and Zhang, Y., 1979, Insectivores and anagalids of Chijiang basin, Jiangxi; *in* Mesozoic and Cenozoic red beds of south China: Beijing, Science Press, pp. 354–359.

Ding, S., Zheng, J., Zhang, Y. and Tong, Y., 1977, The age and characteristics of the Liuniu and the Dongjun faunas, Base basin of Guangxi: Vertebrata PalAsiatica, v. 15, pp. 35–45.

Dong, G., 1987, Further discussions upon the character and age of the Xiaolongtan mammalian fauna, Kaiyuan, Yunnan Province: Vertebrata PalAsiatica, v. 25, pp. 116–123.

Dong, Z., 1972, An ichthyosaur fossil from the Qomolangma Feng region: Memoirs of the Institute of Vertebrate Paleontology and Paleoanthropology, Academia Sinica, v. 9, pp. 7–10.

——, 1973, Dinosaurs from Wuerho: Memoirs of the Institute of Vertebrate Paleontology and Paleoanthropology, Academia Sinica, v. 11, pp. 45–52.

——, 1974, Some fossil crocodiles of Sinkiang: Vertebrata PalAsiatica, v. 12, pp. 187–189.

——, 1977, On the dinosaurian remains from Turpan, Xinjiang: Vertebrata PalAsiatica, v. 15, pp. 59–66.

——, 1978, A new genus of Pachycephalosauria from Laiyang, Shantung: Vertebrata PalAsiatica, v. 16, pp. 225–228.

——, 1979, The Cretaceous dinosaur fossils in southern China; *in* Mesozoic and Cenozoic redbeds in southern China: Beijing, Science Press, pp. 342–350.

——, 1980a, The distribution of Triassic marine reptiles in China: Revista Italiana Paleontologia Stratigrafia, v. 85, pp. 1231–1238.

——, 1980b, On the dinosaurian fauna and their stratigraphical distribution in China: Journal of Stratigraphy, v. 4, pp. 24–38.

——, 1984, A new prosauropod from Ziliujing Formation of Sichuan basin: Vertebrata PalAsiatica, v. 22, pp. 310–313.

——, 1990, On remains of the sauropods from Kelamaili region, Junggar basin, Xinjiang, China: Vertebrata PalAsiatica, v. 28, pp. 43–58.

——, 1992, Dinosaurian faunas of China. Berlin, Springer-Verlag.

——, 1993, The field activities of the Sino-Canadian dinosaur project in China, 1987–1990: Canadian Journal of Earth Sciences, v. 30, pp. 1997–2001.

——, 1995, The dinosaur complexes of China and their biochronology; *in* Sun, A. and Wang, Y., eds., Sixth symposium on Mesozoic terrestrial ecosystems and biota, short papers: Beijing, China Ocean Press, pp. 91–96.

——, 1997a, The geology of the Mazongshan area; *in* Dong, Z., ed., Sino-Japanese Silk Road dinosaur expedition: Beijing, China Ocean Press, pp. 5–10.

——, 1997b, On the crocodiles from Mazongshan area, Gansu Province; *in* Dong, Z., ed., Sino-Japanese Silk Road dinosaur expedition: Beijing, China Ocean Press, pp. 11–12.

——, 1997c, On small theropods from Mazongshan area, Gansu Province, China; *in* Dong, Z., ed., Sino-Japanese Silk Road dinosaur expedition: Beijing, China Ocean Press, pp. 13–18.

——, 1997d, On the sauropods from Mazongshan area, Gansu Province, China; *in* Dong, Z., ed., Sino-Japanese Silk Road dinosaur expedition: Beijing, China Ocean Press, pp. 19–23.

——, 1997e, A small ornithopod from Mazongshan area, Gansu Province, China; *in* Dong, Z., ed., Sino-Japanese Silk Road dinosaur expedition: Beijing, China Ocean Press, pp. 24–26.

——, 1997f, The geology of the Turpan basin; *in* Dong, Z., ed., Sino-Japanese Silk Road dinosaur expedition: Beijing, China Ocean Press, pp. 96–101.

——, 1997g, A gigantic sauropod (*Hudiesaurus sinojapanorum* gen. et sp. nov.) from the Turpan basin, China; *in* Dong, Z., ed., Sino-Japanese Silk Road dinosaur expedition: Beijing, China Ocean Press, pp. 102–110.

——, 1997h, On a large claw of sauropod from the Upper Cretaceous in Turpan basin, Xinjiang, China; *in* Dong, Z., ed., Sino-Japanese Silk Road dinosaur expedition: Beijing, China Ocean Press, pp. 111–112.

Dong, Z. and Azuma, Y., 1997, On a primitive Neoceratopsian from the Early Cretaceous of China; *in* Dong, Z., ed., Sino-Japanese Silk Road dinosaur expedition: Beijing, China Ocean Press, pp. 68–89.

Dong, Z. and Currie, P. J., 1996, On the discovery of an oviraptorid skeleton on a nest of eggs at Bayan Mandahu, Inner Mongolia, People's Republic of China: Canadian Journal of Earth Sciences, v. 33, pp. 631–636.

Dong, Z. and Yu, H., 1997, A new segnosaur from Mazongshan area, Gansu Province, China; *in* Dong, Z., ed., Sino-Japanese Silk Road dinosaur expedition: Beijing, China Ocean Press, pp. 90–95.

Dong, Z., Currie, P. J., and Russell, D. A., 1989, The 1988 field program of the dinosaur project: Vertebrata PalAsiatica, v. 27, pp. 233–236.

Dong, Z., Zhou, S., and Zhang, Y., 1983, The dinosaur remains from Sichuan basin, China: Palaeontologia Sinica, Series C, v. 23.

Du, H., Ma, A., Cheng, J., and Wu, W., 1992, Palaeogene biostratigraphic characteristics of red basin western Henan and its adjacent areas, with an outline of Palaeogene palaeobiogeographic provinces of China: Acta Geologica Sinica, v. 5, pp. 101–117.

Eberth, D. A., Russell, D. A., Braman, D. R. and Deino, A. L., 1993, The age of the dinosaur-bearing sediments at Tebch, Inner Mongolia, People's Republic of China: Canadian Journal of Earth Sciences, v. 30, pp. 2101–2106.

Efremov, I. A., 1937, On the stratigraphic subdivision of the continental Permian and Triassic of the USSR on the basis of the fauna of early Tetrapoda: Doklady Akademiya Nauk SSSR, v. 16, pp. 121–126.

——, 1940, Preliminary description of new Permian and Triassic terrestrial vertebrates from the USSR: Trudy Paleontologicheskaya Instituta Akademia Nauk SSSR, v. 10.

Endo, R., 1934, Geology and mineral resources of Manchuria. Tokyo and Osaka, Sanseido.

——, 1939, A new genus of Thecodontia from the *Lycoptera* beds in Manchuria: Bulletin Central National Museum of Manchuria, v. 2, pp. 1–14.

Erben, H. K., Ashraf, A. R., Böhm, H., Hahn, G., Hambach, U., Krumsiek, K., Stets, J., Thein, J. and Wurster, P., 1995, Die Kreide/Tertiär-Grenze im Nanxiong-Becken (Kontinentalfazies, Südostchina). Stuttgart, Franz Steiner Verlag.

Estes, R., 1983, Sauria terrestria, Amphisbaenia: Handbook of Paleoherpetology, part 10A.

Evans, S. E. and Milner, A. R., 1989, *Fulengia*, a supposed early lizard reinterpreted as a prosauropod dinosaur: Palaeontology, v. 32, pp. 223–230.

Fahlbusch, V., 1987, The Neogene mammalian faunas of Ertemte and Har Obo in Inner Mongolia (Nei Monggol), China-5. The genus *Microtoscoptes* (Rodentia, Cricetidae): Senckenbergeana Lethaia, v. 67, pp. 345–373.

Fahlbusch, V., Qiu, Z., and Storch, G., 1983, Neogene mammalian faunas of Ertemte and Har Obo in Nei Mongol, China—report on field work in 1980 and preliminary results: Scientia Sinica B, v. 26, pp. 205–224.

Fang, Q., 1987, A new species of Middle Jurassic turtle from Sichuan: Acta Herpetologica Sinica, v. 6, pp. 65–69.

Feduccia, A., 1996, The origin and evolution of birds: Cambridge, Harvard University Press.

Flerov, K. K., 1957, The Dinocerata of Mongolia: Trudy Paleontologicheskaya Instituta Akademia Nauk SSSR, v. 86, pp. 1-82.

Flynn, L. J., Tedford, R. H., and Qiu, Z., 1991, Enrichment and stability in the Pliocene mammalian fauna of north China: Paleobiology, v. 17, pp. 246–265.

Flynn, L. J., Wu, W., and Downs, W. R. III, 1997, Dating vertebrate microfaunas in the late Neogene record of northern China: Paleogeography, Palaeoclimatology, Palaeoecology, v. 133, pp. 227–242.

Futakami, M., Matsukawa, M., Chen, P., Cao, Z. and Chen, J., 1995, Barremian ammonites from the Longzhaogou Group in eastern Helongjiang, northeast China: Journal of the Geological Society of Japan, v. 101, pp. 79–85.

Galton, P., 1980, Avian-like tibiotarsi of the pterodactyloids (Reptilia: Pterosauria) from the Upper Jurassic of East Africa: Paläontologische Zeitschrift, v. 54, pp. 331–342.

Galton, P. M., 1990, Basal Sauropodomorpha—prosauropods; *in* Weishampel, D. B., Dodson, P. and Osmólska, H., eds., The Dinosauria: Berkeley, University of California Press, pp. 320–344.

Gao, K., 1986, A new spadefoot toad from the Miocene of Lingu, Shandong with a restudy of Bufo linguensis Young, 1977: Vertebrata PalAsiatica, v. 24, pp. 63–74.

Gao, K. and Hou, L., 1996, Systematics and taxonomic diversity of squamates from the Upper Cretaceous Djadochta Formation, Bayan Mandahu, Gobi Desert, People's Republic of China: Canadian Journal of Earth Sciences, v. 33, pp. 578–598.

Gao, Y., 1976, Eocene vertebrate localities and horizons of Wucheng and Xichuan basins, Henan: Vertebrata PalAsiatica, v. 14, pp. 26–34.

Gaudry, A., 1872, Sur les ossements d'animaux quaternaires que M. l'Abbe David a recueillis en Chine: Bulletin de la Societe Géologique de France (2), v. 29, pp. 177–179.

Gilmore, C. W., 1931, Fossil turtles of Mongolia: Bulletin of the American Museum of Natural History, v. 59, pp. 213–258.

——, 1943, Fossil lizards of Mongolia: Bulletin of the American Museum of Natural History, v. 81, pp. 361–384.

Gingerich, P. D., 1985, South American mammals in the Paleocene of North America; *in* Stehli, F.G. and Webb, S.D., eds., The great American biotic interchange: New York, Plenum Press, pp. 51–65.

Golonka, J., Ross, M. I. and Scotese, C. R., 1994, Phanerozoic paleogeographic and paleoclimatic modeling maps: Canadian Society of Petroleum Geologists, Memoir 17, pp. 1–47.

Grabau, A.W., 1923, Cretaceous Mollusca from north China: Geological Survey of China, Bulletin 5, pp. 183–197.

——, 1928, Stratigraphy of China, Part 2, Mesozoic. Beijing, Geological Survey of China.

Grady, W., 1993, The dinosaur project. Edmonton and Toronto, The Ex Terra Foundation and MacFarlane, Walter & Ross.

Granger, W. and Gregory, W. K., 1923, *Protoceratops andrewsi*, a pre-ceratopsian dinosaur from Mongolia: American Museum Novitates, no. 72.

——, A revision of the Mongolian titanotheres: Bulletin of the American Museum of Natural History, v. 80, pp. 349–389.

Gromova, V., 1954, Gigantic rhinoceroses: Trudy Paleontologicheskaya Instituta Akademia Nauk SSSR, v. 71.

Gu, Y., 1980, A Pliocene macaque's tooth from Hubei: Vertebrata PalAsiatica, v. 18, pp. 324–326.

Gu, Z., 1992, On the boundary between the nonmarine Jurassic and Cretaceous strata in China, with special reference to the occurrence of marine fossils in eastern northeast China; *in* Mateer, N.J. and Chen, P., Aspects of nonmarine Cretaceous geology: Beijing, China Ocean Press, pp. 31–40.

Gu, Z. and Chen, P., 1987, Paleogene palaeogeography and distribution of vertebrate fossils in China: Acta Palaeontologica Sinica, v. 26, pp. 210–221.

Guan, J., 1988, The Miocene strata and mammals from Tongxin, Ningxia and Guanghe, Gansu: Memoirs of Beijing Natural History Museum, no. 42.

——, 1991, The character analysis and phylogeny discussion on the shovel tusk mastodonts: Memoirs of Beijing Natural History Museum, no. 50.

Guan, J., Zhang, B., and Ma, Z., 1981, Surveys of the fossiliferous Miocene deposits in the Tongxin District, Ningxia: Memoirs of the Beijing Natural History Museum, no. 13.

Halstead, L. B. and Turner, S., 1973, Silurian and Devonian ostracoderms; *in* Hallam, A., ed., Atlas of Palaeobiogeography: Amsterdam, Elsevier, pp. 67–79.

Halstead, L. B., Liu, Y., and Pan, K., 1979, Agnathans from the Devonian of China: Nature, v. 282, pp. 831–833.

Handbook of Chinese Vertebrate Fossils Editorial Group, 1979, Handbook of Chinese vertebrate fossils (revised edition). Beijing, Science Press.

Hao, Y. and Guan, S., 1984, The Lower-Upper Cretaceous and Cretaceous-Tertiary boundaries in China: Bulletin of the Geological Society of Denmark, v. 33, pp. 129–138.

Harland, W. B., Armstrong, R. L., Cox, A. V., Craig, L. E., Smith, A. G., and Smith, D. V., 1990, A geologic time scale 1989. Cambridge, Cambridge University Press.

Hartenberger, J.-L., 1996, Les débuts de la radiation adaptative des Rodentia (Mammalia). Compte Rendus Academie Science Paris Série IIa, v. 323, pp. 631–637.

Haubold, H., 1989, Die Dinosaurien. Wittenberg, A. Ziemsen Verlag.

Haughton, S. H., 1924, On reptilian remains from the Karroo beds of East Africa: Quarterly Journal of the Geological Society of London, v. 80, pp. 1–11.

——, 1932, On a collection of Karoo vertebrate fossils from Tangyanika Territory: Quarterly Journal of the Geological Society of London, v. 88, pp. 634–668.

He, X. and Cai, K., 1984, The tritylodont remains from Dashanpu, Zigong, China: Journal of the Chengdu College of Geology, Supplement 2, pp. 33–45.

He, X., Li, K., Cai, K., and Gao, Y., 1984, *Omeisaurus tianfuensis*—a new species of *Omeisaurus* from Dashanpu, Zigong, Sichuan: Journal of the Chengdu College of Geology, Supplement 2, pp. 3–32.

Hennig, E., 1922, Die Säugerzähne des württembergischen Rhät-Lias-Bonebeds: Neues Jahrbuch für Mineralogie, Geologie und Paläontologie, Beilage-Band, v. 46, pp. 181–267.

Holbrook, L. T. and Lucas, S. G., 1997, A new genus of rhinocerotoid from the Eocene of Utah and the status of North American "*Forstercooperia*": Journal of Vertebrate Paleontology, v. 17, pp. 384–396.

Hopson, J. A., 1975, The evolution of cranial display structures in hadrosaurian dinosaurs: Paleobiology, v. 1, pp. 21–43.

Hopson, J. A. and Kitching, J. W., 1972, A revised classification of cynodonts (Reptilia, Therapsida): Palaeontologia Africana, v. 14, pp. 71–85.

Hotton, N. and Kitching, J. W., 1963, Speculations on Upper Beaufort deposition: South African Journal of Science, v. 59, pp. 254–258.

Hou, H. and Wang, S., 1985, Devonian palaeogeography of China: Acta Palaeontologica Sinica, v. 24, pp. 186–197.

Hou, L., 1974, Paleocene lizards from Anhui, China: Vertebrata PalAsiatica, v. 12, pp. 193–200.

——, 1976, New materials of Paleocene lizards of Anhui: Vertebrata PalAsiatica, v. 14, pp. 45–52.

——, 1985, Fossil birds from Zhoukoudian loc. 1; *in* Multidisciplinary Study of the Peking man site at Zhoukoudian: Beijing, Science Press, pp. 114–118.

——, 1995, Morphological comparisons between *Confuciusornis* and *Archaeopteryx*; *in* Sun, A. and Wang, Y., eds., Sixth symposium on Mesozoic terrestrial ecosystems and biota, short papers: Beijing, China Ocean Press, pp. 193–201.

——, 1998, Mesozoic birds of China. Taiwan, Feng-huang-ku Bird Park.

Hou, L., Ye, H., and Zhao, X., 1975, Fossil reptiles from Fusui, Kwangshi: Vertebrata PalAsiatica, v. 13, pp. 24–33.

Hou, L., Martin, L. D., Zhou, Z., and Feduccia, A., 1996, Early adaptive radiation of birds: Evidence from fossils from northeastern China: Science, v. 274, pp. 1164–1167.

Hou, L., Zhou, G., Gu, Y., and Zhang, H., 1993, *Confuciusornis sanctus*, a new Late Jurassic sauriurine bird from China. Chinese Science Bulletin, 40: pp. 726–729.

Hou, L., Zhou, Z., Martin, L. D. and Feduccia, A., 1995, A beaked bird from the Jurassic of China: Nature, v. 377, pp. 616–618.

Hou, L., Martin, L.D., Zhou, Z., Feduccia, A. and Zhang, F., 1999, A diapsid skull in a new species of the primitive bird *Confuciusornis*: Nature, v. 399, pp. 679–682.

Hsieh, C., 1973, Atlas of China. New York, McGraw-Hill Book Company.

Hsü, K. J., 1989, Origin of sedimentary basins of China; *in* Zhu, X., ed., Chinese sedimentary basins: Amsterdam, Elsevier, pp. 207–227.

Hsü, K. J., Sun, S., Li, J., Chen, H., Pen, H., and Sengör, A.M.C., 1988, Mesozoic overthrust tectonics in south China: Geology, v. 16, pp. 418–427.

Hu, C., 1962, Cenozoic mammalian fossil localities in Kansu and Ningxia: Vertebrata PalAsiatica, v. 6, pp. 162–172.

Hu, C. and Qi, T., 1978, Gongwangling Pleistocene mammalian fauna of Lantian, Shaanxi: Palaeontologia Sinica C, no. 21.

Hu, S., 1973, A new hadrosaur from the Cretaceous of Chucheng, Shantung: Acta Geologica Sinica, v. 2, pp. 179–202.

——, 1993, A new Theropoda (*Dilophosaurus sinensis* sp. nov.) from Yunnan, China: Vertebrata PalAsiatica, v. 31, pp. 65–69.

Hu, Y., 1993, Two new genera of Anagalidae (Anagalida, Mammalia) from the Paleocene of Qianshan, Anhui and the phylogeny of anagalids: Vertebrata PalAsiatica, v. 31, pp. 153–182.

Hu, Y., Wang, Y., Luo, Z., and Li, C., 1998, A new symmetrodont mammal from China and its implications for mammalian evolution: Nature, v. 390, pp. 137–142.

Huang, R., 1988, Charophytes of Nanxiong basin, Guangdong, and its Cretaceous-Tertiary boundary: Acta Palaeontologia Sinica, v. 27, pp. 457–474.

Huang, W. and Hou, L., 1984, Vertebrate fossils from Yunshui cave, Beijing: Vertebrata PalAsiatica, v. 22, pp. 117–122.

Huang, X., 1977, *Archaeolambda* fossils from Anhui: Vertebrata PalAsiatica, v. 15, pp. 249–260.

——, 1982, Preliminary observation on the Oligocene deposits and mammalian fauna from Alashan Zuogi, Nei Monggol: Vertebrata PalAsiatica, v. 22, pp. 305–309.

——, 1985, Middle Oligocene ctenodactylids (Rodentia, Mammalia) of Ulantatal, Nei Monggol: Vertebrata PalAsiatica, v. 23, pp. 27–38.

Huang, X. and Qi, T., 1982, Notes on late Eocene tapiroids from the Lunan basin, eastern Yunnan: Vertebrata PalAsiatica, v. 20, pp. 315–326.

Huene, F. von, 1925, Die sudafrikanische Karoo-Formation als geologisches und faunitisches Lebensbied: Fortschritt Geologie-Paläontologie, v. 12, p. 124.

——, 1942, Die Anomodontien des Ruhuhu-Gebeits in der Tübingen Sammlung: Palaeontographica Abteilung A, v. 94, pp. 154–184.

——, 1958, The first Chinese labyrinthodonts: Vertebrata PalAsiatica, v. 2, pp. 101–102.

——, 1960, Ein grosser Pseudosuchier aus der Orenburger Trias: Palaeontographica Abteilung A, v. 114, pp. 105–111.

Ivakhnenko, M. F., 1987, Permian parareptiles of the USSR: Trudy Paleontologicheskaya Instituta Akademiya Nauk SSSR, v. 223, pp. 1–59.

Janus, C. G. and Brashler, W., 1975, The search for Peking Man. New York, Macmillan Publishing Co.

Janvier, P., 1975, Anatomie et position systematique des Galéaspides, des Céphalaspides du Dévonien inferieur de Yunnan (Chine): Bulletin Museum National d'Histoire Naturelle, v. 278, pp. 1–16.

——, 1996, Early vertebrates. Oxford, Clarendon Press.

Janvier, P. and Blieck, A., 1979, New data on the internal anatomy of the Heterostraci (Agnatha), with general remarks on the phylogeny of the Craniata: Zoologica Scripta, v. 8, pp. 287–296.

Jarvik, E., 1980, Basic structure and evolution of vertebrates. New York, Academic Press.

——, 1981, Lungfishes, tetrapods, paleontology and plesiomorphy: Systematic Zoology, v. 30, pp. 378–384.

Jenkins, F. A., Jr. and Schaff, C. R., 1988, The Early Cretaceous mammal *Gobiconodon* (Mammalia, Triconodonta) from the Cloverly Formation in Montana: Journal of Vertebrate Paleontology, v. 8, pp. 1–24.

Jerzykiewicz, T. and Russell, D. A., 1991, Late Mesozoic stratigraphy and vertebrates of the Gobi basin: Cretaceous Research, v. 12, pp. 345–377.

Jerzykiewicz, T., Currie, P. J., Johnston, P. A., Koster, E. H., and Gradzinski, R., 1989, Upper Cretaceous dinosaur-bearing eolianites in the Mongolian basin: 20th International Geological Congress, Washington, D.C., Abstracts, v. 2, pp. 122–123.

Jerzykiewicz, T., Currie, P. J., Eberth, D. A., Johnston, P. A., Koster, E. H., and Zheng, J., 1993, Djadokhta Formation correlative strata in Chinese Inner

Mongolia: An overview of the stratigraphy, sedimentary geology, and paleontology and comparisons with the type locality in the pre-Altai Gobi: Canadian Journal of Earth Sciences, v. 30, pp. 2180–2195.

Jessen, H., 1975, A new choanate fish, *Powicthys thorsteinssoni* n.g., n.sp., from the early Lower Devonian of the Canadian Arctic Archipelago: Colloques International Centre National Recherches Scientifique, v. 218, pp. 213–222.

——, 1980, Lower Devonian Porolepiformes from the Canadian Arctic with special reference to *Powicthys thorsteinssoni* Jessen: Palaeontographica A, v. 167, pp. 180–214.

Ji, H., 1975, The lower Pleistocene mammalian fossils from Lantian district, Shaanxi: Vertebrata PalAsiatica, v. 13, pp. 222–230.

Ji, Q., 1989, The Dapoushang section—an excellent section for the Devonian-Carboniferous boundary stratotype in China. Beijing, Science Press.

Ji, Q. and Ji, S., 1996, On discovery of the earliest bird fossil in China and the origin of birds: Chinese Geology 1996, no. 10, pp. 30–33.

——, 1997, *Protoarchaeopteryx*, a new genus of Archaeopterygidae in China: Chinese Geology 1997, no. 3, pp. 38–41.

Ji, Q., Currie, P. J., Norell, M. A., and Ji, S., 1998, Two feathered dinosaurs from northeastern China: Nature, v. 393, pp. 753–761.

Ji, S. and Pan, J., 1997, The macropetalichthyids (Placodermi) from Guangxi and Hunan, China: Vertebrata PalAsiatica, v. 35, pp. 18–34.

Jia, L., 1980, Early man in China. Beijing, Foreign Languages Press.

Jia, L. and Huang, W., 1990, The story of Peking Man. Beijing, Foreign Languages Press.

Jin, F., 1991, A new genus and species of Hiodontidae from Xintai, Shandong: Vertebrata PalAsiatica, v. 29, pp. 46–54.

——, 1995, Late Mesozoic acipenseriforms (Osteichthyes: Actinopterygii) in Central Asia and their biogeographical implications; *in* Sun, A. and Wang, Y., eds., Sixth symposium on Mesozoic terrestrial ecosystems and biota, short papers: Beijing, China Ocean Press, pp. 15–21.

——, 1996, New advances in the late Mesozoic stratigraphic research of western Liaoning, China: Vertebrata PalAsiatica, v. 34, pp. 102–122.

Johnson, M. E., 1985, A. W. Grabau and the fruition of a new life in China: Journal of Geological Education, v. 33, pp. 106–111.

Kalandadze, N. N., 1970, New Triassic kannemeyeriids from the southern pre-Urals; *in* Materialy po evolyutsii nazemnykh pozvonochnykh: Moscow, Izdatelstvo "Nauka," pp. 51–57.

Kalandadze, N. N., 1975, First discovery of a lystrosaur in the territory of the European part of the USSR: Paleontological Journal, v. 4, pp. 140–142.

Kemp, T. S., 1975, Vertebrate localities in the Karoo system of the Luangwa Valley, Zambia: Nature, v. 254, p. 415.

Kermack, D. M. and Kermack, K. A., 1984, The evolution of mammalian characters. London, Croom Helm.

Kermack, K. A., Mussett, F. and Rigney, H. W., 1973, The lower jaw of *Morganucodon*: Zoological Journal of the Linnaean Society of London, v. 53, pp. 87–175.

——, 1981, The skull of *Morganucodon*: Zoological Journal of the Linnaean Society of London, v. 71, pp. 1–158.

Keyser, A .A., 1973, A new Triassic vertebrate fauna from South West Africa: Palaeontologia Africana, v. 16, pp. 1–15.

Keyser, A. A. and Cruickshank, A. R. I., 1979, The origins and classification of Triassic dicynodonts: Transactions Geological Society South Africa, v. 82, pp. 81–108.

Keyser, A. W. and Smith, R. M. H., 1977, Vertebrate biozonation of the Beaufort Group with special reference to the western Karoo basin: Annals of the Geological Survey of South Africa, v. 12, pp. 1–35.

King, G. M., 1981, The functional anatomy of a Permian dicynodont: Philosophical Transactions Royal Society of London B, v. 291, pp. 243–322.

——, 1988, Anomodontia: Encyclopedia of Paleoherpetology, Part 17c.

——, 1990, The dicynodonts a study in paleobiology. London, Chapman and Hall.

Kitching, J. W., 1970, A short review of the Beaufort zoning in South Africa: Proceedings Second Gondwana Symposium, Johannesburg, pp. 309–312.

——, 1995, Biostratigraphy of the *Dicynodon* Assemblage Zone: South African Committee for Stratigraphy Biostratigraphic Series, no. 1, pp. 29–34.

Koken, E., 1885, Über fossile Säugethiere aus China, nach dem Sammlungen des Herrn Ferdinand Freiherrn von Richthofen bearbeitet: Geologisches Paläontologisches Abhandlungen, v. 3, pp. 31–113.

Korth, W. W., 1994, The Tertiary record of rodents in North America. New York, Plenum Press.

Kühne, W. G., 1956, The Liassic therapsid *Oligokyphus*. London, British Museum (Natural History).

Kukla, G., 1987, Loess stratigraphy in central China: Quaternary Science Reviews, v. 6, pp. 191–219.

Kurochkin, E.N., 1976, A survey of the Paleogene birds of Asia: Smithsonian Contributions to Paleobiology, no. 27, pp. 75–86.

——, 1985, Lower Cretaceous birds from Mongolia and their evolutionary significance: Acta XVIII Congressus Internationalis Ornithologici, v. 1, pp. 191–199.

——, 1995, Synopsis of Mesozoic birds and early evolution of Class Aves: Archaeopteryx, v. 13, pp. 47–66.

Kurtén, B., 1952, The Chinese *Hipparion* faunas: Commentationes Biologicae, v. 13, pp. 1–182.

Kurzanov, S. M. and Bannikov, A. F., 1983, A new sauropod from the Upper Cretaceous of Mongolia: Paleontological Journal, 1983, pp. 91–97.

Laurin, M., 1996, A reappraisal of *Utegenia*, a Permo-Carboniferous seymouriamorph (Tetrapoda: Batrachosauria) from Kazakhstan: Journal of Vertebrate Paleontology, v. 16, pp. 374–383.

Lee, J. S., 1939, The geology of China. London, Thomas Murby & Co.

Li, C., 1962, A Tertiary beaver from Changpei, Hopei Province: Vertebrata PalAsiatica, v. 6, pp. 72–79.

——, 1963, Paramyid and sciuravids from north China: Vertebrata PalAsiatica, v. 7, pp. 151–160.

——, 1977a, Paleocene eurymyloids (Anagalida, Mammalia) of Qianshan, Anhui: Vertebrata PalAsiatica, v. 15, pp. 103–118.

——, 1977b, A new Miocene cricetodont rodent of Fangshan, Nanking: Vertebrata PalAsiatica, v. 15, pp. 67–75.

Li, C. and Qiu, Z., 1980, Early Miocene mammalian fossils of Xining basin, Qinghai: Vertebrata PalAsiatica, v. 18, pp. 210–218.

Li, C. and Ting, S., 1983, The Paleogene mammals of China: Bulletin Carnegie Museum of Natural History 21.

——, 1985, Possible phylogenetic relationship of Asiatic eurymylids and rodents, with comments on Mimotonids; *in* Luckett, W. P. and Hartenberger, J.-L., eds., Evolutionary relationships among rodents: New York, Plenum Press, pp. 35–58.

Li, C. and Yan, D., 1979, The systematic position of eurymylids (Mammalia) and the origin of Rodentia: 12th Annual Conference and 3rd National Congress of the Paleontological Society of China Abstracts, pp. 155–156.

Li, C., Wu, W., and Qiu, Z., 1984, Chinese Neogene—subdivision and correlation: Vertebrata PalAsiatica, v. 22, pp. 163–178.

Li, C., Chiu, C., Yan, D., and Hsieh, S., 1979, Notes on some early Eocene mammalian fossils of Hengtung, Hunan: Vertebrata PalAsiatica, v. 17, pp. 71–82.

Li, C., Wilson, R. W., Dawson, M. R., and Krishtalka, L., 1987, The origin of rodents and lagomorphs: Current Mammalogy, v. 1, pp. 97–108.

Li, C., Lin, Y., Gu, Y., Hou, L., Wu, W., and Qiu, Z., 1983, The Aragonian vertebrate fauna of Xiacacewan, Jiangsu—1. A brief introduction to the fossil localities and preliminary report on the new material: Vertebrata PalAsiatica, v. 21, pp. 313–327.

Li, J., 1976, Fossils of Sebecosuchia discovered from Nanxiong, Guangdong: Vertebrata PalAsiatica, v. 14, pp. 169–173.

——, 1980, *Kannemeyeria* fossil from Inner Mongolia: Vertebrata PalAsiatica, v. 18, pp. 94–99.

——, 1983, Tooth replacement in a new genus of procolophonid from the Early Triassic of China: Palaeontology, v. 26, pp. 567–583.

——, 1984, A new species of *Planocrania* from Hengdong, Hunan: Vertebrata PalAsiatica, v. 22, pp. 123–133.

——, 1985, Revision of *Edentosuchus tienshanensis* Young: Vertebrata PalAsiatica, v. 23, pp. 196–206.

——, 1988, *Lystrosaurus* of Xinjiang China: Vertebrata PalAsiatica, v. 26, pp. 241–249.

Li, J. and Cheng, Z., 1995a, The first discovery of bolosaurs from Upper Permian of China: Vertebrata PalAsiatica, v. 33, pp. 17–23.

——, 1995b, A new Late Permian vertebrate fauna from Dashankou, Gansu with comments on Permian and Triassic vertebrate assemblage zones of China; *in* Sun, A. and Wang, Y., eds., Sixth symposium on Mesozoic terrestrial ecosystems and biota, short papers: Beijing, China Ocean Press, pp. 33–37.

——, 1997a, A captorhinid from the Upper Permian of Nei Mongol, China; *in* Tong, Y., et al., eds., Evidence for evolution—essays in honor of Prof. Chung-chien Young on the hundredth anniversary of his birth: Beijing, China Ocean Press, pp. 119–124.

——, 1997b, First discovery of eotitanosuchian (Therapsida, Synapsida) of China: Vertebrata PalAsiatica, v. 35, pp. 268–282.

——, 1999, New anthracosaur and temnospondyl amphibians from Gansu, China—the fifth report on Late Permian Dashankou lower tetrapod fauna: Vertebrata Palasiatica, v. 37, pp. 234–247.

Li, J. and Wang, B., 1987, A new species of *Alligator* from Shanwang, Shandong: Vertebrata PalAsiatica, v. 25, pp. 199–207.

Li, K., He, X., and Cai, K., 1996, The Jurassic dinosaurs and their burial environments of Sichuan basin. Beijing, Geological Publishing House [30th IGC Field Trip Guide T115].

Li, M., 1989, Sporo-pollen from Shanghu Formation of early Paleocene in Nanxiong basin, Guangdong: Acta Palaeontologia Sinica, v. 28, pp. 741–750.

Li, W. and Liu, Z., 1994, The Cretaceous palynofloras and their bearing on stratigraphic correlation in China: Cretaceous Research, v. 15, pp. 333–365.

Li, Z., Powell, C. M., and Trench, A., 1993, Palaeozoic global reconstructions; *in* Long, J.A., ed., Palaeozoic vertebrate biostratigraphy and biogeography: London, Bethaven Press, pp. 25–53.

Linnaeus, C., 1758, Systema Naturae per regna tria naturae, secundum classes, ordines, genera, species cum characteribus, differentiis, synonymis, locis. Volume 1: Stockholm, Laurentii Salvii.

Liu, D., Li, C., and Zhai, R., 1978, Pliocene vertebrates from Lantian, Shaanxi Province: Professional Papers in Stratigraphy and Paleontology, no. 7, pp. 149–200.

Liu, H. 1961, A new amiid from Inner Mongolia, China: Vertebrata PalAsiatica, v. 5, pp. 125–129.

Liu, H. and Ma, T., 1973, A new palaeoniscoid fish from the Chichitsao Series (Permian) of Sinkiang: Memoirs of the Institute of Vertebrate Paleontology and Paleoanthropology, Academia Sinica, v. 10, pp. 6–14.

Liu, H. and Su, D., 1983, Fossil amiids (Pisces) of China and their biostratigraphic significance: Palaeontologica Polonica, v. 28, pp. 181–194.

Liu, T. and Ding, M., 1983, Discussion on the age of "Yuanmo Man": Acta Anthropologica Sinica, v. 2, pp. 40–48.

——, 1984, A tentative chronological correlation of early fossil horizons in China with the loess-deep sea records: Acta Anthropologica Sinica, v. 3, pp. 93–101.

Liu, T. and Li, X., 1984, Mammoths in China; *in* Martin, P.S. and Klein, R.G., eds., Quaternary extinctions a prehistoric revolution: Tucson, The University of Arizona Press, pp. 517–527.

Liu, T. and Li, Y., 1963, New species of *Listriodon* from Miocene of Lantien, Shensi, China: Vertebrata PalAsiatica, v. 7, pp. 291–304.

Liu, T., An, Z., Yuan, B., and Han, J., 1985, The loess-paleosol sequence in China and climatic history: Episodes, v. 8, pp. 21–28.

Liu, X. and Wang S., 1985, Jurassic fish faunas; *in* Wang, S., ed., The Jurassic System of China: Beijing, Geological Publishing House, pp. 282–286.

Liu, Y., 1963, On the Antiarchi from Chutsing, Yunnan: Vertebrata PalAsiatica, v. 7, pp. 39–45.

———, 1965, New Devonian agnathans of Yunnan: Vertebrata PalAsiatica, v. 9, pp. 125–134.

———, 1975, Lower Devonian agnathans of Yunnan and Sichuan: Vertebrata PalAsiatica, v. 13, pp. 202–216.

———, 1979, On the arctolepid Arthrodira from Lower Devonian of Yunnan: Vertebrata PalAsiatica, v. 17, pp. 23–34.

Long, J. A., 1993, Morphological characteristics of Palaeozoic vertebrates used in biostratigraphy; *in* Long, J. A., ed., Palaeozoic vertebrate biostratigraphy and biogeography: London, Belhaven Press, pp. 3–24.

———, 1995, The rise of fishes. Baltimore, The Johns Hopkins University Press.

Long, J. A. and Burrett, C. F., 1989, Tubular phosphatic microproblematica from the Early Ordovician of China: Lethaia, v. 22, pp. 439–446.

Lu, J., 1997, A new Iguanodontidae (*Probactrosaurus mazongshanensis* sp. nov.) from Mazongshan area, Gansu Province, China; *in* Dong, Z., ed., Sino-Japanese Silk Road dinosaur expedition: Beijing, China Ocean Press, pp. 27–47.

Lucas, S.G., 1982, The phylogeny and composition of the order Pantodonta (Mammalia, Eutheria): Third North American Paleontological Convention Proceedings, v. 2, pp. 337–342.

———, 1986, Pyrothere systematics and a Caribbean route for land-mammal dispersal during the Paleocene: Revista Geologica de America Central, v. 5, pp. 1–35.

———, 1990, Toward a vertebrate biochronology of the Triassic: Albertiana, no. 8, pp. 36–41.

———, 1991, Dinosaurs and Mesozoic biochronology: Modern Geology, v. 16, pp. 127–138.

———, 1993a, The *Shansiodon* biochron, nonmarine Middle Triassic of Pangea: Albertiana, v. 11, pp. 40–42.

———, 1993b, Vertebrate biochronology of the Triassic of China: New Mexico Museum of Natural History and Science, Bulletin 3, pp. 301–306.

———, 1993c, Vertebrate biochronology of the Jurassic-Cretaceous boundary, North American Western Interior: Modern Geology, v. 18, pp. 371–390.

——, 1993d, Pantodonts, tillodonts, uintatheres, and pyrotheres are not ungulates; *in* Szalay, F. S., Novacek, M. J. and McKenna, M. C., eds., Mammaly phylogeny: placentals: New York, Springer-Verlag, pp. 182–194.

——, 1996a, Vertebrate biochronology of the Jurassic of China: Museum of Northern Arizona Bulletin 60, pp. 23–33.

——, 1996b, Fossil mammals and the age of the Changxindian Formation, Northeastern China: Palaeovertebrata, v. 25, pp. 133–140.

——, 1996c, Vertebrate biochronology of the Mesozoic of China: Memoirs of the Beijing Natural History Museum, no. 55, pp. 109–148.

——, 1998a, Toward a tetrapod biochronology of the Permian: New Mexico Museum of Natural History and Science, Bulletin 12, pp. 71–91.

——, 1998b, Global Triassic tetrapod biostratigraphy and biochronology: Palaeogeography, Palaeoclimatology, Palaeoecology, v. 143, pp. 347–384.

——, 1998c, Fossil mammals and the Paleocene/Eocene Series boundary in Europe, North America, and Asia; *in* Aubry, M.-P., Lucas, S. G. and Berggren, W. A., eds., Late Paleocene-early Eocene climatic and biotic events in the marine and terrestrial records: New York, Columbia University Press, pp. 451–500.

Lucas, S. G. and Bendukidze, O. G., 1997, Proboscidea (Mammalia) from the early Miocene of Kazakhstan: Neues Jahrbuch für Geologie und Paläontologie Monatshefte 1997, pp. 659–673.

Lucas, S. G., and González-León, C., 1995, Ichthyosaurs from the Upper Triassic of Sonora, Mexico and the biochronology of Triassic ichthyosaurs: Geological Society of America Special Paper, no. 301, pp. 17–20.

Lucas, S. G. and Hunt, A. P., 1993a, *Fukangolepis* from the Triassic of China is not an aetosaur: Journal of Vertebrate Paleontology, v. 13, pp. 145–147.

——, 1993b, A review of Triassic labyrinthodont amphibians from China; Geobios, v. 26, pp. 121–128.

Lucas, S. G. and Luo, Z., 1993, *Adelobasileus* from the Upper Triassic of West Texas: The oldest mammal: Journal of Vertebrate Paleontology, v. 13, pp. 309–334.

Lucas, S. G. and Sobus, J. C., 1989, The systematics of indricotheres; *in* Prothero, D.R. and Schoch, R.M., eds., The evolution of perissodactyls: New York, Oxford University Press, pp. 358–378.

Lucas, S. G. and Williamson, 1995, Systematic position and biochronological significance of *Yuodon* and *Palasiodon*, supposed Paleocene "condylanths" from China: Neues Jahrbuch für Geologie und Paläontologie Monatshefte, v. 1995, pp. 73–81.

Lucas, S. G., Schoch, R. M., and Manning, E., 1981, The systematics of *Forstercooperia*, a middle to late Eocene hyracodontid (Perissodactyla, Rhinocerotoidea) from Asia and western North America: Journal of Paleontology, v. 55, pp. 826–841.

Luo, Z., 1999, A refugium for relicts: Nature, v. 400, pp. 23–24.

Luo, Z. and Sun, A., 1993, *Oligokyphus* (Cynodontia: Tritylodontidae) from the lower Lufeng Formation (Lower Jurassic) of Yunnan, China: Journal of Vertebrate Paleontology, v. 13, pp. 477–482.

Luo, Z. and Wu, X., 1994, The small vertebrate fauna of the lower Lufeng Formation, Yunnan; *in* Fraser, N.C. and Sues, H.-D., eds., In the shadow of dinosaurs. Early Mesozoic tetrapods: Cambridge, Cambridge University Press, pp. 251–270.

——, 1995, Correlation of vertebrate assemblage of the lower Lufeng Formation, Yunnan, China; *in* Sun, A. and Wang, Y., eds., Sixth symposium on Mesozoic terrestrial ecosystems and biota, short papers: Beijing, China Ocean Press, pp. 83–88.

Luo, Z., Lucas, S.G., Li, J. and Zhen, S., 1995, A new specimen of *Morganucodon oehleri* (Mammalia, Triconodonta) from the Liassic lower Lufeng Formation of Yunnan, China: Neues Jahrbuch für Geologie und Paläontologie Monatshefte, v. 1995, pp. 671–680.

Lydekker, R., 1881, Observations of ossiferous beds of Hundes in Tibet: Records of the Geological Survey of India, v. 14, pp. 178–184.

——, 1883, Note on the probable occurrence of Siwalik strata in China and Japan: Records of the Geological Survey of India, v. 16, pp. 58–61.

——, 1891, On a collection of mammalian bones from Mongolia: Records of the Geological Survey of India, v. 24, pp. 207–211.

——, 1901, On the skull of a chiru-like antelope from the ossiferous deposits of the Hundes (Tibet): Quarterly Journal of the Geological Society of London, v. 57, pp. 289–292.

Ma, A. and Cheng, J., 1991, On biostratigraphical subdivision of Yuhuangding Formation in Liguanqiao basin of eastern Qinling region: Scientia Geologica Sinica, v. 1991 (1), pp. 21–29.

Ma, F., 1980, A new genus of Lycopteridae from Ningxia, China: Vertebrata PalAsiatica, v. 18, pp. 286–295.

MacFadden, B. J., 1984, Systematics and phylogeny of *Hipparion, Neohipparion, Nannipus,* and *Cormohipparion* (Mammalia, Equidae) from the Miocene and Pliocene of the New World: Bulletin of the American Museum of Natural History, v. 179, pp. 1–196.

MacFadden, B. J., 1992, Fossil horses. Cambridge, Cambridge University Press.

Madden, C. T. and Van Couvering, J. A., 1976, The proboscidean datum event: Early Miocene migration from Africa: Geological Society of America, Abstracts with Programs, p. 992.

Mader, B. J. and Bradley, R. L., 1989, A redescription and revised diagnosis of the syntypes of the Mongolian tyrannosaur *Alectrosaurus olseni*: Journal of Vertebrate Paleontology, v. 9, pp. 41–55.

Maleev, E., 1955a, Gigantic carnivorous dinosaurs of Mongolia: Doklady Akademii Nauk SSSR, v. 104, pp. 634–637.

——, 1955b, New carnivorous dinosaurs from the Upper Cretaceous of Mongolia: Doklady Akademii Nauk SSSR, v. 104, pp. 779–782.

——, 1974, Giant carnosaurs of the family Tyrannosauridae: Trudy Sovmestnaya—Mongolskaya Paleontologicheskii Ekspeditsiya, v. 1, pp. 132–190.

Martin, L. D., Zhou, Z., Hou, L., and Feduccia, A. 1998. *Confuciusornis sanctus* compared to *Archaeopteryx lithographica*. Naturwissenschaften, v. 85, pp. 286–289.

Martin, P. S., 1984, Prehistoric overkill: the global model; *in* Martin, P. S. and Klein, R. G., eds., Quarternary extinctions a prehistoric revolution: Tucson, The University of Arizona Press, pp. 354–403.

Maryanska, T., 1977, Ankylosauridae (Dinosauria) from Mongolia: Paleontologia Polonica, v. 37, pp. 85–151.

Maryanska, T. and Osmólska, H., 1975, Protoceratopsidae (Dinosauria) of Asia: Palaeontologia Polonica, no. 33, pp. 133–181.

Mateer, N. J., 1989, Upper Cretaceous reptilian eggs from Zhejiang Province, China; *in* Gillette, D. D. and Lockley, M. G., eds., Dinosaur tracks and traces: New York, Cambridge University Press, pp. 116–118.

Mateer, N. J. and Chen, P., 1992, A review of the nonmarine Cretaceous-Tertiary transition in China: Cretaceous Research, v. 13, pp. 81–90.

Mateer, N. J. and Lucas, S. G., 1985, Swedish vertebrate palaeontology in China: A history of the Lagrelius Collection: Bulletin of the Geological Institutions of the University of Uppsala, New Series, v. 11, pp. 1–24.

Mateer, N. J. and McIntosh, J. S., 1985, A new reconstruction of the skull of *Euhelopus zdanskyi* (Saurischia: Sauropoda): Bulletin of the Geological Institutions of the University of Uppsala, New Series, v. 11, pp. 125–132.

Matsukawa, M. and Obata, I., 1994, Dinosaurs and sedimentary environments in the Japanese Cretaceous: A contribution to dinosaur facies in Asia based on molluscan paleontology and stratigraphy: Cretaceous Research, v. 15, pp. 101–125.

Matsukawa, M., Futakami, M., Lockley, M. G., Chen, P., Chen, J., Cao, Z., and Bolotsky, U. L., 1995, Dinosaur footprints from the Lower Cretaceous of eastern Manchuria, northeastern China: Implications for the recognition of an ornithopod ichnofacies in East Asia: Palaios, v. 10, pp. 3–15.

Matthew, W. D., 1915a, A revision of the lower Eocene Wasatch and Wind River faunas. Part IV. Entelonychia, Primates, Insectivora (part): Bulletin of the American Museum of Natural History, v. 34, pp. 429–483.

——, 1915b, Climate and evolution: Annals of the New York Academy of Science, v. 24, pp. 171–318.

Matthew, W.D. and Granger, W., 1923, The fauna of the Houldjin Gravels: American Museum Novitates, no. 97, pp. 1–6.

——, 1924, New Carnivora from the Tertiary of Mongolia: American Museum Novitates, no. 104, pp. 1–9.

——, 1925, Fauna and correlation of the Gashato Formation of Mongolia: American Museum Novitates, no. 189, pp. 1–12.

Matthew, W. D., Granger, W. and Simpson, G. G., 1929, Additions to the fauna of the Gashato Formation of Mongolia: American Museum Novitates, no. 376, pp. 1–12.

Mazin, J. M. and Sander, P. M., 1993, Palaeobiogeography of the Early and Late Triassic Ichthyopterygia: Paleontologia Lombarda Nuova Serie, v. 2, pp. 93–107.

McIntosh, J. S., 1990, Sauropoda; *in* Weishampel, D.B., Dodson, P. and Osmolska, H., eds., The Dinosauria: Berkeley, University of California Press, pp. 345–401.

McKenna, M. C., 1962, Studies of the natural history of the Mongolian People's Republic and adjacent areas, made by the American Museum of Natural History: The Mongolia Society Newsletter, v. 1, no. 3, pp. 31–35.

———, 1975, Fossil mammals and early Eocene North Atlantic continuity: Annals of the Missouri Botanical Garden, v. 62, pp. 335–353.

McKenna, M. C. and Bell, S. K., 1997, Classification of mammals above the species level. New York, Columbia University Press.

McKenna, M. C., Xue, X., and Zhou, M., 1984, *Prosarcodon lonanensis*, a new Paleocene micropternodontid palaeoryctoid insectivore from Asia: American Museum Novitates, no. 2780, pp. 1–17.

McKenna, M. C., Chow, M., Ting, S., and Luo, Z., 1989, *Radinskya yupingae*, a perissodactyl-like mammal from the late Paleocene of China; *in* Prothero, D. R. and Schoch, R.M., eds., The evolution of perissodactyls: New York, Oxford University Press, pp. 24–36.

Mendrez, C. H., 1972, On the skull of *Regisaurus jacobi*, a new genus and species of Bauriamorpha Watson and Romer 1956 (= Scaloposauria Boonstra 1953), from the *Lystrosaurus*-zone of South Africa; *in* Joysey, V.A. and Kemp, T.S., eds., Studies in vertebrate evolution: Edinburgh, Oliver and Boyd, pp. 191–219.

Meng, J., Wyss, A. R., Dawson, M. R., and Zhai, R., 1994, Primitive fossil rodent from Inner Mongolia and its implications for mammalian phylogeny: Nature, v. 370, pp. 134–136.

Metcalfe, I., 1988, Origin and assembly of south-east Asian continental terranes; *in* Audley-Charles, M. G. and Hallam, A., eds., Gondwana and Tethys: London, Geological Society Special Paper 37, pp. 101–118.

———, 1994, Late Palaeozoic and Mesozoic palaeogeography of eastern Pangea and Tethys: Canadian Society of Petroleum Geologists, Memoir 17, pp. 97–111.

———, 1996, Pre-Cretaceous evolution of SE Asian terranes; *in* Hall, R. and Blundell, D., eds., Tectonic evolution of southeast Asia: London, Geological Society Special Publication 106, pp. 97–122.

Meyerhoff, A. A., Kamen-Kaye, M., Chen, C., and Taner, I., 1991, China-stratigraphy, paleogeography and tectonics. Dordrecht, Kluwer Academic Publishers.

Miao, D., 1982, Early Tertiary fossil mammals from the Shinao basin, Panxian County, Guizhou Province: Acta Palaeontologia Sinica, v. 21, pp. 526–536.

——, 1988, Skull morphology of *Lambdopsalis bulla* (Mammalia, Multitubercu-
lata) and its implications to mammalian evolution: Contributions to Geology,
University of Wyoming, Special Paper 4, p. 104.

——, 1996, Minchen Chow (Zhou Min-zhen), 1918–1996: Society of Vertebrate
Paleontology New Bulletin, no. 167, pp. 80–83.

Mikhailov, K., Sabath, K., and Kurzanov, S., 1994, Eggs and nests from the Cretaceous of Mongolia; *in* Carpenter, K., Hirrsch, K. F., and Horner, J. R., eds.,
Dinosaur eggs and babies: Cambridge, Cambridge University Press, pp. 88–115.

Milner, A. R., 1993, Biogeography of Palaeozoic tetrapods; *in* Long, J. A., ed.,
Palaeozoic vertebrate biostratigraphy and biogeography: London, Belhaven
Press, pp. 324–353.

Molnar, R. E., Kurzanov, S. M. and Dong, Z., 1990, Carnosauria; *in* Weishampel,
D. B., Dodson, P. and Osmólska, H., eds., The Dinosauria: Berkeley, University of California Press, pp. 169–209.

Mook, C. C., 1940, A new fossil crocodilian from Mongolia: American Museum
Novitates, no. 1097.

Morita, G., 1939, The Fushin series of Manchoukou, a preliminary note: Japanese
Journal of Geology and Geography, v. 17, pp. 14–15.

Morris, F. K., 1936, Central Asia in Cretaceous time: Geological Society of America Bulletin, v. 47, pp. 1477–1534.

Moy-Thomas, J. A. and Miles, R. S., 1971, Palaeozoic fishes. Philadelphia, W. B.
Saunders Company.

Mu, E., Boucot, A. J., Chen, X., and Rong, J., 1986, Correlation of the Silurian
rocks of China: Geological Society of America Special Paper 202.

Muizon, C. de and Marshall, L. G., 1992, *Alcidedorbignya inopinata* (Mammalia:
Pantodonta) from the early Paleocene of Bolivia: Phylogenetic and paleobiogeographic implications: Journal of Paleontology, v. 66, pp. 499-511.

Needham, J., 1959, Science and civilization in China, v. 3, mathematics and the
sciences of the heavens and the earth. Cambridge, Cambridge University Press.

Newton, E. T., 1893, On some new fossil reptiles from the Elgin Sandstones: Philosophical Transactions of the Royal Society Series B, v. 184, pp. 431–503.

Nicholls, E. L. and Russell, A. P., 1981, A new specimen of *Struthiomimus altus*
from Alberta, with comments on the classificatory characters of the Upper Cretaceous ornithomimids: Canadian Journal of Earth Sciences, v. 18, pp. 518–
526.

Nie, S., 1991, Paleoclimatic and paleomagnetic constraints on the Paleozoic reconstructions of south China, North China and Tarim: Tectonophysics, v. 196, pp.
279–308.

Nowinski, A., 1971, *Nemegtosaurus mongoliensis* n. gen. n. sp. (Sauropoda) from the
uppermost Cretaceous of Mongolia: Palaeontologia Polonica, v. 25, pp. 57–81.

Ochev, V. G. and Shishkin, M. A., 1989, On the principles of global correlation of
the continental Triassic on the tetrapods: Acta Palaeontologica Polonica, v. 34,
pp. 149–173.

Olson, E. C. and Chudinov, P., 1992, Upper Permian terrestrial vertebrates of the USA and Russia: 1991: International Geology Review, v. 34, pp. 1143–1160.

Olshevsky, G., 1991, A revision of the parainfraclass Archosauria Cope, 1869, excluding the advanced Crocodylia: Mesozoic Meanderings, no. 2.

Osborn, H. F., 1894, A division of the eutherian mammals into the Mesoplacentalia and Cenoplacentalia: Transactions of the New York Academy of Sciences, v. 13, pp. 234–237.

——, 1910, The age of mammals in Europe, Asia and North America. New York, MacMillan.

——, 1924, Sauropoda and Theropoda of the Lower Cretaceous of Mongolia: American Museum Novitates, no. 144p.

Osborn, H. F., and Earle, C., 1895, Fossil mammals of the Puerco beds, collection of 1892: Bulletin of the American Museum of Natuaral History, v. 7, pp. 1–70.

Osborn, H. F. and Granger, W., 1932, Coryphodonts and uintatheres from the Mongolian expedition of 1930: American Museum Novitates, no. 552, pp. 1–16.

Owen, R., 1870, On fossil remains of mammals found in China: Quarterly Journal of the Geological Society of London, v. 26, pp. 417–434.

Padian, K. and Chiappe, L.M., 1998, The origin and early evolution of birds: Biological Reviews, v. 73, pp. 1–42.

Pan, J., 1984, The phylogenetic position of the eugaleaspida in China: Proceedings of the Linnaean Society of South Wales, v. 107, pp. 309–319.

——, 1986, Notes on Silurian vertebrates in China: Bulletin Chinese Academy of Geological Sciences, no. 15, pp. 161–190.

——, 1992, New galeaspids (Agnatha) from the Silurian and Devonian of China. Beijing, Geological Publishing House.

Pan, J. and Dineley, D. L., 1988, A review of early (Silurian and Devonian) vertebrate biogeography and biostratigraphy of China: Proceedings of the Royal Society of London B, v. 235, pp. 29–61.

Pan, J. and Lu, L., 1997, *Grammaspis*, a new antiarch fish (placoderm) from Early Devonian of Jiangyou, Sichuan Province; *in* Tong, Y. et al., eds., Evidence for evolution– essays in honor of Professor Chungchien Young on the hundreth anniversary of his birth: Beijing, China Ocean Press, pp. 97–103.

Pan, J. and Wang, S., 1980, New findings of Galeaspiformes in south China: Acta Palaeontologica Sinica, v. 19, pp. 1–7.

——, 1981, New discoveries of polybranchiaspids of Yunnan Province: Vertebrata PalAsiatica, v. 19, pp. 113–121.

Pan, J., Wang, S. and Liu, Y., 1975, The Lower Devonian Agnatha and Pisces of south China: Professional Papers on Stratigraphy and Palaeontology 1, pp. 135–169.

Pan, J., Wang, S., Gao, L. and Hou, J., 1978, Devonian continental and sedimentary formations of south China; *in* Symposium on the Devonian System in China: Beijing, Geological Publishing House, pp. 240–269.

Pan, J. and Zeng, X., 1985, Dayongaspidae, a new family of Polybranchiaspi-formes (Agnatha) from Early Silurian of Hunan, China: Vertebrata PalAsiatica, v. 23, pp. 207–213.

Pan, J., Hou, F., Cao, J., Gu, Q., Liu, S., Wang, J., Gao, L., and Liu, C., 1987, Continental Devonian of Ningxia and its biotas. Beijing, Geological Publishing House.

Pan, K., 1964, Some Devonian and Carboniferous fishes from South China: Acta Palaentologica Sinica, v. 12, pp. 139–168.

Parrish, J. M., 1992, Phylogeny of the Erythrosuchidae (Reptilia: Archosauri-formes): Journal of Vertebrate Paleontology, v. 12, pp. 93–102.

Pei, W., 1957, The zoogeographical divisions of Quaternary mammalian faunas in China: Vertebrata PalAsiatica, v. 1, pp. 9–23.

Peng, G., 1995, A new protosuchian from the Late Jurassic of Sichuan, China; *in* Sun, A. and Wang, Y., eds., Sixth symposium on Mesozoic terrestrial ecosystems and biota, short papers: Beijing, China Ocean Press, pp. 63–68.

——, 1996, Late Jurassic protosuchian *Sichuanosuchus huidongensis* (Archosauria: Crocodyliformes) from Zigong, Sichuan, China: Vertebrata PalAsiatica, v. 34, pp. 269–278.

Peng, J., 1991, A new genus of Proterosuchia from Lower Triassic of Shaanxi, China: Vertebrata PalAsiatica, v. 29, pp. 95–107.

Peng, J. and Brinkman, D. B., 1993, New material of *Xinjiangchelys* (Reptilia: Testudines) from the Late Jurassic Qigu Formation (Shishigou Group) of the Pingtengshan locality, Junggar basin, Xinjiang: Canadian Journal of Earch Sciences, v. 30, pp. 2013–2026.

Ponomarenko, A. C. and Popov, Y. A., 1980, Paleobiocenoses of Early Cretaceous Mongolian lakes: Paleontological Journal, v. 1980, pp. 3–13.

Poplin, C., Wang, N., Richter, M., and Smith, M., 1991, An enigmatic actinopterygian (Pisces: Osteichthyes) from the Upper Permian of China: Zoological Journal of the Linnaean Society, v. 103, pp. 1–20.

Preston, D. J., 1986, Dinosaurs in the attic. New York, Ballatine Books.

Qi, T., 1975, An early Oligocene mammalian fauna of Ningxia: Vertebrata PalAsiatica, v. 13, pp. 217–224.

——, 1979, A general account of the early Tertiary mammalian fauna of Shara Murun area, Inner Mongolia: Second Congress on Stratigraphy, Beijing, 1979, pp. 1–9.

——, 1987, The middle Eocene Arshanto fauna (Mammalia) of Inner Mongolia: Annals of the Carnegie Museum, v. 56, pp. 1–73.

Qiu, D., 1990, The distribution of ancient seas, marine sedimentary facies and land areas during the Triassic Period in China: ESCAP Atlas of Stratigraphy 9 (United Nations Mineral Resource Development Series 59), pp. 2–10.

Qiu, Z., 1977, New genera of Pseudictopidae (Anagalida, Mammalia) from Middle-Upper Paleocene of Qianshan, Anhui: Acta Palaeontologia Sinica, v. 16, pp. 128–148.

——, 1985, The Neogene mammalian fauna of Ertemte and Hau Obo in Inner Mongolia (Nei Monggol), China—3. Jumping mice, Rodentia: Lophocricetinae: Senckenbergea Lethaea, v. 66, pp. 39–67.

——, 1987, Die Hyaeniden aus dem Ruscinium und Villafranchium Chinas: Müncher Geowissenschaftliche Abhandlungen A, v. 9, pp. 109.

——, 1990, The Chinese Neogene mammalian biochronology—its correlation with the European Neogene mammalian zonation; *in* Lindsay, E. H., Fahlbusch, V. and Mein, P., eds., European Neogene mammal chronology: New York, Plenum Press, pp. 527–556.

——, 1996, Middle Miocene micromammalian fauna from Tunggur, Nei Mongol. Beijing, Science Press.

Qiu, Z. and Gu, Z., 1988, A new locality yielding mid-Tertiary mammals near Lanzhou, Gansu: Vertebrata PalAsiatica, v. 26, pp. 198–213.

Qiu, Z. and Lin, Y., 1986, The Aragonian vertebrate fauna of Xiacaowan, Jiangsu—5. Sciuridae (Rodentia, Mammalia): Vertebrata PalAsiatica, v. 24, pp. 195–209.

Qiu, Z. and Qiu, Z., 1990, Neogene local mammalian faunas—succession and ages: Journal of Stratigraphy, v. 14, pp. 241–245.

——, 1995, Chronological sequence and subdivision of Chinese Neogene mammalian faunas: Palaeogeography, Palaeoclimatology, Palaeoecology, v. 116, pp. 41–70.

Qiu, Z., Huang, W., and Guo, Z., 1987, The Chinese hipparionine fossils: Palaeontologia Sinica, v. 25.

Qiu, Z., Li, C. and Wang, S., 1981, Miocene mammalian fossils from Xining basin, Qinghai: Vertebrata PalAsiatica, v. 19, pp. 156–173.

Qiu, Z., Ye, J. and Cao, J., 1988, A new species of *Percrocuta* from Tongxin, Ningxia: Vertebrata PalAsiatica, v. 26, pp. 116–127.

Radinsky, L. B., 1964, Notes on Eocene and Oligocene fossil localities in Inner Mongolia: American Novitates, no. 2180, pp. 1–11.

——, 1965, Early Tertiary Tapiroidea of Asia: Bulletin of the American Museum of Natural History, v. 129, pp. 181–263.

Radinsky, L.B., 1967, A review of the rhinocerotoid family Hyracodontidae (Perissodactylal: Bulletin of the American Museum of Natural History, v. 136, pp. 1–46.

Ren, M., Liu, Z., Jin, J., Deng, X., Wang, F., Peng, B., Wang, X. and Wang, Z., 1981, Evolution of limestone caves in relation to the life of early man at Zhoukoudian, Beijing: Scientia Sinica, v. 24, pp. 843–850.

Rensberger, J.M. and Li, C., 1986, A new prosciurine rodent from Shantung Province, China: Journal of Paleontology, v. 60, pp. 763–771.

Repetski, J.E., 1978, A fish from the Upper Cambrian of North America: Science, v. 200, pp. 529–531.

Reshetov, V.Y., 1979, Early Tertiary tapiroids of Mongolia and the USSR: Trudy Sovmestnaya Sovietskovo-Mongolskaya Paleontologichesky Ekspeditsiy, v. 11.

Riabinin, A.N., 1930, *Mandshurosaurus amurensis* nov. gen., nov. sp., a hadrosaurian dinosaur from the Upper Cretaceous Amur River: Monograf Russkovo Paleontologicheskovo Obshchestva, v. 2, pp. 1–36.

Rich, P., Zhang, Y., Chow, M., Wang, B., Komarower, P., Fan, J., Sloss, R., Moody, J.K.M. and Dawson, J., 1994, A Chinese-English and English-Chinese dictionary of vertebrate palaeontology terms. Melbourne, Monash University.

Richthofen, F. von, 1877–1912, China, Ergebnisse eineger Reisen und darauf gegrundeter Studien. Berlin, 5 volumes.

Rieppel, O., 1998, The systematic status of *Hanosaurus hupehensis* (Reptilia, Sauropterygia) from the Triassic of China: Journal of Vertebrate Paleontology, v. 18, pp. 545–557.

——, 1999, The sauropterygian genera *Chinchenia, Kwangsisaurus,* and *Sanchiaosaurus* from the Lower and Middle Triassic of China: Journal of Vertebrate Paleontology, v. 19, pp. 321–337.

Rigby, J.K., Jr., Snee, L.W., Unruh, D.M., Harlan, S.S., Guan, J., Li, F., Rigby, J.K., Sr. and Kowalis, B.J., 1993, ^{40}Ar / ^{39}Ar and U-Pb dates for dinosaur extinction, Nanxiong basin Guongdong Province, People's Republic of China: Geological Society of America, Abstracts with Programs, v. 25, no. 6, pp. A–296.

Romer, A.S., 1956, Osteology of the reptiles. Chicago, University of Chicago Press.

——, 1966, Vertebrate paleontology. Chicago, University of Chicago Press.

——, 1973, Permian reptiles, *in* Hallam, A., ed., Atlas of palaeobiogeography: Amsterdam, Elsevier, pp. 159–167.

Rosen, D.E., Forey, P.L., Gardiner, B.G. and Patterson, C., 1981, Lungfishes, tetrapods, paleontology, and plesiomorphy: Bulletin of the American Museum of Natural History, v. 167, pp. 163–275.

Rozhdestvensky, A. K., 1977, The study of dinosaurs in Asia: Journal of the Paleontological Society of India, v. 20, pp. 102–119.

Russell, D. A., 1972, Ostrich dinosaurs from the Late Cretaceous of western Canada: Canadian Journal of Earth Sciences, v. 9, pp. 375–402.

——, 1993, The role of Central Asia in dinosaurian biogeography: Canadian Journal of Earth Sciences, v. 30, pp. 2002–2012.

Russell, D. A. and Zhao, X., 1996, New psittacosaur occurrences in Inner Mongolia: Canadian Journal of Earth Sciences, v. 33, pp. 637–648.

Russell, D. A., Russell, D. E., and Sweet, A. R., 1993, The end of the dinosaurian era in the Nanxiong basin: Vertebrata PalAsiatica, v. 31, pp. 139–145.

Russell, D. E. and Zhai, R., 1987, The Paleogene of Asia: Mammals and stratigraphy: Memoire du Museum National d'Histoire Naturelle C, v. 52.

Samoilov, V. S., Ivanov, V. G. and Smirnov, V. N., 1988, Late Mesozoic riftogenic magmatism in the northeastern part of the Gobi Desert (Mongolia): Soviet Geology and Geophysics, v. 29, pp. 10–16.

Savage, D. E., and Russell, D. E., 1983, Mammalian paleofaunas of the world. London, Addison-Wesley.

Schafer, E. H., 1967, Ancient China. New York, Time-Life Books.

Schlosser, M., 1903, Die fossilen Säugethiere Chinas nebst einer Odontographie der recenten Antilopen: Abhandlungen Bayerische Akademie von Wissenschaften, v. 22.

——, 1924, Tertiary vertebrates from Mongolia: Palaeontologia Sinica C, v. 1.

Schoch, R. M., 1989, A review of the tapiroids; *in* Prothero, D.R. and Schoch, R.M., eds., The evolution of perissodactyls: New York, Oxford University Press, pp. 298–320.

Sereno, P., 1991, *Lesothosaurus*, "fabrosaurids" and the early evolution of Ornithischia: Journal of Vertebrate Paleontology, v. 11, pp. 168–197.

Sereno, P. and Rao, C., 1992, Early evolution of avian flight and perching: new evidence from the Lower Cretaceous of China: Science, v. 255, pp. 845–848.

Sha, J., Fürsich, F. T., and Grant-Mackie, J. A., 1994, A revised Early Cretaceous age for the Longzhaogou and Jixi Groups of eastern Heilongjiang, China, Previously considered Jurassic: Palaeogeographic implications: Newsletters in Stratigraphy, v. 31, pp. 101–114.

Shapiro, H. L., 1971, The strange unfinished saga of Peking Man: Natural History, v. 80, pp. 8–10, 12, 14, 16–18, 74, 76–83.

Shen, Y. and Mateer, N. J., 1992, An outline of the Cretaceous System in northern Xinjiang, western China; *in* Mateer, N. J. and Chen, P., eds., Aspects of nonmarine Cretaceous geology: Beijing, China Ocean Press, pp. 49–77.

Shishkin, M. A., Ochev, V. G., and Tverdokhlebov, P. V., eds., 1995, Bio-stratigraphy of the continental Triassic of the southern pre-Urals. Moscow, Nauka.

Shu, D., Luo, H., Conway Morris, S., Zhang, X., Hu, S., Chen, L., Han, J., Zhu, M., Li, Y., and Chen, L., 1999, Lower Cambrian vertebrates from south China: Nature, v. 402, pp. 42–46.

Sigogneau-Russell, D., 1981, Presence d'un champsosauride dan le Cretace superieur de Chine: Comptes Rendus de l'Academie des Sciences Paris, v. 292, pp. 1–4.

Sigogneau-Russel, D. and Sun, A., 1979, A brief review of Chinese synapsids: Geobios, v. 14, pp. 215–219.

Simmons, D. J., 1965, The non-therapsid reptiles of the Lufeng basin, Yunnan, China: Fieldiana: Geology, v. 15(1), pp. 1–96.

Simons, E. L. and Chopra, S. R. K., 1969, *Gigantopithecus* (Pongidae, Hominoidea). A new species from northern India: Postilla, no. 138.

Simons, E. L. and Ettel, P. C., 1970, *Gigantopithecus*: Scientific American, v. 222, no. 1, pp. 76–85.

Sloan, R. E., 1987, Paleocene and latest Cretaceous mammal ages, biozones, magnetozones, rates of sedimentation, and evolution: Geological Society of America Special Paper 209, pp. 165–200.

Smith, H. M. R., 1993, Changing fluvial environments across the Permian-Triassic boundary in the main Karoo basin, South Africa and possible causes of the tetrapod extinctions: New Mexico Museum of Natural History and Science, Bulletin 3, p. 132.

Smith, P. E., Evensen, N. M., York, D., Chang, M., Jin, F., Li, J., Cumbaa, S., and Russell, D., 1995, Dates and rates in ancient lakes: ^{40}Ar-^{39}Ar evidence for an Early Cretaceous age for the Jehol Group, northeast China: Canadian Journal for Earth Sciences, v. 32, pp. 1426–1431.

Song, C. and Zhang, M., 1991, Discovery of *Chirodipterus* (Dipnoi) from lower Upper Devonian of Hunan, south China; *in* Chang, M., Liu, Y. and Zhang, G., eds., Early vertebrates and related problems of evolutionary biology: Beijing, Science Press, pp. 465–576.

Stensiö, E. A., 1935, *Sinamia zdanskyi*, a new amiid from the Lower Cretaceous of Shantung, China: Palaeontologia Sinica, C, v. 3, pp. 1–48.

Storch, G., 1987, The Neogene mammalian faunas of Ertemte and Hau Obo in Inner Mongolia (Nei Mongol), China—7. Muridae (Rodentia): Senckenbergeana Lethaea, v. 67, pp. 401–431.

Storch, G. and Qiu, Z., 1983, The Neogene mammalian faunas of Ertemte and Hau Obo in Inner Mongolia (Nei Mongol), China—2. Moles, Insectivora: Senckenbergeana Lethaea, v. 63, pp. 89–127.

Storrs, G. W., 1991, Anatomy and relationships of *Corosaurus alcovensis* (Diapsida: Sauropterygia) and the Triassic Alcova Limestone of Wyoming: Peabody Museum of Natural History Yale University, Bulletin 44, p. 151.

Su, T., 1978, A new Triassic palaeoniscoid fish from Fukang, Sinkiang: Institute of Vertebrate Paleontology and Paleoanthropology Memoirs, v. 13, pp. 55–59.

——, 1985, On late Mesozoic fish fauna from Xinjiang, China: Institute of Vertebrate Paleontology and Paleoanthropology Memoirs, v. 17, pp. 61–136.

Sues, H.-D., 1985, First record of tritylodontid *Oligokyphus* (Synapsida) from the Lower Jurassic of western North America: Journal of Vertebrate Paleontology, v. 5, pp. 328–335.

——, 1986, Relationships and biostratigraphic significance of the Tritylodontidae (Synapsida) from the Kayenta Formation of northeastern Arizona; *in* Padian, K., ed., The beginning of the age of dinosaurs: Cambridge, Cambridge University Press, pp. 275–278.

Sues, H.-D. and Norman, D. B., 1990, Hypsilophodontidae, Tenontosaurus, Dryosauridae; *in* Weishampel, D. B., Dodson, P., and Osmolska, H., eds., The Dinosauria: Berkeley, University of California Press, pp. 498–509.

Sun, A., 1962, Discovery of neorhachitomous vertebrae from Lufeng, Yunnan: Vertebrata PalAsiatica, v. 6, pp. 109–110.

——, 1963, The Chinese kannemeyerids: Palaentologia Sinica, New Series C, v. 17, p. 109.

——, 1964, Preliminary report on a new species of *Lystrosaurus* of Sinkiang: Vertebrata PalAsiatica, v. 8, pp. 216–217.

——, 1972, Permo-Triassic reptiles of Sinkiang: Scientia Sinica, v. 16, pp. 152–156.

——, 1973, A new species of *Dicynodon* from Sinkiang: Vertebrata PalAsiatica, v. 11, pp. 52–58.

———, 1978, Two new genera of Dicynodontidae: Memoirs of the Institute of Vertebrate Paleontology and Paleoanthropology, v. 13, pp. 19–25.

———, 1984, Skull morphology of the tritylodont genus *Bienotheroides* of Sichuan: Scientia Sinica B, v. 27, pp. 970–984.

———, 1986, New material of *Bienotheroides* (tritylodont reptile) from Shaximiao Formation of Sichuan: Vertebrata PalAsiatica, v. 24, pp. 165–170.

———, 1989, Before dinosaurs land vertebrates of China 200 million years ago. Beijing, China Ocean Press.

———, 1991, A review of Chinese therocephalian reptiles: Vertebrata PalAsiatica, v. 29, pp. 85–94.

Sun, A. and Cui, G., 1986, A brief introduction to the lower Lufeng saurischian fauna (Lower Jurassic: Lufeng, Yunnan, People's Republic of China); *in* Padian, K., ed., The beginning of the age of dinosaurs: Cambridge, Cambridge University Press, pp. 275–278.

———, 1989, Tritylodont reptile from Xinjiang: Vertebrata PalAsiatica, v. 27, pp. 1–8.

Sun, A. and Li, Y., 1985, The postcranial skeleton of the last tritylodont *Bienotheroides*: Vertebrata PalAsiatica, v. 23, pp. 135–151.

Sun, A., Li, J., Ye, X., Dong, Z., and Hou, L., 1992, The Chinese fossil reptiles and their kins. Beijing, Science Press.

Sushkin, P. P., 1926, Notes on the pre-Jurassic Tetrapoda from Russia: 1. *Dicynodon amalitzkii*: Palaeontologica Hungarica, v. 1, pp. 323–327.

Swisher, C. C., III, Wang, Y., Wang, X., Xu, X., and Wang, Y., 1999, Cretaceous age for the feathered dinosaurs of Liaoning, China: Nature, v. 400, pp. 58–61.

Sych, L., 1971, Mixodontia, a new order of Mammalia from the Paleocene of Mongolia: Palaeontologia Polonica, v. 25, pp. 147–158.

Szalay, F. S., 1982, A critique of some recently proposed Paleogene primate taxa and suggested relationships: Folia Primatologia, v. 37, pp. 153–162.

Széchenyi, B., 1899, Wissenschaftliche Ergebnisse der Reise des Grafen Béla Széchenyi in Ostasien 1877–1880. Dritter Band. Die Berarbeitung des gesamme Hen Materials. Vienna, Ed. Hölzel, p. 222.

Tan, H. C., 1923, New research on the Mesozoic and early Tertiary in Shantung: Bulletin Geological Survey of China, v. 5, pp. 148–181.

Tang, Y., 1978, New materials of Oligocene mammalian fossils from Qujing basin, Yunnan: Professional Papers on Stratigraphy and Paleontology, v. 7, pp. 75–79.

———, 1980, Note on a small collection of early Pleistocene mammalian fossils from northern Hebei: Vertebrata PalAsiatica, v. 18, pp. 314–323.

Tang, Y. and Ji, H., 1983, A Pliocene-Pleistocene transitional fauna from Yuxian, northern Hebei: Vertebrata PalAsiatica, v. 21, pp. 245–254.

Tang, Y. and Qiu, Z., 1979, Vertebrate faunas of Baise, Guangxi; *in* Mesozoic and Cenozoic red beds of south China: Beijing, Science Press, pp. 407–415.

Tang, Y. and Yan, D., 1976, Notes on some mammalian fossils from the Paleocene of Qianshan and Xuancheng, Anhui: Vertebrata PalAsiatica, v. 14, pp. 91–99.

Taquet, P., 1991, The status of *Tsintaosaurus spinorhinus* Young, 1958 (Dinosauria); *in* Kielan-Jaworowska, Z., Heintz, N., and Nakrem, H. A., eds., Fifth symposium on Mesozoic terrestrial ecosystems and biota extended abstracts: Oslo, University of Oslo, p. 63.

Tarlo, L .B., 1967, Agnatha; *in* Harland, W. B., ed., The fossil record: London, Geological Society of London, pp. 629–636.

Tassy, P., 1990, The "proboscidean datum event:" how many proboscideans and how many events?; *in* Lindsay, E. H., Fahlbusch, V., and Mein, P., eds., European Neogene mammal chronology: New York, Plenum Press, pp. 237–252.

——, 1996, The earliest gomphotheres; *in* Shoshani, J. and Tassy, P., eds., The Proboscidea: Evolution and paleoecology of elephants and their relatives: Oxford, Oxford University Press, pp. 89–91.

Taylor, D. W., 1960, Late Cenozoic molluscan faunas from the High Plains: U.S. Geological Survey, Professional Paper 337.

Tedford, R. H., 1970, Principles and practices of mammalian geochronology in North America: Proceedings North American Paleontological Convention 1, pp. 666–703.

Teilhard de Chardin, P., 1926, Description de mammifères Tertiaires de Chine et de Mongolies: Annales de Paleontologie, v. 15, pp. 1–51.

Teilhard de Chardin, P. and Piveteau, J., 1930, Les mammifères fossiles de Nihowan (China): Annales de Paleontologie, v. 19.

Thanh, T.-D. and Janvier, P., 1987, Les vertebres Devoniens du Vietnam: Annales de Paleontologie, v. 73, pp. 165–194.

——, 1990, Les vertebres du Devonien inferieur du Bac Bo oriental (provinces de Bac Thai et Langson, Viet Nam): Bulletin du Museum d'Histoire Naturelle, 4th series, v. 12, pp. 143–223.

Thorne, A. G. and Wolpoff, M. H., 1981, Regional continuity in Australasian Pleistocene hominid evolution: American Journal of Physical Anthropology, v. 55, pp. 337–349.

Thulborn, R., 1992, Nest of the dinosaur *Protoceratops*: Lethaia, v. 25, pp. 145–149.

Ting, S., 1993, A preliminary report on an early Eocene mammalian fauna from Hengdong, Hunan Province, China: Kaupia-Darmstädter Beiträge zur Naturgeschichte, v. 3, pp. 201–207.

——, 1998, Paleocene and early Eocene land mammal ages of Asia: Bulletin of the Carnegie Museum of Natural History, v. 34, pp. 124–147.

Tobien, H., Chen, G. and Li, Y., 1984, The mastodonts (Proboscidea, Mammalia) of China: evolution, palaeobiogeography, palaeoecology; *in* Whyte, R.O. et al., eds., The evolution of the east Asian environment Volume II Paleobotany, Palaeozoology and Palaeoanthroplogy: Hong Kong, University of Hong Kong, pp. 689–696.

——, 1986, Mastodonts (Proboscidea, Mammalia) from the late Neogene and early Pleistocene of the People's Republic of China. Part 1. Historical account; the genera *Gomphotherium, Choerolophodon, Synconolophus, Amebelodon, Platy-*

belodon, Sinomastodon: Mainzer Geowissenschaftliche Mitteilungen, v. 15, pp. 119–181.

——, 1988, Mastodonts (Proboscidea, Mammalia) from the late Neogene and early Pleistocene of the People's Republic of China. Part 2. The genera *Tetralophodon, Anancus, Stegotetrabelodon, Zygolophodon, Mamut, Stegolophodon*. Some generalities on the Chinese mastodonts: Mainzer Geowissenschaftliche Mitteilungen, v. 17, pp. 95–220.

Tong, Y., 1978, Late Paleocene mammals of Turfan basin, Sinkiang: Memoirs Institute of Vertebrate Paleontology and Paleoanthropology, v. 13, pp. 82–101.

——, 1979, A late Paleocene primate from south China: Vertebrata PalAsiatica, v. 17, pp. 65–70.

——, 1982, Some taligrades from the Upper Paleocene of the Nanxiong basin, Guangdong: Vertebrata PalAsiatica, v. 20, pp. 26–34.

——, 1987, A new species of *Sinolagomys* (Lagomorpha, Ochotonidae) from Xinjiang: Vertebrata PalAsiatica, v. 27, pp. 103–116.

——, 1992, *Pappocricetodon*, a pre-Oligocene cricetid genus (Rodentia) from central China: Vertebrata PalAsiatica, v. 30, pp. 1–16.

Tong, Y. and Lei, Y., 1984, Fossil tapiroids from the Upper Eocene of Xichuan, Henan: Vertebrata PalAsiatica, v. 23, pp. 269–280.

Tong, Y. and Wang, J., 1980, Subdivision of the Upper Cretaceous and lower Tertiary of the Tantou basin, the Lushi basin and the Lingbao basin of W. Henan: Vertebrata PalAsiatica, v. 18, pp. 21–27.

Tong, Y., Zheng, S., and Qiu, Z., 1995, Cenozoic mammal ages of China: Vertebrata PalAsiatica, v. 33, pp. 290–314.

Trofimov, B., 1978, First triconodonts [Mammalia, Triconodonta] from Mongolia: Doklady Akademii Nauk SSSR, v. 24, pp. 213–216.

Turner, S., 1986, Vertebrate fauna of the Silverband Formation, Grampians, western Victoria: Proceedings of the Royal Society of Victoria, v. 98, pp. 53–62.

Turner, S. and Tarling, D. H., 1982, Thelodont and other agnathan distributions as tests of lower Paleozoic continental reconstructions: Palaeogeography, Palaeoclimatology, Palaeoecology, v. 39, pp. 295–311.

Turner, S., Wang, S., and Young, G.C., 1995, Lower Devonian microvertebrates from Longmenshan, Sichuan, China: Preliminary report: Geobios Mémoire Spécial no. 19, pp. 383–387.

Upchurch, P., 1998, The phylogenetic relationships of sauropod dinosaurs: Zoological Journal of the Linnaean Society, v. 124, pp. 43–103.

Van Valen, L, 1988, Paleocene dinosaurs or Cretaceous ungulates in South America: Evolutionary Monographs, no. 10.

Veevers, J. J., 1988, Gondwana facies started when Gondwanaland merged in Pangea: Geology, v. 16, pp. 732–734.

Von Koenigswald, G. H. R., 1983, The significance of hitherto undescribed Miocene hominoids from the Siwaliks of Pakistan in the Senckenberg Museum,

Frankfurt; *in* Ciochon, R. L. and Corruccini, R. S., eds., New interpretations of ape and human ancestry: New York, Plenum Press, pp. 517–526.

Vorobyeva, E. I. and Minikh, M. G., 1968, Oput primeneniya biometrii k izucheniyu zubnykh plastinok tseratodontid [Experimental application of biometry to the study of ceratodontid toothplates]: Paleontologicheskiy Zhurnal, v. 1988, pp. 76–87.

Vyushkov, B. P., 1969, New dicynodonts from the Triassic of southern Cisuralia: Paleontological Journal, v. 3, pp. 237–242.

Wall, W. P., 1980, Cranial evidence for a proboscis in *Cadurcodon* and a review of snout structure in the family Amynodontidae (Perissodactyla, Rhinocerotoidea): Journal of Paleontology, v. 54, pp. 968–977.

Wang, B., 1975, Paleocene mammals of Chaling basin, Hunan: Vertebrata PalAsiatica, v. 13, pp. 154–162.

——, 1976, Late Paleocene mesonychid from Nanxiong basin, Guangdong: Vertebrata PalAsiatica, v. 14, pp. 259–262.

——, 1987, Discovery of Aplodontidae (Rodentia, Mammalia) from middle Oligocene of Nei Monggol, China: Vertebrata PalAsiatica, v. 25, pp. 32–45.

——, 1991, Discovery of *Yindirtemis* (Ctenodactylidae, Rodentia, Mammalia) from late Oligocene of Nei Mongol, China: Vertebrata PalAsiatica, v. 29, pp. 298–302.

——, 1997, The mid-Tertiary Ctenodactylidae (Rodentia, Mammalia) of eastern and central Asia. Bulletin of the American Museum of Natural History, no. 234, p. 88.

Wang, B. and Dawson, M. R., 1994, A primitive cricetid (Mammalia: Rodentia) from the middle Eocene of Jiangsu Province, China: Annals of Car-negie Museum, v. 63, pp. 239–256.

Wang, B. and Ding, S., 1979, *Bemalambda* of Chijiang basin, Jiangxi; *in* Mesozoic and Cenozoic red beds of south China: Beijing, Science Press, pp. 351–353.

Wang, B. and Emry, R. J., 1991, Eomyidae (Rodentia: Mammalia) from the Oligocene of Nei Monggol, China: Journal of Vertebrate Paleontology, v. 11, pp. 370–377.

Wang, B. and Li, C., 1990, First Paleogene mammalian fauna from northeast China: Vertebrata PalAsiatica, v. 28, pp. 165–205.

Wang, B. and Zhou, S., 1982, Late Eocene mammals from Pingchangguan basin, Henan: Vertebrata PalAsiatica, v. 38, pp. 203–215.

Wang, B. and Zhang, Y., 1983, New finds of fossils from Paleogene of Qujing, Yunnan: Vertebrata PalAsiatica, v. 21, pp. 119–128.

Wang, B., Chang, J., Meng, X., and Chen, J., 1981, Stratigraphy of the upper middle Oligocene of Qianlishan District, Nei Mongol (Inner Mongolia): Vertebrata PalAsiatica, v. 19, pp. 26–34.

Wang, H., ed., 1985, Atlas of the palaeogeography of China. Beijing, Cartographic Publishing House.

Wang, J., 1982, New materials of Dinichthyidae: Vertebrata PalAsiatica, v. 20, pp. 180–186.

——, 1991, New material of *Hunanolepis* from the Middle Devonian of Hunan; *in* Chang, M., Liu, Y. and Zhang, G., eds., Early vertebrates and related problems of evolutionary biology: Beijing, Science Press, pp. 213–247.

Wang, K., 1959, Ueber eine neue Fossile Reptiform von Provinz Hupeh, China: Acta Palaeontologica Sinica, v. 7, pp. 367–378.

Wang, N., 1977a, A new pholidophorid fish from Hengnan, Hunan: Vertebrata PalAsiatica, v. 15, pp. 177–183.

——, 1977b, Jurassic fishes from Lingling-Hengyang, Hunan, and its stratigraphic significance: Vertebrata PalAsiatica, v. 15, pp. 233–243.

——, 1984, Thelodont, acanthodian, and chondrichtyan fossils from the Lower Devonian of southwest China: Proceedings of the Linnean Society of New South Wales, v. 107, pp. 419-441.

——, 1991, Two new Silurian galeaspids (jawless craniates) from Zhejiang Province, China, with a discussion of galeaspid-gnathastome relationships; *in* Chang, M., Liu, Y. and Zhang, G., eds., Early vertebrates and related problems of evolutionary biology: Beijing, Science Press, pp. 41–65.

——, 1997, Restudy of thelodont microfossils from the lower part of the Cuifengshan Group of Qujing, eastern Yunnan, China: Vertebrata PalAsiatica, v. 35, pp. 1–17.

Wang, N. and Dong, Z., 1989, Discovery of Late Silurian microfossils of Agnatha and fishes from Yunnan, China: Acta Palaeontologica Sinica, v. 8, pp. 192–206.

Wang, N. and Wang, J., 1999, Discovery of placoderm inferognathal from China: Vertebrata PalAsiatica, v. 37, pp. 249–256.

Wang, N., Zhang, S., Wang, J., and Zhu, M., 1998, Early Silurian chondrichtyan microfossils from Bachu County, Xinjiang: Vertebrata PalAsiatica, v. 36, pp. 257–267.

Wang, S., 1989, Biostratigraphy of vertebrate microfossils; *in* Ji, Q. et al., eds., the Dapoushang section—an excellent section for the Devonian-Carboniferous boundary stratotype in China: Beijing, Science Press, pp. 36–38, 103–107.

——, 1991, Lower Devonian vertebrate paleocommunities from south China; *in* Chang, M., Liu, Y. and Zhang, G., eds., Early vertebrates and related problems of evolutionary biology: Beijing, Science Press, pp. 487–497.

——, 1993, Vertebrate biostratigraphy of the middle Paleozoic of China; *in* Long, J.A., ed., Palaeozoic vertebrate biostratigraphy and biogeography: London, Belhaven Press, pp. 252–276.

Wang, S. and Turner, S., 1985, Vertebrate microfossils of the Devonian-Carboniferous boundary, Muhua Section, Guizhou Province: Vertebrata PalAsiatica, v. 23, pp. 223–234.

Wang, S., Dong, Z,. and Turner, S., 1986, Middle Devonian Turiniidae (Thelodonti, Agnatha) from western Yunnan, China: Alcheringa, v. 10, pp. 315–325.

Wang, X., 1996, Mesozoic and Cenozoic paleobiocoenoses in Shandong Province. Beijing, Geological Publishing House [30th International geological Congress Field Trip Guide T320].

Wang, X., Wang, Y, Wang, Y., Xu, X., Tang, Z., Zhang, F., Hu, Y., Gu, G. and Hao, Z., 1998, Stratigraphic sequence and vertebrate-bearing beds of the lower part of the Yixian Formation in Sihetun and neighboring area, western Liaoning, China. Vertebrata PalAsiatica v. 36, pp. 81–101.

Wang, Y. and Sun, D., 1983, A survey of the Jurassic System of China: Canadian Journal of Earth Sciences, v. 20, pp. 1646–1656.

Wang, Y., Hu, Y., Chow, M., and Li, C., 1998, Chinese Paleocene mammal faunas and their correlation: Bulletin of the Carnegie Museum of Natural History, v. 34, pp. 89–123.

Wang, Y., Hu, Y., Zhou, M., and Li, C., 1995, Mesozoic mammal localities in western Liaoning, northeast China; in Sun, A. and Wang, Y., eds., Sixth symposium on Mesozoic terrestrial ecosystems and biota, short papers: Beijing, China Ocean Press, pp. 221–227.

Wang, Y., Xue, X., Yue, L., Zhao, J., and Liu, S., 1979, Discovery of Dali fossil man and its preliminary study: Kexue Tongba, v. 24, pp. 303–306.

Warren, A., Jupp, R., and Bolton, B., 1986, Earliest tetrapod trackway: Alcheringa, v. 10, pp. 183–186.

Watabe, M., 1992, Phylogeny of Chinese *Hipparion* (Perissodactyla, Mammalia): Their relationships with the western Old World and North American hipparionines: Paleontologia i Evolució, v. 24-25, pp. 155–172.

Watson, D. M. S., 1914, The zones of the Beaufort beds of the Karoo system in South Africa: Geological Magazine, New Series, v. 1, pp. 203–208.

Webb, S. D. and Taylor, B. E., 1980, The phylogeny of hornless ruminants and a description for the cranium of *Archaeomeryx*: Bulletin of the American Museum of Natural History, v. 167, pp. 117–158.

Weidenreich, F., 1943, The skull of *Sinanthropus pekinensis*; a comparative study on a primitive hominid skull: Palaeontologia Sinica, New Series D, no. 10.

Weishampel, D. B., 1990, Dinosaurian distribution; in Weishampel, D.B., Dodson, P., and Osmólska, H., eds., The Dinosauria: Berkeley, University of California Press, pp. 63–139.

Weishampel, D. B. and Horner, J. R., 1986, The hadrosaurid dinosaurs from the Iren Dabasu fauna (People's Republic of China, Late Cretaceous): Journal of Vertebrate Paleontology, v. 6, pp. 38–45.

——, 1990, Hadrosauridae; in Weishampel, D.B., Dodson, P. and Osmolska, H., eds., The Dinosauria: Berkeley, University of California Press, pp. 534–561.

Welles, S. P., 1984, *Dilophosaurus wetherilli* (Dinosauria, Theropoda) osteology and comparisons: Palaeontographica A, v. 185, pp. 86–180.

Wellnhofer, P., 1991, The illustrated encyclopedia of pterosaurs. New York, Crescent Books.

Werneburg, R., 1988, Paläobiogeographie der labyrinthodonten Amphibien im Oberkarbon und Rotliegenden Mitteleuropas: Zeitschrift für Geologischen Wissenschaft, v. 16, pp. 929–932.

Wetmore, A., 1934, Fossil birds from Mongolia and China: American Museum Novitates, no. 711.

Williams, H. S., 1901, The discrimination of time-values in geology: Journal of Geology, v. 9, pp. 570–585.

Wilson, J. A., and Sereno, P. C., 1998, Early evolution and higher-level phylogeny of sauropod dinosaurs: Society of Vertebrae Paleontology Memoir 5.

Wiman, C., 1929, Die Kreide-Dinosaurier aus Shantung: Palaeontologia Sinica, Series C, v. 6, pp. 1–67.

———, 1930, Fossil Schildkröten aus China: Palaeontologia Sinica, Series C, v. 6, pp. 5–53.

Wolfe, D. G. and Kirkland, J. I., 1998, *Zuniceratops christopheri* n. gen & n. sp., a ceratopsian dinosaur from the Moreno Hill Formation (Cretaceous, Turonian) of west-central New Mexico: New Mexico Museum of Natural History and Science, Bulletin 14, pp. 303–317.

Wood, A. E., 1974, Early Tertiary vertebrate faunas, Vieja Group, Trans-Pecos Texas: Rodentia: Texas Memorial Museum Bulletin, v. 21.

Woodburne, M. O. and Bernor, R. L., 1980, On superspecific groups of some Old World hipparionine horses: Journal of Paleontology, v. 54, pp. 1319–1348.

Wu, R., 1983, Hominid fossils from China and their bearing on human evolution: Canadian Journal of Anthropology, v. 3, pp. 207–214.

———, 1985, The cranium of *Ramapithecus* and *Sivapithecus* from Lufeng, China; *in* Andrews, P. and Franzen, J.L., eds., The early evolution of man: Senckenberg, Courier Forschungs Institut, pp. 41–48.

———, 1987, A revision of the classification of the Lufeng great apes: Acta Anthropologica Sinica, v. 3, pp. 193–200.

Wu, R. and Lin, S., 1983, Peking man: Scientific American, v. 248(6), pp. 86–94.

Wu, R. and Pan, Y., 1984, A late Miocene gibbon-like primate from Lufeng, Yunnan Province: Acta Anthropologica Sinica, v. 3, pp. 193–200.

———, 1985, A new adapid primate from the Lufeng Miocene, Yunnan: Acta Anthropologica Sinica, v. 4, pp. 1–6.

Wu, R., Xu, Q., and Lu, Q., 1983, Morphological features of *Ramapithecus* and *Sivapithecus* and their phylogenetic relationships—morphology and comparison of the cranium: Acta Anthropologica Sinica, v. 2, pp. 1–10.

Wu, X., 1981, The discovery of a new thecodont from north-east Shensi: Vertebrata PalAsiatica, v. 19, pp. 122–132.

———, 1990, The evolution of humankind in China: Acta Anthropologica Sinica, v. 9, pp. 312-321.

———, 1994, Late Triassic-Early Jurassic sphenodontians from China and the phylogeny of the Sphenodontia; *in* Fraser, N. C. and Sues, H.-D., eds., In the shadow of the dinosaurs: Cambridge, Cambridge University Press, pp. 38–69.

Wu, X. and Chatterjee, S., 1993, *Dibothrosuchus elaphros*, a crocodylomorph from the Lower Jurassic of China and the phylogeny of the Sphenosuchia: Journal of Vertebrate Paleontology, v. 13, pp. 58–89.

Wu, X. and Poirier, F.E., 1995, Human evolution in China. New York, Oxford University Press.

Wu, X. and Sues, H.-D., 1995, Protosuchians (Archosauria: Crocodyliformes) from China; *in* Sun, A. and Wang, Y., eds., Sixth symposium on Mesozoic terrestrial ecosystems and biota, short papers: Beijing, China Ocean Press, pp. 57–62.

——, 1996, Reassessment of *Platyognathus hsui* Young, 1944 (Archosauria: Crocodyliformes) from the lower Lufeng Formation (Lower Jurassic) of Yunnan, China: Journal of Vertebrate Paleontology, v. 16, pp. 42–48.

Xu, Q., 1977, Two new genera of old Ungulata from the Paleocene of Qianshan basin, Anhui: Vertebrata PalAsiatica, v. 15, pp. 119–125.

Xu, Q., Tian, M., Li, L., and Liu, J., 1996, A brief introduction to the Quaternary geology and paleoanthropology of the Zhoukoudian site, Beijing. Beijing Geological Publishing House [30th IGC Field Trip Guide T203].

Xu, X., 1997, A new psittacosaur (*Psittacosaurus mazongshanensis* sp. nov.) from Mazongshan area, Gansu Province, China; *in* Dong, Z., ed., Sino-Japanese Silk Road dinosaur expedition: Beijing, China Ocean Press, pp. 48–67.

Xu, X. and Wang, X., 1998, New psittacosaur (Ornithischia, Ceratopsia) occurrence from the Yixian Formation of Liaoning, China and its stratigraphical significance: Vertebrata PalAsiatica, v. 36, pp. 147–158.

——, 1965, A new genus of amynodont from the Eocene of Lantian, Shensi: Vertebrata PalAsiatica, v. 9, pp. 83–88.

——, 1966, Amynodonts of Inner Mongolia: Vertebrata PalAsiatica, v. 10, pp. 123–190.

——, 1976, Some new forms of Coryphodontidae from the Eocene of Sichuan basin in Honan: Vertebrata PalAsiatica, v. 14, pp. 185–193.

——, 1980, New material of fossil *Manteodon youngi* from Yichang, Hubei: Vertebrata PalAsiatica, v. 18, pp. 296–298.

Xu, Y., Yan, D., Zhou, S., Han, S. and Zhong, Y., 1979, Subdivision of the red beds of Liguangiao basin with description of fossil mammals therefrom: Mesozoic and Cenozoic red beds of south China, pp. 416–432.

Xu, Z., 1990, Mesozoic volcanism and volcanogenic iron-ore deposits in eastern China: Geological Society of America Special Paper 237.

Xue, X., 1981, An early Pleistocene mammalian fauna and its stratigraphy of the River You, Weinan, Shensi: Vertebrata PalAsiatica, v. 19, pp. 35–44.

——, 1982, Notes on *Megaloceros luochuanensis* (sp. nov.) from Heimugou, Luochuan, Shaanxi Province (in addition, the other mammalian fossils from the same district): Vertebrata PalAsiatica, v. 20, pp. 228–235.

——, 1984, The Quaternary mammalian fossils in the loess area of China; *in* Sasijama, S. and Wang, Y., eds., The recent research of loess in China: Kyoto, Kyoto Institute of Natural History, pp. 112–159.

Xue, X. and Zhang, Y., 1991, Quaternary mammalian fossils and fossil human beings; *in* Zhang, Z. and Shao, S., eds., The Quaternary of China: Beijing, China Ocean Press, pp. 307–374.

Xue, X., Zhang, Y., Bi, Y., Yue, L., and Chen, D., 1996, The development and environmental changes of the intermontane basins in the eastern part of Qinling Mountains. Beijing, Geological Publishing House.

Yan, D., 1979, Einiger der fossilen miozänen Säugetiere der Kreis von Fangxian in der Proviz Hupei: Vertebrata PalAsiatica, v. 17, pp. 189–199.

Yan, D. and Tang, Y., 1976, Mesonychids from the Paleocene of Anhui: Vertebrata PalAsiatica, v. 14, pp. 252–258.

Yan, D., Qiu, Z., and Men, Z., 1983, Miocene stratigraphy and mammals of Shanwang, Shandong: Vertebrata PalAsiatica, v. 21, pp. 210–222.

Yang, S., Pan, K., and Hou, H., 1981, The Devonian system in China: Geological Magazine, v. 118, pp. 113–224.

Yang, Z., Cheng, Y., and Wang, H., 1986, The geology of China. Clarendon Press, Oxford.

Ye, H., 1959, A new dicynodont from the *Sinokannemeyeria* fauna from Shansi: Vertebrata PalAsiatica, v. 3, pp. 187–204.

——, 1963, Fossil turtles of China: Palaeontologica Sinica, Series C, v. 18, pp. 1–112.

——, 1982, Middle Jurassic turtles from Sichuan, S.W. China: Vertebrata PalAsiatica, v. 20, pp. 282–290.

——, 1990, Fossil turtles from Dashanpu, Zigong, Sichuan: Vertebrata PalAsiatica, v. 28, pp. 304–311.

——, 1994, Fossil and recent turtles of China. Beijing, Science Press.

Ye, J. and Jia, H., 1986, *Platybelodon* (Proboscidea, Mammalia) from the Middle Miocene of Tongxin, Ningxia: Vertebrata PalAsiatica, v. 24, pp. 139–151.

Young, C. C., 1927, Fossile Nägetiere aus Nord-China. Palaeontologia Sinica C, v. 5.

——, 1931, On some new dinosaurs from Western Suiyan, Inner Mongolia: Bulletin Geological Survey of China, v. 2, pp. 159–266.

——, 1935a, On a new nodosaurid from Ninghsia: Palaeontologia Sinica, Series C, v. 11, pp. 1–34.

——, 1935b, On two skeletons of Dicynodontia: Bulletin Geological Society of China, v. 14, pp. 483–517.

——, 1936a, A Miocene fossil frog from Shantung: Bulletin of the Geological Society of China, v. 15, pp. 189–193.

——, 1936b, On a new *Chasmatosaurus* from Sinkiang: Bulletin of the Geological Society of China, v. 15, pp. 291–320.

——, 1939a, On a new Sauropoda, with notes on other fragmentary reptiles from Szechuan: Bulletin Geological Society of China, v. 19, pp. 279–315.

——, 1939b, Additional Dicynodontia remains from Sinkiang: Bulletin Geological Society of China, v. 19, pp. 111–136.

——, 1943, Notes on some fossil footprints in China: Bulletin Geological Society of China, v. 23, pp. 151–154.

——, 1944, On a supposed new pseudosuchian from Upper Triassic saurischian-bearing beds of Lufeng, Yunnan, China: American Museum Novitates, no. 1264, pp. 1–4.

——, 1947, Mammal-like reptiles from Lufeng, Yunnan: Proceedings Zoological Society London, v. 117, pp. 537–597.

——, 1948, Fossil crocodiles in China, with notes on dinosaurian remains associated with the Kansu crocodiles: Bulletin Geological Society of China, v. 28, pp. 255–288.

——, 1951a, The Lufeng saurischian fauna in China: Palaeontologia Sinica Series C, v. 13, pp. 1–96.

——, 1951b, Main vertebrate horizons in China, their geological and geographical distribution, faunistic character and correlation: Report 18th Session International Geological Congress, part II, pp. 66–73.

——, 1952, On a new therocephalian from Sinkiang, China: Acta Scientia Sinica, v. 2, pp. 152–165.

——, 1957, *Neoprocolophon asiticus* [sic], a new cotylosaurian reptile in China: Vertebrata PalAsiatica, v. 1, pp. 1–8.

——, 1958a, On a new locality of *Yabeinosaurus tenuis* Endo and Shikama: Vertebrata PalAsiatica, v. 2, pp. 153–156.

——, 1958b, The dinosaurian remains of Laiyang, Shantung: Palaeontologia Sinica Series C, v. 16, pp. 53–159.

——, 1958c, New sauropods from China: Vertebrata PalAsiatica, v. 2, pp. 1–28.

——, 1958d, On the new Pachypleurosauroidea from Keichow, southwest China: Vertebrata PalAsiatica, v. 2, pp. 69–81.

——, 1960, Fossil footprints in China: Vertebrata PalAsiatica, v. 4, pp. 53–66.

——, 1964a, The pseudosuchians in China: Palaeontologia Sinica Series C, v. 19, pp. 105–205.

——, 1964b, New fossil crocodiles from China: Vertebrata PalAsiatica, v. 8, pp. 189–208.

——, 1964c, On a new pterosaurian from Sinkiang, China: Vertebrata PalAsiatica, v. 8, pp. 221–256.

——, 1965a, On the first occurrence of fossil salamanders from the Upper Miocene of Shantung, China: Acta Palaeontologia Sinica, v. 13, pp. 455–459.

——, 1965b, Fossil eggs from Nanhsiung, Kwangtung and Kanehou, Kiangsi: Vertebrata PalAsiatica, v. 9, pp. 141–170.

——, 1965c, On the new nothosaurs from Hupeh and Kweichow, China: Vertebrata Palasiatica, v. 9, pp. 315–356.

——, 1966, Two footprints from the Jiaoping Coal Mine of Tungcuan, Shensi: Vertebrata PalAsiatica, v. 10, pp. 68–71.

——, 1973a, A new fossil crocodile from Wuerho: Memoirs Institute Vertebrate Paleontology and Paleoanthropology, Academia Sinica, v. 11, pp. 37–45.

——, 1973b, Pterosaurs from Wuerho: Memoirs Institute Vertebrate Paleontology and Paleoanthropology Academia Sinica, v. 11, pp. 1–7.

——, 1973c, On the occurrence of *Vjushkovia* in Sinkiang: Memoirs Institute Vertebrate Paleontology and Paleoanthropology Academia Sinica, v. 10, pp. 38–52.

——, 1974a, New materials of Therapsida from Lufeng, Yunnan: Vertebrata PalAsiatica, v. 12, pp. 111–116.

——, 1974b, A new genus of Traversodontidae in Jiyuan, Honan: Vertebrata PalAsiatica, v. 12, pp. 203–211.

——, 1977, On some Salientia and Chiroptera from Shanwang, Lingu, Shandong: Vertebrata PalAsiatica, v. 15, pp. 76–80.

——, 1978, A Late Triassic vertebrate fauna from Fukang, Sinkiang: Institute of Vertebrate Paleontology and Paleoanthropology, Memoir 13, pp. 60–67.

——, 1979a, Note on an egg from Ninghsia: Vertebrata PalAsiatica, v. 17, pp. 35–36.

——, 1979b, A new Late Permian fauna from Jiyuan, Honan: Vertebrata PalAsiatica, v. 17, pp. 99–113.

——, 1982, On a primitive crocodile from Lufeng, Yunnan; *in* Selected works of Yang Zhungjian: Beijing, Science Press, pp. 26–28.

Young, C. C. and Bien, M. N., 1936, Cenozoic geology of the Kaolan-Yungteng area of central Kansu: Bulletin of the Geological Survey of China, v. 16, pp. 221–260.

Young, C. C. and Chao, H., 1972, *Mamenchisaurus hochuanensis* sp. nov.: Institute of Vertebrate Paleontology and Paleoanthropology, Monograph 8, pp. 1–30.

Young, C. C. and Chow, M., 1953, New Mesozoic reptiles from Sichuan: Acta Palaeontologia Sinica, v. 1, pp. 216–243.

Young, C. C. and Dong, Z., 1972, Aquatic reptiles from Triassic of China: Memoirs Institute Vertebrate Paleontology and Paleoanthropology, v. 9, pp. 1–34.

Young, C. C. and Tchang, T. L., 1936, Fossil fishes from the Shanwang Series of Shantung: Bulletin of the Geological Society of China, v. 15, pp. 197–205.

Young, C. C. and Ye, H., 1963, On a new pareiasaur from the Upper Permian of Shansi, China: Vertebrata PalAsiatica, v. 7, pp. 195–212.

Young, C. C., Liu, D., and Zhang, M., 1982, Ichthyosauria from Xizang Zhong; *in* Monograph on Mount Xixia Bongma scientific expedition, 1964: Beijing, Science Press, pp. 350–355.

Young, G. C., 1981, Biogeography of Devonian vertebrates: Alcheringa, v. 5, pp. 225–243.

——, 1990, Devonian vertebrate distribution patterns and cladistic analysis of palaeogeographic hypotheses; *in* McKerrow, W. C. and Scotese, C. R., eds., Palaeozoic palaeogeography and biogeography: Geological Society of London Memoir, v. 12, pp. 243–255.

Young, G., 1993, Vertebrate faunal provinces in the middle Paleozoic; *in* Long, J. A., ed., Palaeozoic vertebrate biostratigraphy and biogeography: London, Belhaven Press, pp. 293–323.

Yu, W., Boucot, A. J., Rong, J., and Yang, X., 1984, Silurian and Devonian biogeography of China: Geological Society of America Bulletin, v. 95, pp. 265–279.

Yuan, P. L. and Young, C. C., 1934a, On the discovery of a new dicynodon in Sinkiang: Bulletin of the Geological Society of China, v. 13, pp. 563–574.

——, 1934b, On the occurrence of *Lystrosaurus* in Sinkiang: Bulletin of the Geological Society of China, v. 13, pp. 575–580.

Yue, L. and Xue, X., 1996, The mammalian faunas in North Chinese loess and their position in magnetostratigraphy: Vertebrata PalAsiatica, v. 34, pp. 305–311.

Zangerl, R., 1981, Chondrichthyes I Paleozoic Elasmobranchii: Handbook of Paleoichthyology, v. 3A.

Zdansky, O., 1927, Preliminary notice on two teeth of a hominid from a cave in Chihli (China): Bulletin of the Geological Society of China, v. 5, pp. 281–284.

——, 1928, Die Säugetiere der Quartärfauna von Chouk'-ou-Tien: Palaeontologia Sinica C, v. 5, pp. 1–146.

——, 1930, Die Alttertiären Säugetiere Chinas nebst stratigraphischen Bemerkungen: Palaeontologia Sinica C, v. 6.

Zeng, D. and Zhong, J., 1979, Fossil dinosaur eggs from the western part of the Donting basin, Hunan Province: Vertebrata PalAsiatica, v. 17, pp. 131–136.

Zhai, R., 1977, Supplementary remarks on the age of Changxindian Formation: Vertebrata PalAsiatica, v. 15, pp. 173–176.

——, 1978a, Two new early Eocene mammals from Sinkiang: Memoirs Institute of Vertebrate Paleontology and Paleoanthropology, v. 13, pp. 102–106.

——, 1978b, More fossil evidence favoring an early Eocene connection between Asia and Neoarctic: Memoirs Institute of Vertebrate Paleontology and Paleoanthropology, v. 13, pp. 107–115.

——, 1978c, Late Oligocene mammals from the Taoshuyuanzi Formation of eastern Turfan basin: Memoirs Institute of Vertebrate Paleontology and Paleoanthropology, v. 13, pp. 126–131.

Zhang, F., 1975, A new thecodont *Lotosaurus*, from Middle Triassic of Hunan: Vertebrata PalAsiatica, v. 13, pp. 144–147.

Zhang, F., Li, Y. and Wang, X., 1984, A new occurrence of Permian seymouriamorphs in Xinjiang, China: Vertebrata PalAsiatica, v. 22, pp. 294–304.

Zhang, G., 1978, The antiarchs from the Early Devonian of Yunnan: Vertebrata PalAsiatica, v. 16, pp. 147–186.

Zhang, M., 1982, The braincase of *Youngolepis*, a Lower Devonian crossopterygian from Yunnan, south-western China. Stockholm, Swedish Museum of Natural History.

——, 1991, Head exoskeleton and shoulder girdle of *Youngolepis*; *in* Chang, M., Liu, Y., and Zhang, G., eds., Early vertebrates and related problems of evolutionary biology: Beijing, Science Press, pp. 355–378.

Zhang, M. and Yu, X., 1981, A new crossopterygian, *Youngolepis praecursor*, gen. et sp. nov., from Lower Devonian of E. Yunnan, China: Scientia Sinica, v. 24, pp. 89–97.

——, 1987, A *nomen novum* for *Diabolichthys* Chang et Yu, 1981: Vertebrata PalAsiatica, v. 25, p. 79.

Zhang, X., Zhou, G., Hu, Y. and Lan, Y., 1981, Stratigraphy of *Ramapithecus*-bearing Pliocene of Lufeng, Yunnan. Memoirs of the Beijing Natural History Museum, no. 10.

Zhang, Y. and Tong, Y., 1981, New anagaloid mammals from the Paleocene of south China: Vertebrata PalAsiatica, v. 19, pp. 133–144.

Zhang, Y., Huang, W., Tang, T., Ji, H., You, Y., Tong, Y., Ding, S., Huang, X., and Zheng, J., 1978, The Cenozoic of the Lantian region, Shaanxi Province. Beijing, Academia Sinica.

Zhao, X. and Currie, P.J., 1993, A large crested theropod from the Jurassic of Xinjiang, People's Republic of China: Canadian Journal of Earth Sciences, v. 30, pp. 2027–2036.

Zhao, X., Cheng, Z. and Xu, X., 1999, The earliest ceratopsian from the Tuchengzi Formation of Liaoning, China: Journal of Vertebrate Paleontology, v. 19, pp. 681–691.

Zhao, Z., 1975, The microstructure of fossil dinosaur eggs from Nanxiong County, Guangdong Province, and issues in their classification: Vertebrata PalAsiatica, v. 13, pp. 105–117.

——, 1979a, Progress in the research of dinosaur eggs; *in* Mesozoic and Cenozoic red beds of south China: Beijing, Science Press, pp. 330–340.

——, 1979b, Discovery of the dinosaurian eggshells from Alxa, Ningxia and its stratigraphic significance: Vertebrata PalAsiatica, v. 14, pp. 42–44.

——, 1994, Dinosaur eggs in China: On the structure and evolution of eggshells; *in* Carpenter, K., Hirsch, K. F. and Horner, J. R., eds., Dinosaur eggs and babies: Cambridge, Cambridge University Press, pp. 184–203.

Zhao, Z. and Ding, S., 1976, The discovery of fossil dinosaur eggs in Alxa Left Banner, Ningxia Huizu Autonomous Region, and their significance: Vertebrata PalAsiatica, v. 14, pp. 42–44.

Zhao, Z. and Li, Z., 1988, A new structural type of the dinosaur eggs from Anlu County, Hubei Province: Vertebrata PalAsiatica, v. 26, pp. 107–115.

Zhao, Z., Ye, J., Li, H., Zhao, Z,. and Van, Z., 1991, Extinctions of the dinosaurs across the Cretaceous-Tertiary boundary in Nanxiong basin, Guangdong Province: Vertebrata PalAsiatica, v. 29, pp. 1–20.

Zhen, S., Li, J., Han, Z., and Yang, X., 1996, The study of dinosaur footprints in China. Chongqing, Sichuan Scientific and Technological Publishing House.

Zhen, S., Zhen, B., Mateer, N. J., and Lucas, S.G., 1985, The Mesozoic reptiles of China: Bulletin of the Geological Institutions of the University of Uppsala, New Series, v. 11, pp. 133–150.

Zhen, S., Li, J., Rao, C., Mateer, N. J. and Lockley, M. G., 1989, A review of dinosaur footprints in China; *in* Gillette, D. D. and Lockley, M. G., eds., Dinosaur tracks and traces: Cambridge, Cambridge University Press, pp. 187–197.

Zhen, S., Li, J., Zhang, B., Chen, W., and Zhu, S., 1994, Dinosaur and bird footprints from the Lower Cretaceous of Emei County, Sichuan, China: Memoirs of Beijing Natural History Museum, no. 54, pp. 105–120.

Zheng, J., 1978, Description of some late Eocene mammals from Lian-Kan Formation of Turfan basin, Sinkiang: Memoirs Institute of Vertebrate Paleontology and Paleoanthropology, v. 13, pp. 116–125.

Zheng, S. and Han, D., 1991, Quaternary mammals of China; *in* Liu, T., ed., Quaternary geology and environments in China: Beijing, Science Press, pp. 101–114.

Zheng, J. and Chi, H., 1978, Some of the latest Eocene Condylarthra mammals from Guangsi, south China: Vertebrata PalAsiatica, v. 16, pp. 97–101.

Zheng, J., Tung, Y., and Chi, H., 1975, Discovery of Miacidae (Carnivora) in Yuan-shui basin, Kiangsi Province: Vertebrata PalAsiatica, v. 13, pp. 96–104.

Zheng, J., Tang, Y., Zhai, R., Ding, S., and Huang, X., 1978, Early Tertiary strata of Lunan basin Yunnan: Professional Papers on Stratigraphy and Paleontology, v. 7, pp. 22–29.

Zhou, J., 1992, Initial study of the bird fossils of the early Mesozoic in Liaoning: Science Bulletin 1992, pp. 435–437.

Zhou, M., 1964, Mammals of "Lantian Man" locality at Lantian, Shaanxi: Vertebrata PalAsiatica, v. 8, pp. 301–307.

——, 1965, The characters and geological age of the fauna in association with the Lantian Man: Kexue Tongba, v. 6, pp. 482–487.

——, 1978, Tertiary mammalian faunas of the Lantian district, Shensi: Professional Papers in Stratigraphy and Paleontology, v. 7, pp. 98–108.

Zhou, M., Rich, P. V., and Qi, T., 1982, A late Eocene–early Oligocene bird and mammal from Usu (Wusu) Xinjiang (Sinkiang), northwestern China. Memoirs of Beijing Natural History Museum, no. 16.

Zhou, M., Zhang, Y., and You, Y., 1978, Notes on some mastodonts from Yunnan: Professional Papers in Stratigraphy and Paleontology, v. 7, pp. 68–74.

Zhou, X., Sun, Y., and Wang, H., 1990, Gulonghan cave site, an Upper Paleolithic site at Dalian City. Beijing, Beijing Science and Technology Press.

Zhou, X., Zhai, R., Gingerich, P. D. and Chen, L., 1995, Skull of a new mesonychid (Mammalia, Mesonychia) from the late Paleocene of China: Journal of Vertebrate Paleontology, v. 15, pp. 387–400.

Zhou, Z. and Chen, P., eds., 1992, Biostratigraphy and geological evolution of Tarim. Beijing, Science Press.

Zhou, Z., Jin, F., and Zhang, J., 1992. Preliminary report on a Mesozoic bird from Liaoning, China. Chinese Science Bulletin 37, pp. 1365–1368.

Zhu, M., 1991, New information on *Diandongpetalichthys* (Placodermi: Petalichthyida); *in* Chang, M., Liu, Y. and Zhang, G., eds., Early vertebrates and related problems of evolutionary biology: Beijing, Science Press, pp. 179–194.

——, 1998, Early Silurian sinacanths (Chondrichthyes) from China: Palaeontology, v. 41, pp. 157–171.

Zhu, M. and Fan, J., 1995, *Youngolepis* from the Xishancun Formation (early Lochkovian) of Qujing, China: Geobios Mémoire Spécial, no. 19, pp. 293–299.

Zhu, M. and Wang, J., 1996, A new macropetalichthyid from China, with special reference to the historical zoogeography of the Macropetalichthyidae (Placodermi): Vertebrata PalAsiatica, v. 34, pp. 253–268.

Zhu, Y., 1989, The discovery of dicynodonts in Daqingshan Mountain, Nei Mongol: Vertebrata PalAsiatica, v. 27, pp. 9–27.

Zidek, J., 1976, Some fishes of the Wild Cow Formation (Pennsylvanian), Manzanita Mountains, New Mexico: New Mexico Bureau of Mines and Mineral Resources Circular, no. 35.

Ziegler, A. M., 1990, Phytogeographic patterns and continental configurations during the Permian Period; *in* McKerrow, W. S. and Scotese, C. R., eds., Paleozoic Paleogeography and biogeography: London Geological Society Memoir 12, pp. 363–379.

Index

Printed in the USA
CPSIA information can be obtained
at www.ICGtesting.com
JSHW051456221024
72172JS00010B/88

9 780231 084833